Advances in Physical Geochemistry

Volume 2

Advances in Physical Geochemistry

Edited by
Surendra K. Saxena

With Contributions by
P. M. Bell   J. G. Blencoe   S. Carbonin
A. Cundari   A. Dal Negro   J. Ganguly
D. R. Gaskell   S. Ghose   B. S. Hemingway
R. Kretz   H. K. Mao   G. A. Merkel
G. M. Molin   E. M. Piccirillo
S. K. Saxena   M. K. Seil   V. M. Shmonov
K. I. Shmulovich   R. G. J. Strens
T. Yagi   V. A. Zharikov

With 113 Illustrations

Springer-Verlag
New York   Heidelberg   Berlin

Series Editor
Surendra K. Saxena
Department of Geology
Brooklyn College
City University of New York
Brooklyn, NY 11210
U.S.A.

Library of Congress Cataloging in Publication Data
Main entry under title:
Advances in physical geochemistry.
  (Advances in physical geochemistry; v. 2)
  Bibliography: p.
  Includes index.
  1. Mineralogical chemistry. 2. Chemistry,
Physical and theoretical. I. Saxena, Surendra
Kumar, 1936–   . II. Series.
QE371.A33   549'.6   82-3258
             AACR2

Printed in the United States of America

9 8 7 6 5 4 3 2 1

ISBN 0-387-90644-4 Springer-Verlag New York Heidelberg Berlin
ISBN 3-540-90644-4 Springer-Verlag Berlin Heidelberg New York

Preface

The second volume of this series consists of three parts. Part I focuses on the research on intracrystalline reactions. This work, which began nearly two decades ago, is critically reviewed by Ghose and Ganguly in Chapter 1. Besides the review, the authors include some of their previously unpublished work to demonstrate how future research could aid in obtaining data on thermodynamics of solid solutions and in understanding the cooling history of igneous and metamorphic rocks. The latter is also the theme adopted by Kretz in the second chapter, which examines the redistribution of Fe and Mg in coexisting silicates during cooling. Chapter 3 contains new data on Fe–Mg distribution in clinopyroxenes. Dal Negro and his co-authors have selected a series of clinopyroxenes from volcanic rocks and present site occupancy data on several clinopyroxenes of intermediate compositions. The data set has not been published before and is the first of its kind.

Part II of this book begins with a chapter on melts by Gaskell, who explores the relationship between density and structure of silicate melts. This is followed by the synthesis of data generated in the U.S.S.R. by Shmulovich and his co-authors on fluids. Blencoe, Merkel and Seil present a thorough analysis of the phase equilibrium data on feldspars coexisting with fluids in the third chapter in this part.

The last part of the book contains two chapters on thermodynamic methods of calculating phase equilibria, one chapter on the critical analysis of Gibbs free energy of formation of substances in laterites and bauxites, and two chapters on crystal-chemistry presenting new data on pyroxenes and perovskite.

This volume, with the reviews of many current topics and the new crystal-chemical and thermodynamic data, should be very useful to geochemists, geophysicists, and other material scientists. The help of Alex Navrotsky, D. Kerrick, R. Kretz, E. Busenberg, R. C. Newton, J. Ganguly, S. Ghose, J. Blencoe, G. Rossi, L. Ungeretti, and R. F. Mueller in reviewing one or more chapters is gratefully acknowledged.

S. K. Saxena

Contents

Contributors

BELL, P. M.
Geophysical Laboratory, 2801 Upton Street N.W., Washington, D.C. 20008, U.S.A.

BLENCOE, J. G.
Department of Geosciences, The Pennsylvania State University, University Park, Pennsylvania 16802, U.S.A.

CARBONIN, S.
Instituto di Mineralogia e Petrologia, University of Padova, Corso Garibaldi, 37, 35100 Padova, Italy

CUNDARI, A.
Department of Geology, University of Melbourne, Melbourne, Parkville, Victoria, 3052, Australia

DAL NEGRO, A.
Instituto di Mineralogia e Petrologia, University of Padova, Corso Garibaldi, 37, 35100 Padova, Italy

GANGULY, J.
Department of Geosciences, University of Arizona, Tucson, Arizona 85721, U.S.A.

GASKELL, D. R.
Laboratory for Research on the Structure of Matter and Department of Materials Science and Engineering, University of Pennsylvania, Philadelphia, Pennsylvania 19104, U.S.A.

GHOSE, S.
Department of Geological Sciences, University of Washington, Seattle, Washington 98195, U.S.A.

HEMINGWAY, B. S.
U.S. Geological Survey, Reston, Virginia 22092, U.S.A.

KRETZ, R.   Department of Geology, University of Ottawa, Ot-
                             tawa, Canada

MAO, H. K.   Geophysical Laboratory, 2801 Upton Street N.W.,
                             Washington D.C. 20008, U.S.A.

MERKEL, G. A.   Ceramics Research, Corning Glass Works, Corning,
                             New York 14831, U.S.A.

MOLIN, G. M.   Instituto di Mineralogia e Petrologia, University of
                             Padova, Corso Garibaldi, 37, 35100 Padova, Italy

PICCIRILLO, E. M.   Instituto di Mineralogia e Petrografia, University of
                             Bari, Piazza Umberto, 1, 70121 Bari, Italy

SAXENA, S. K.   Department of Geology, Brooklyn College, Brook-
                             lyn, New York 11210, U.S.A.

SEIL, M. K.   Department of Geosciences, The Pennsylvania State
                             University, University Park, Pennsylvania 16802,
                             U.S.A.

SHMONOV, V. M.   Institute of Experimental Mineralogy, Academy of
                             Sciences of the USSR, 142432, Chernogolovka,
                             U.S.S.R.

SHMULOVICH, K. I.   Institute of Experimental Mineralogy, Academy of
                             Sciences of the USSR, 142432, Chernogolovka,
                             U.S.S.R.

STRENS, R. G. J.   School of Physics, University of New Castle, New
                             Castle Upon Tyne NE1 7IR, UK

YAGI, T.   Institute of Solid State Physics, University of Tokyo,
                             Roppongi, Minato—Ku, Tokyo 106 Japan

ZHARIKOV, V. A.   Institute of Experimental Mineralogy, Academy of
                             Sciences of the USSR, 142432, Chernogolovka,
                             U.S.S.R.

I. Ferromagnesian Silicates: Order–Disorder, Kinetics, and Phase Equilibria

Chapter 1
Mg–Fe Order–Disorder in Ferromagnesian Silicates

S. Ghose and J. Ganguly

Introduction

Intracrystalline distribution in silicates has been a subject of interest among geochemists since the classical work of Goldschmidt (1954).[1] The phenomenon of strong Fe^{2+}–Mg ordering in two of the common rock-forming ferromagnesian silicates, pyroxenes and amphiboles, was discovered in the early sixties through single-crystal X-ray diffraction study. Soon thereafter, the possibility of rapid determination of intracrystalline Fe^{2+}–Mg distributions through the newly discovered technique of Mössbauer resonance spectroscopy attracted the attention of a number of mineralogists and petrologists, because such distributions are related to the thermodynamic mixing properties of the Fe^{2+} and Mg-end member components, which are necessary for phase equilibrium calculations, and also to the cooling history of rocks. The purpose of this work (Parts I and II) is to critically review and synthesize the various contributions made in this field in the last two decades. We also include some of our own results, which are presented here for the first time. The crystalchemical details, which are necessary to understand the atomic forces governing the Fe^{2+}–Mg distribution in ferromagnesian silicates, as well as the experimental techniques commonly utilized to determine Fe^{2+}–Mg distribution, are reviewed in Part I (Ghose). Part II (Ganguly) deals with the thermodynamics and kinetics of Fe^{2+}–Mg order–disorder and their application to geologic problems. Hopefully, this review will stimulate further interest in the subject and provide directions for future work.

---

[1] Goldschmidt, V. M. (1954) *Geochemistry*. Clarendon Press, Oxford.

I. Crystal Chemistry (S. Ghose)

Rationale for Mg–Fe Ordering

In ferromagnesian silicates with nonequivalent divalent cation sites, Fe^{2+} – Mg^{2+} ordering is the rule rather than the exception. Intracrystalline site partitioning of Fe^{2+} and Mg^{2+} depends on the intrinsic differences of the electronic structures of these two ions and the stereochemical differences of the crystallographic sites. In contrast to Mg^{2+}, Fe^{2+} possesses six d-electrons and is a transition metal ion with distinct electronic and magnetic properties. The electronic configurations of the high-spin and low-spin Fe^{2+} are shown in Fig. 1. Under conditions prevalent in the Earth's crust and the upper mantle, Fe^{2+} in oxides and silicates is known to exist in the high-spin state only. The charge density distribution within the Fe^{2+} ion is pronouncedly anisotropic due to the six d-electrons present. As a result, within a crystal where the Fe^{2+} ion is surrounded by a regular octahedron of six ligands such as oxygen, the energy levels of the five d-orbitals are no longer degenerate and split up into two levels e_g and t_{2g} containing two and four d-electrons, respectively. The energy differences between these two levels is called the crystal field splitting, Δ_0. The three t_{2g} orbitals are lowered by $\frac{2}{5}\Delta_0$ below and the e_g orbitals are raised by $\frac{3}{5}\Delta_0$ above the baricenter. Each electron occurring in a t_{2g} orbital stabilizes a transition metal ion by $\frac{2}{5}\Delta_0$, whereas for each electron occurring in an e_g orbital, the stability is decreased by $\frac{3}{5}\Delta_0$. The net gain in stabilization energy is called the crystal field stabilization energy (CFSE) (see Burns (1970)). If the coordination octahedron is distorted from the regular octahedral symmetry, the twofold and threefold degeneracy within the e_g and t_{2g}

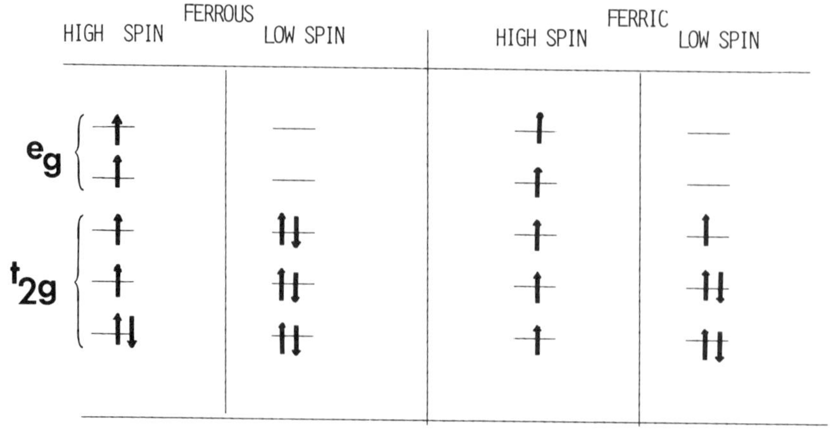

Fig. 1. Electronic configuration of the Fe^{2+} ion in octahedral coordination.

levels respectively, may be lifted, such that their respective mean energy is unaffected. If two octahedral sites are distorted to different degrees from octahedral symmetry, the Fe^{2+} ion is expected to be stabilized in the more distorted site due to a net gain in CFSE (Burns, 1970; Walsh et al., 1974). The crystal field theory is based strictly on electrostatic considerations, where Fe^{2+} and the ligand ions (O^{2-}) are considered to be point charges. The observed crystal field parameters, however, cannot be accounted for by point charge calculations; this discrepancy is due to the neglect of the covalency effects (Varret, 1976). From superhyperfine splitting of the electron spin resonance spectra of transition metal ions in silicates and magnetic spin density studies of transition metal compounds by polarized neutrons, there is abundant evidence that the d-electrons in transition metal ions are considerably delocalized and spend a considerable part of their time on ligand orbitals (Tofield, 1975, 1976; Forsyth, 1980). On the other hand, MgO and Mg–O bonds are considered to be completely ionic or nearly so. Hence, covalency effects in Fe–O bonds as opposed to the Mg–O bonds have to be seriously considered in ferromagnesian silicates. The Mössbauer isomer shifts of Fe^{2+} in different crystallographic sites offer a measure of the s-electron density at the nucleus, i.e., a degree of the covalency of the Fe–O bonds in these sites. Distinct differences in the isomer-shift values for Fe^{2+} in different crystallographic sites in a crystal indicate different degrees of covalency in the Fe–O bonds at these sites. This difference in the degree of covalency is a strong stabilization factor in the site partitioning of Fe^{2+} as opposed to Mg^{2+} in ferromagnesian silicates (Ghose, 1961, 1965a; Burnham et al., 1971; Hafner and Ghose, 1971; Ghose and Wan, 1974; Ghose et al., 1975, 1976). Finally the small ionic size difference between Mg^{2+} (0.72 Å) and Fe^{2+} (0.78 Å) (Shannon and Prewitt, 1969) may be a factor in Fe–Mg site partitioning in ferromagnesian silicates (Ghose, 1962).

Crystal Chemistry of Ferromagnesian Pyroxenes, $(Mg, Fe)_2Si_2O_6$

General Considerations

The most common ferromagnesian pyroxenes are the orthopyroxenes $(Mg, Fe)_2Si_2O_6$, which crystallize in the space group $Pbca$, and span the entire range of Fe–Mg compositions. Magnesium-rich pyroxenes, containing up to 13 mol% $Fe_2Si_2O_6$, crystallize in the space group $Pbcn$ at high temperatures; these are known as protopyroxenes. The low calcium clinopyroxenes (clinohypersthene and pigeonite with up to 15 mol% $Ca_2Si_2O_6$ component) are metastable at room temperature. They crystallize in the space group $P2_1/c$ at low temperature, and show a rapid reversible phase transition at high temperature to $C2/c$. The high calcium pyroxenes crystallize in the space group

Table 1. Unit cell dimensions and space groups of ferromagnesian pyroxenes.

Name	Chemical Composition	a (Å)	b (Å)	c (Å)	β (°)	Space Group
Orthoenstatite[a]	$Mg_2Si_2O_6$	18.225(2)	8.814(1)	5.174(1)	—	$Pbca$
Hypersthene[b]	$Fe_{0.79}Mg_{1.21}Si_2O_6$	18.313(3)	8.912(2)	5.210(1)	—	$Pbca$
Orthoferrosilite[c]	$Fe_2Si_2O_6$	18.431(4)	9.080(2)	5.238(1)	—	$Pbca$
Protoenstatite[d]	$Mg_2Si_2O_6$	9.252(2)	8.740(1)	5.316(1)	—	$Pbcn$
Low clinoenstatite[a]	$Mg_2Si_2O_6$	9.607(2)	8.817(1)	5.169( )	108.34	$P2_1/c$
Low clinohypersthene[e]	$Ca_{0.03}Mg_{0.62}Fe_{1.34}Si_2O_6$, RT	9.691(3)	8.993(3)	5.231(2)	108.61(2)	$P2_1/c$
High clinohypersthene[e]	$Ca_{0.03}Mg_{0.62}Fe_{1.34}Si_2O_6$, 760°C	9.851(4)	9.045(5)	5.326(3)	110.05(3)	$C2/c$
Low clinoferrosilite[c]	$Fe_2Si_2O_6$	9.709(1)	9.087(1)	5.228(1)	108.43	$P2_1/c$
Low pigeonite[f]	$(Ca_{0.18}Mg_{0.78}Fe_{1.04})Si_2O_6$, 24°C	9.706(2)	8.950(1)	5.246(1)	108.59	$P2_1/c$
High pigeonite[f]	$(Ca_{0.18}Mg_{0.78}Fe_{1.04})Si_2O_6$, 960°C	9.858(4)	9.053(3)	5.329(3)	109.42(1)	$C2/c$

[a] Stephenson, Sclar, and Smith (1966).
[b] Ghose et al. (1975).
[c] Burnham (1966).
[d] Smyth (1971).
[e] Smyth (1974).
[f] Brown et al. (1972).

$C2/c$. The unit cell dimensions of $Mg_2Si_2O_6$ and $Fe_2Si_2O_6$ polymorphs and low calcium pyroxenes are shown in Table 1.

The crystal structures of all pyroxene polymorphs are characterized by corner-sharing single silicate chains and bands of (Fe, Mg)–O octahedra. The silicate chains are packed together in slabs parallel to the (100) plane. The apical oxygens of the silicate tetrahedra point in opposite directions normal to the (100) plane. The surface layers of these silicate chain slabs consist of oxygen atoms in imperfect hexagonal close packing. Between two such surface layers belonging to two adjacent silicate slabs occur octahedral holes, which contain the (Mg, Fe) ions. Successive silicate slabs may be related to each other by a glide plane parallel to (100), the glide component being $\frac{1}{3}c$ parallel to [001]. The glide component may operate in the $+$ or $-$ z-direction. Since the glide plane runs through the (Mg, Fe)–O octahedra, successive silicate chains may be stacked in two ways. In protoenstatite the displacement is alternatively in $+$ and $-$ directions $(+ - + -)$; in clinohypersthene and pigeonite (both $C2/c$ and $P2_1/c$ space groups) always in the same direction $(+ + + +$ or $- - - -)$; and in orthopyroxene in alternate pairs, two in one way and two in the other way $(+ + - -)$ (Brown, Morimoto, and Smith, 1961). High-resolution electron microscopy of orthopyroxenes indicate that fine lamellae of clinopyroxene quite commonly occur as mistakes in the stacking sequence of successive silicate layers parallel to (100) (Iijima and Buseck, 1975; Buseck and Iijima, 1975).

Following Thompson (1970), Papike et al. (1973) have considered the polymorphism in pyroxenes in terms of the stacking sequences of the close-packed oxygen layers, which result in two different types of tetrahedral rotations. In O and S rotations, the tetrahedra in the chains rotate so that the triangular faces are *oppositely* (O) or *similarly* (S) directed to the triangular faces of the octahedral strip (both parallel to the (100) plane) to which they are linked. The *Pbca* structure has only O-rotations, whereas the $P2_1/c$ structure has both O- and S-rotations. The basic structural unit of pyroxenes can be considered as an I-beam in allusion to the appearance of the tetrahedral–octahedral–tetrahedral sandwich unit when projected down the c-axis (Fig. 2). The four known pyroxene polymorphs can be schematically represented by means of these I-beam diagrams (Fig. 3). The "tilt" of the octahedral faces are indicated as $+$ or $-$; A and B in $P2_1/c$ and *Pbca* structures indicate that the tetrahedral chains above and below the octahedral strip are symmetrically different. The O and S indicate the O- and S-rotations of the tetrahedra with respect to the octahedral strip. The stacking sequence of the silicate layers with respect to the octahedral cations (called octahedral stacking sequence by Papike et al. (1973)) within each polymorph can be determined by noting the "tilt" "$+$" or "$-$" of the octahedra within the I-beams along the a-axis.

In all pyroxene polymorphs, there are two distinct octahedral sites, M1 and M2, which occur in the ratio $1:1$. Their multiplicities per unit-cell and point symmetries within the different polymorphs are shown in Table 2. The M2 site is more distorted than M1, and strongly prefers Fe^{2+}. The strong preference

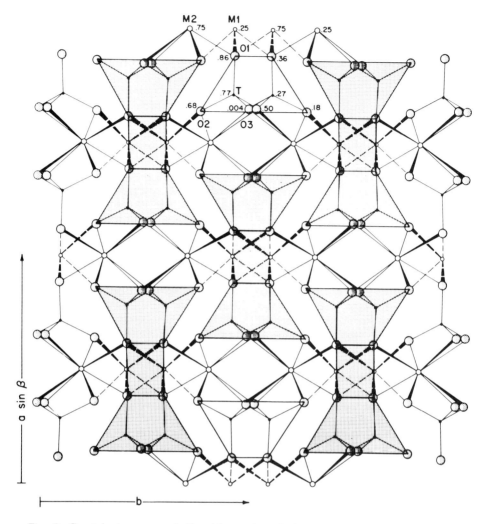

Fig. 2. Crystal structure of diopside projected down the *c* axis, showing the tetrahedral–octahedral–tetrahedral sandwich unit (I-beam) (after Cameron and Papike, 1981).

of Fe^{2+} for the distorted octahedral M2 site in orthopyroxene and pigeonite were first observed by Ghose (1960, 1962, 1965a) and Morimoto *et al.* (1960) respectively.

Orthopyroxenes

The orthopyroxene crystal structure is characterized by two crystallographically distinct single silicate chains, A and B, the latter being more distorted

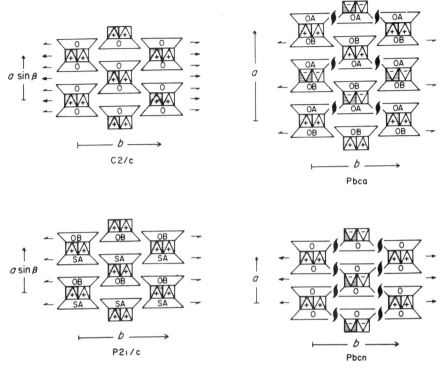

Fig. 3. I-beam representation of the four pyroxene polymorphs (after *Papike et al.* 1973).

Table 2. Multiplicities and site symmetries of the octahedral cation sites in pyroxene space groups.

Space Group	Site	Multiplicity	Point Symmetry
Pbca	M1	8	1
	M2	8	1
Pbcn	M1	4	2
	M2	4	2
$P2_1/c$	M1	4	1
	M2	4	1
$C2/c$	M1	4	2
	M2	4	2

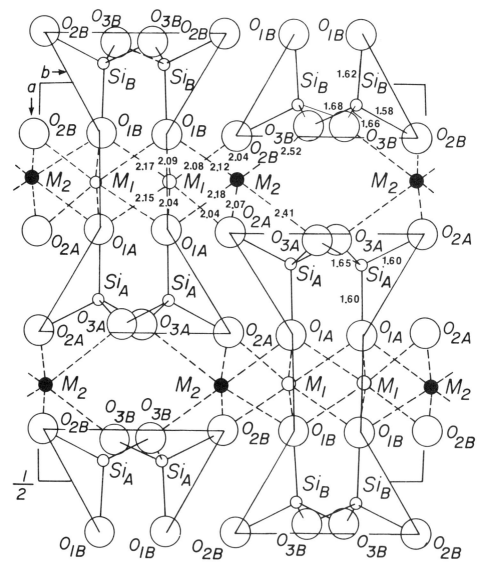

Fig. 4. Projection of the orthopyroxene structure down the *c* axis (modified from Ghose, 1965a).

than the former (Fig. 4). The M1 and M2 octahedra form a zigzag edge-sharing two-octahedral band running parallel to the *c*-axis. The M1 octahedron is nearly regular; the average M1–O distance ranges from 2.075 Å in enstatite to 2.135 Å in ferrosilite. The M1 site occurs in the interior of the double octahedral band and corresponds to the Mg-position in diopside. The M1 site can accept small divalent or trivalent cations such as Mg^{2+}, Fe^{2+}, Ni^{2+}, Co^{2+}, Al^{3+}, Fe^{3+}, etc. Four of the oxygen atoms (01A, 01B) bonded to

M1 are charge balanced according to Pauling's electrostatic valence rule (Pauling, 1960), because each of them is bonded to one silicon and three divalent cations each. The two other oxygen atoms (O2B) are slightly charge deficient because each of them is bonded to one silicon and two divalent cations, one M1 and one M2. The M2 position lies on the exterior of the octahedral band and corresponds to the Ca position in diopside. It plays a special role in the pyroxene structure insofar as it links one I-beam with two adjacent I-beams (Fig. 4). In the M2 octahedron four oxygen atoms are in their regular octahedral positions (O1A, O1B, O2A, O2B) with average M-O distances of 2.042 Å in enstatite and 2.076 Å in ferrosilite; two others (O3A, O3B) are much farther away (M2–O3A 2.290 Å (En)-2.460 Å (Fs); M2-O3B 2.447 Å (En) 2.600 Å (Fs)) (Table 3). The latter two oxygen atoms are further shared between two silicon atoms each; hence, these two oxygen atoms are overbonded. In all orthopyroxene structures the two charge deficient oxygen atoms, O2A and O2B, are very closely bonded to M2. The M2-O2A and M2-O2B distances are 2.032 and 1.992 Å in enstatite and 2.024 and 1.994 Å in ferrosilite respectively. These distances are very similar in clinoenstatite and clinoferrosilite and are apparently independent of chemical substitution of Mg^{2+} by Fe^{2+} and temperature (Smyth, 1973). They are considerably shorter

Table 3. Bond lengths (Å) within the M1 and M2 polyhedra in orthopyroxenes.

M1 Site Occupancy	Enstatite[a] Mg^{2+}:1.00	Hypersthene[b] Fe^{2+}:0.190(1) Mg:0.810	Eulite[c] Fe:0.743(3) Mg:0.257	Ferrosilite[d] Fe^{2+}:1.00
M1-O1A	2.151(1)	2.156(1)	2.178(2)	2.195(3)
-O1A[a]	2.028(1)	2.040(1)	2.076(2)	2.085(3)
-O1B	2.171(1)	2.177(1)	2.186(2)	2.194(3)
-O1B[a]	2.065(1)	2.072(1)	2.106(2)	2.124(3)
-O2A	2.006(1)	2.034(1)	2.086(2)	2.090(3)
-O2B	2.047(1)	2.073(1)	2.106(2)	2.124(3)
Mean	2.078	2.092	2.123	2.135

M2 Site Occupancy	Mg^{2+}:1.00	Fe^{2+}:0.604 (1) Mg^{2+}:0.396	Fe^{2+}:0.957 Ca:0.043	Fe^{2+}:1.00
M2-O1A	2.089(1)	2.141(2)	2.161(2)	2.158(3)
-O1B	2.056(1)	2.102(1)	2.130(2)	2.129(3)
-O2A	2.032(1)	2.043(1)	2.035(2)	2.024(3)
-O2B	1.992(1)	1.990(1)	1.997(2)	1.994(3)
-O3A	2.290(1)	2.353(1)	2.444(2)	2.460(3)
-O3B	2.447(1)	2.510(1)	2.576(2)	2.600(3)
Mean	2.151	2.190	2.224	2.228

[a] Ghose *et al.* (1980), synthetic.
[b] Ghose *et al.* (1975) and Ghose *et al.* (1982), synthetic $Fe_{0.79}Mg_{1.21}Si_2O_6$.
[c] Burnham *et al.* (1971).
[d] Sueno *et al.* (1976), synthetic $Fe_2Si_2O_6$.

than the sum of the ionic radii of Mg^{2+} and O^{2-} (2.12 Å) and Fe^{2+} and O^{2-} (2.18 Å) (Shannon and Prewitt, 1970). Ghose (1965a) considered these bonds to be highly covalent. Note that the degree of deviation of the observed Fe^{2+} –O bond length from the sum of the corresponding ionic radii is considerably more than that of the Mg^{2+} –O bond length at the M2 site. The higher degree of covalent bonding of Fe^{2+} at the M2 site compared to the M1 site has been confirmed by the smaller Mössbauer isomer shift of Fe^{2+} at the M2 site (Evans, Ghose, and Hafner, 1967; Burnham et al., 1971). Ghose (1965a) advanced the idea that the greater degree of covalency of the Fe^{2+} –O bond at the M2 site is the driving force for the strong Fe^{2+} site preference for the M2 site in orthopyroxenes. Burns (1970) determined the difference of the crystal field stabilization energies (CFSE) for Fe^{2+} at the M1 and M2 sites from optical absorption spectra to be 0.2–0.3 kcal/mole. However, the Gibbs free energy difference of intracrystalline Fe^{2+} –Mg^{2+} exchange has been estimated to be 3.6 and 2.5 kcal/mole based on the ideal and regular solution models for Fe–Mg mixing at the M1 and M2 sites respectively (Virgo and Hafner, 1969; Saxena and Ghose, 1971). Hence, the difference in CFSE determined by Burns (1970) is completely inadequate to explain the observed site preference of Fe^{2+} in orthopyroxenes (Ghose, 1970; O'Nions and Smith, 1973). The inadequacy of the crystal field theory stems from the fact that it ignores completely the possibility of covalent bonding between the ligands and the transitional metal ions. O'Nions and Smith (1973) have presented a schematic molecular orbital model for Fe^{2+} at the M1 and M2 sites. However, what is urgently needed is an ab initio molecular orbital calculation based on a large enough cluster of M1 and M2 octahedra and silicate tetrahedra to determine the relative energies of Mg^{2+} and Fe^{2+} at the M1 and M2 sites.

Low Calcium Clinopyroxenes: The $P2_1/c$ to $C2/c$ Transition

The high calcium pyroxenes (diopside, augite) crystallize in the space group $C2/c$, whereas the low calcium clinopyroxenes (clinohypersthene, pigeonite) crystallize in the space group $P2_1/c$ at low temperature and $C2/c$ at high temperature (Morimoto, 1956; Bown and Gay, 1957). In high calcium ($C2/c$) clinopyroxenes, there is only one crystallographically distinct silicate chain and the M2 site is eight-coordinated, which is mostly occupied by calcium. In low calcium $P2_1/c$ clinopyroxenes, the twofold rotation axis is lost and, therefore, the silicate chains on either side of the octahedral cations are no longer equivalent. Like in orthopyroxenes, the more distorted silicate chain is called B and the less distorted one A. The M2 site is six-coordinated and is distorted in the same way as in orthopyroxenes (Table 4).

The $P2_1/c$ to $C2/c$ transition temperature in low Ca pyroxenes depends on the Ca and Fe content, being about 1000°C for the magnesium-rich pigeonites and about 500°C for very iron-rich pigeonites (Prewitt et al., 1971). The

Table 4. Bond lengths (Å) within the M1 and M2 polyhedra in low-Ca clinopyroxenes.

Site Occupancy (M1)	Clinoenstatite[a] Mg:1.00	Clinohypersthene[b] Mg:0.497(4) Fe:0.503	Clinoferrosilite[c] Fe:1.00
M1-O1A	2.142(1)	2.155(3)	2.199(7)
-O1A'	2.032(1)	2.074(2)	2.102(5)
-O1B	2.178(1)	2.187(3)	2.200(6)
-O1B'	2.067(1)	2.088(2)	2.126(5)
-O2A	2.006(1)	2.065(2)	2.082(6)
-O2B	2.042(1)	2.094(2)	2.113(6)
Mean	2.078	2.111	2.137
Site Occupancy (M2)	Mg:1.00	Mg:0.134 Fe:0.834 Ca:0.032	Fe:1.00
M2-O1A	2.090(1)	2.156(3)	2.159(7)
-O1B	2.053(1)	2.121(2)	2.136(6)
-O2A	2.034(1)	2.029(2)	2.032(5)
-O2B	1.987(1)	2.002(2)	1.985(5)
-O3A	2.279(1)	2.394(3)	2.444(5)
-O3B	2.412(1)	2.543(3)	2.587(6)
-O3B'			
Mean of 6	2.143	2.208	2.224

[a] Ohashi and Finger (1976), synthetic $Mg_2Si_2O_6$.
[b] Smyth (1974), $Ca_{0.03}Mg_{0.62}Fe_{1.34}Si_2O_6$.
[c] Burnham (1966), synthetic $Fe_2Si_2O_6$.

reversible structural changes accompanying the phase transition in clinohypersthene (Smyth and Burnham, 1972; Smyth, 1974) and pigeonite (Brown *et al.*, 1972) are primarily caused by the straightening out of the two crystallographically distinct silicate chains, such that they become crystallographically identical above the phase transition (Fig. 5). Simultaneously, the coordination of the M1 and M2 sites becomes more symmetrical, each of them possessing a twofold rotation axis as point symmetry above the transition. Thus, in $C2/c$ pyroxenes, there is only one set of M-O bond distances as compared to two sets in $P2_1/c$ pyroxenes in each of these coordination polyhedra. The characteristics of the M2 coordination remain essentially the same above the transition, namely, four oxygen atoms close to their regular octahedral positions and two farther oxygen atoms shared by two silicon atoms each complete the distorted octahedron (Table 5). Hence, the site preference of Fe^{2+} for the M2 site is not expected to be affected much by the changes in the crystal structure due to the phase transition. In pigeonites, where the M2 site is partially blocked by Ca, the site preference of Fe^{2+} for the M2 site will be a function of the calcium content as well because of the stereochemical effect of the Ca ions on the coordination of the M2 site.

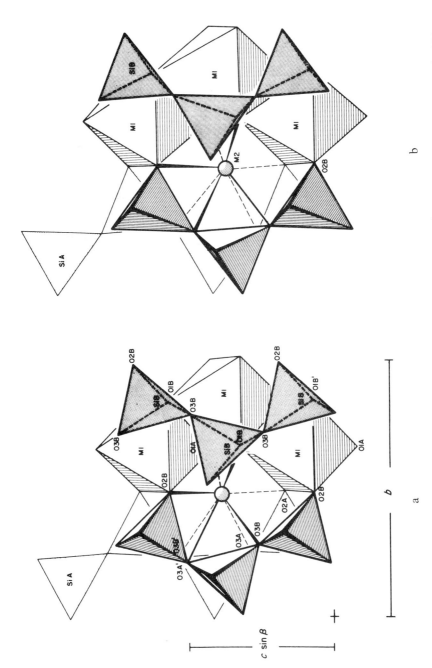

Fig. 5. The crystal structures of pigeonite (a) below and (b) above the $P2_1/c \rightarrow C2/c$ transition (after Brown *et al.* 1972).

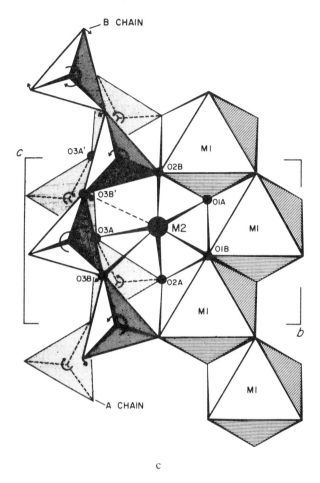

c

Fig. 5. (c) Atomic movements involved in the low to high pigeonite transition.

Protoenstatite

Although no Mg–Fe determination in protoenstatite has yet been undertaken because it is only stable at about 1000°C, it is worthwhile to consider the crystal chemistry of this phase with respect to the possible site preference of Fe^{2+} between the two octahedral sites M1 and M2. Two refinements of the protoenstatite structure have been undertaken at high temperatures (Sadanaga *et al.*, 1969; Smyth, 1971). The accuracy of these determinations is quite low and the bond distances are not reliable. We base our discussion on an accurate determination of the structure of a Li–Sc–Mg protopyroxene with the composition $Li_{0.1}Sc_{0.1}Mg_{1.8}Si_2O_6$, which is stable at room temperature (Smyth and Ito, 1977). The lithium and scandium atoms are ordered in the M2 and M1 sites respectively.

As in the other pyroxene polymorphs, in protoenstatite the M1 and M2 octahedra form an edge-sharing zigzag two-octahedral band (Fig. 6). The M2

Table 5. Bond lengths in low Ca-clinopyroxenes below and above the $P2_1/c$ → $C2/c$ phase transition.

	Mull Pigeonite[a]		Clinohypersthene[b]	
	24°C	960°C	700°C	760°C
M1 Site	Mg: 0.702(5)	Mg: 0.613(14)	Mg: 0.524	Mg: 0.521
Occupancy	Fe: 0.298	Fe: 0.387	Fe: 0.476	Fe: 0.479
M1-O1A	2.159(3)	2.229(7)	2.235(9)	2.257(8)
-O1A′	2.054(3)	2.062(6)	2.039(8)	2.081(8)
-O1B	2.173(3)	2.229(7)	2.217(9)	2.257(8)
-O1B′	2.074(3)	2.062(6)	2.085(8)	2.081(8)
-O2A	2.048(3)	2.078(8)	2.070(8)	2.090(8)
-O2B	2.075(3)	2.078(8)	2.136(8)	2.090(8)
Mean	2.097	2.123	2.134	2.143
	Mg: 0.078	Mg: 0.167	Mg: 0.112	Mg: 0.115
M2 Site	Fe: 0.742	Fe: 0.653	Fe: 0.857	Fe: 0.853
Occupancy	Ca: 0.180	Ca: 0.180	Ca: 0.032	Ca: 0.032
M2-O1A	2.168(3)	2.176(7)	2.108(9)	2.146(8)
-O1B	2.140(3)	2.176(7)	2.225(10)	2.146(8)
-O2A	2.071(3)	2.081(7)	1.987(10)	1.997(8)
-O2B	2.035(4)	2.081(7)	1.990(9)	1.997(8)
-O3A	2.460(4)	2.656(8)	2.504(11)	2.682(9)
-O3B	2.663(4)	3.173(8)	2.833(12)	3.245(9)
-O3B′	2.935(4)	2.656(8)	2.926	2.682(9)
-O3A′	3.407(4)	3.173(8)	—	—
Mean of 6	2.256	2.304	2.275	2.368
Mean of 8	2.485	2.521	2.290	2.275

[a] Brown *et al.* (1972).
[b] Smyth (1974).

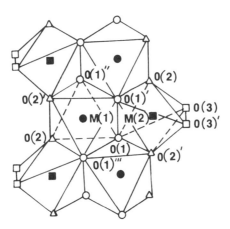

Fig. 6. Octahedral band of M1 and M2 octahedra in protoenstatite (after Smyth, 1971).

Table 6. M-O bond distances (Å) in protoenstatite, $Mg_2Si_2O_6$ at 1100°C and protopyroxene, $Li_{0.1}Sc_{0.1}Mg_{1.8}Si_2O_6$ at RT.

	Protoenstatite[a]	Protopyroxene[b]
M1 Site Occupancy	Mg: 1.0	Mg: 0.9 Sc: 0.1
(×2) M1-O1	2.15(2)	2.085(4)
(×2) -O1′	2.30(2)	2.199(4)
(×2) -O2	2.10(2)	2.003(4)
Mean	2.18	2.096
M2 Site Occupancy	Mg: 1.0	Mg: 0.9 Li: 0.1
(×2) M2-O1	2.07(2)	2.086(4)
(×2) -O2	2.06(2)	2.078(4)
(×2) -O3	2.28(2)	2.393(4)
Mean	2.12	2.186

[a] Protoenstatite, $Mg_2Si_2O_6$: Smyth (1971).
[b] Protopyroxene, $Li_{0.1}Sc_{0.1}Mg_{1.8}Si_2O_6$: Smyth and Ito (1977).

octahedra occur on the exterior of this band and are considerably more distorted than the M1 octahedra (Table 6). The M2 coordination is sixfold (4 + 2), where two farthest oxygen atoms (O3) are bonded to two silicon atoms each and have their charges satisfied. There is a possibility of considerable covalent bonding of M2 with O2, which is slightly charge deficient, being bonded to one silicon and two divalent cations. Furthermore, the M2 octahedron is much larger (average M-O distance = 2.186 Å) than the M1 octahedron (average M-O distance = 2.096 Å). On the basis of these crystal chemical features, we predict that Fe^{2+} will be preferred in the M2 site in protoenstatite, although within its stability field the degree of the Fe^{2+} site preference may not be as strong as in orthopyroxene.

Crystal Chemistry of Ferromagnesian Amphiboles, $(Mg, Fe)_7Si_8O_{22}(OH)_2$, and Other Pyriboles

General Considerations

There are many similarities between pyroxene and amphibole structures. The most common iron-rich ferromagnesian amphiboles cummingtonite and

Table 7. Unit cells and space groups of ferromagnesian amphiboles and new pyriboles.

Name and Chemical Composition	a (Å)	b (Å)	c (Å)	β(°)	Space Group
Anthophyllite[a] (Mg$_{5.53}$Fe$_{1.47}$)Si$_8$O$_{22}$(OH)$_2$	18.560(3)	18.013(2)	5.282(1)	—	Pnma
P-Cummingtonite[b] (Ca$_{0.07}$Mn$_{0.02}$Mg$_{6.00}$Fe$_{0.97}$)Si$_8$O$_{22}$(OH)$_2$	9.485(9)	18.001(13)	5.292(4)	101.927(7)	P2$_1$/m
C-Cummingtonite[c] (Ca$_{0.35}$Mn$_{0.17}$Mg$_{4.05}$Fe$_{2.50}$)(Si$_{7.9}$Al$_{0.1}$)$_8$O$_{22}$(OH)$_2$	9.516	18.139	5.311	102.11(1)	C2/m
Grunerite[d] (Ca$_{0.06}$Mn$_{0.05}$Mg$_{0.77}$Fe$_{6.14}$)Si$_8$O$_{22}$(OH,F)$_2$	9.564(1)	18.393(2)	5.339(1)	101.89(1)	C2/m
Protoamphibole[e] (Na$_{0.03}$Li$_{1.20}$Mg$_{6.44}$)(Si$_{7.84}$Al$_{0.04}$)O$_{21.71}$(F, OH)$_2$	9.330	17.879	5.288	—	Pnmn
Jimthompsonite[f] (Mg, Fe)$_{10}$Si$_{12}$O$_{32}$(OH)$_4$	18.626(1)	27.230(1)	5.297(1)	—	Pbca
Clinojimthompsonite[f] (Mg, Fe)$_{10}$Si$_{12}$O$_{32}$(OH)$_4$	9.874(4)	27.24(3)	5.316(3)	109.47(3)	C2/c
Chesterite[f] (Mg, Fe)$_{17}$Si$_{20}$O$_{54}$(OH)$_6$	18.614(1)	45.306(1)	5.297(1)	—	A2$_1$/m
Monoclinic chesterite (Mg, Fe)$_{17}$Si$_{20}$O$_{54}$(OH)$_6$	9.867	45.310	5.292	109.7	A2/m Am or A2

[a] Finger (1970a).
[b] Rice et al. (1974).
[c] Viswanathan and Ghose (1965).
[d] Finger (1969).
[e] Gibbs (1969).
[f] Veblen and Burnham (1978a).

grunerite are monoclinic, which crystallize in the space group $C2/m$ (or $I2/m$). The magnesium-rich amphiboles may crystallize metastably as a monoclinic phase (space group $P2_1/m$) or stably as an orthorhombic phase, anthophyllite, with the space group $Pnma$. The protoamphibole, with the space group $Pnmn$ is only known as a synthetic product (Gibbs, 1969). The unit cells and space groups of amphiboles and other pyriboles are shown in Table 7.

The structures of all ferromagnesian amphiboles are based on double silicate chains, two of which sandwich between them Fe^{2+} and Mg^{2+} cations in octahedral coordination. The resulting octahedral bands are alternately three and four octahedra wide and the silicate chains occur back to back, both components running parallel to the c-axis. As in pyroxenes, when projected down the c-axis, the tetrahedral–octahedral–tetrahedral unit (T-O-T unit) has the appearance of an I-beam. The different octahedral stacking sequences give rise to the four different amphibole polymorphs (Fig. 7). Note that each pyroxene polymorph has an amphibole analog, the only difference between the corresponding polymorphs being the width of the I-beams; the amphibole I-beams are roughly twice in width compared to those in pyroxenes. The new

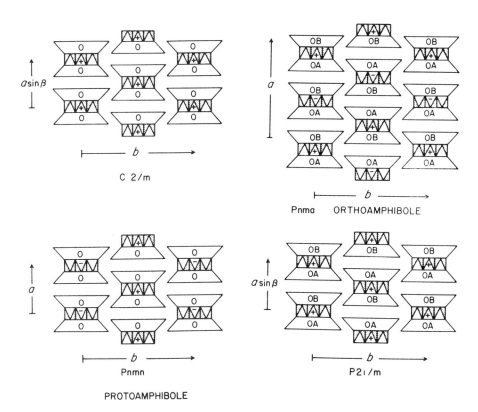

Fig. 7. I-beam representation of four amphibole polymorphs (after Papike and Ross, 1970).

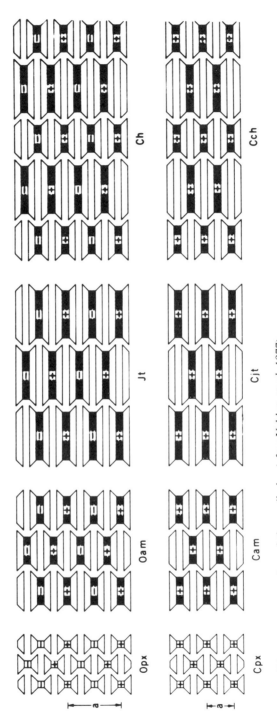

Fig. 8. I-beam representation of the pyriboles (after Veblen *et al.* 1977).

ferromagnesian pyribole structures, jimthompsonite and clinojimthompsonite, based on triple silicate chains and chesterite and its monoclinic analog based on alternating double and triple silicate chains can also be represented by similar I-beam diagrams (Veblen *et al.*, 1977) (Fig. 8). Note that all the orthopyriboles (orthopyroxene, orthoamphibole, jimthompsonite, and

Table 8. Multiplicities and point symmetrics of octahedral cation sites in amphiboles and other pyriboles.

Name	Space Group	Site	Multiplicity	Point Symmetry
Anthophyllite	*Pnma*	M1	8	1
		M2	8	1
		M3	4	*m*
		M4	8	1
P-Cummingtonite	$P2_1/m$	M1	4	1
		M2	4	1
		M3	2	*m*
		M4	4	1
C-Cummingtonite	$C2/m$	M1	4	2
		M2	4	2
		M3	2	$2/m$
		M4	4	2
Protoamphibole	*Pnmn*	M1	4	2
		M2	4	2
		M3	2	$2/m$
		M4	4	2
Jimthompsonite	*Pbca*	M1	8	1
		M2	8	1
		M3	8	1
		M4	8	1
		M5	8	1
Clinojimthompsonite	$C2/c$	M1	4	
		M2	4	2
		M3	4	2
		M4	4	2
		M5	4	2
Chesterite	$A2_1m_a$	MT1	8	1
		MT2	8	1
		MT3	8	1
		MT4	8	1
		MT5	8	1
		MD1	8	1
		MD2	8	1
		MD3	4	*m*
		MD4	8	1

chesterite) have the same octahedral stacking sequence $(+ + - -)$, whereas all the clinopyriboles (clinopyroxene, clinoamphibole, clinojimthompsonite, and the monoclinic analog of chesterite) have the same octahedral stacking sequence $(+ + + +)$ or $(- - - -)$. The chemical and structural analogies of the different pyroxene, amphibole, and other pyribole phases are not fortuitous, but depend strictly on the similarity of the structural components and the way they are linked within each structure. One of the most significant structural consequences of these phenomena is the fact that irrespective of the pyribole, there are two types of octahedral sites within the octahedral band: regular octahedral sites within the interior of the band (M1 in pyroxene; M1, M2, M3 in amphibole; M1, M2, M3, M4 in jimthompsonite) and a highly distorted octahedral site on the exterior of the band (M2 in pyroxene; M4 in amphibole; M5 in jimthompsonite). It is always the highly distorted octahedral site, which strongly prefers Fe^{2+} over Mg^{2+}. The chemical bonding characteristics of the $Mg^{2+}-Fe^{2+}$ cations in this site are closely comparable in all the structural homologs. The multiplicities and site symmetries of M-cation sites in amphibole and other pyriboles are listed in Table 8. In all amphibole polymorphs the M2 and M4 sites have six oxygen atoms each as ligands, whereas the M1 and M3 sites have four oxygen atoms and two (OH) ions each as ligands. The strong site preference of Fe^{2+} for the M4 site in cummingtonite was first determined by Ghose (1961), who predicted a similar site preference in anthophyllite (Ghose, 1965b), subsequently confirmed by Bancroft et al. (1966) and Finger (1970a). The strong site preference of Fe^{2+} for the M5 site in jimthompsonite and clinojimthompsonite and the MD4 and MT5 sites in chesterite were determined by Veblen and Burnham (1978b), who also determined their crystal structures.

Anthophyllite

Anthophyllites are magnesium-rich and may contain up to about 27 mol% of the Fe component (Rabbit, 1948), whereas cummingtonites including "primitive" cummingtonites are mostly Fe-rich, but extend at least to 86 mol% of the Mg component (Rice et al., 1974) with a substantial range of compositional overlap with anthophyllites. In comparison with cummingtonites, anthophyllites usually contain more aluminum and less calcium. The very aluminum-rich orthoamphibole, gedrite, which also contains considerable amounts of sodium in the A-site is not strictly a ferromagnesian amphibole and will not be treated here (Papike and Ross, 1970). The stability relations between the orthoamphiboles and clinoamphiboles have not been clearly established. A low-pressure temperature–composition diagram for the cummingtonite to anthophyllite transformation is given by Mueller (1973), based on the common occurrence of cummingtonite in low to intermediate grade metamorphic iron formation, where anthophyllite is considered to be the high-temperature phase separated from cummingtonite by a narrow region of two-phase assemblage. On the other hand, based on crystallographic studies of exsolution and phase transition in cummingtonites, Ross, Papike and Shaw (1969) and Prewitt, Papike and Ross (1970) postulated that "a $C2/m$ amphibole containing Ca and Mg (or Fe^{2+}) at the M4 site, which is stable at high temperature with respect to anthophyllite, on cooling unmixes sufficient calcic clinoamphibole to place its composition in the stability field of anthophyllite." Because of kinetic barriers quite often inversion is prevented and the $P2_1/m$ cummingtonite as a metastable phase is obtained. Evans et al. (1974) studied a primitive cummingtonite from Ticino, Switzerland which showed partial inversion to anthophyllite and a small amount of exsolution of tremolite. This situation is closely comparable to the case of pigeonite, which on cooling exsolves augite lamellae and then inverts into orthopyroxene (cf. Ghose et al., 1973).

The orthoamphibole structure is illustrated in Fig. 9. As in orthopyroxenes, it is characterized by two crystallographically distinct silicate chains, A and B, the latter being more distorted than the former. The M1, M2 and M3 octahedra have average M-O distances of 2.084, 2.076, and 2.070 Å, respectively. The M4 cation is effectively five-coordinated, where four oxygen atoms are at distances of 2.081 ± 0.085 Å, with a fifth one at 2.387 Å; two further oxygens at 2.865 and 2.867 Å may be considered as part of the coordination sphere (Finger, 1970a) (Table 9). In conformity with the prediction made by Ghose (1965b), in anthophyllite the M1 and M3 sites have nearly the same (Mg, Fe^{2+}) content, whereas the M2 site is enriched in Mg and the M4 site by Fe^{2+} as determined by X-ray diffraction (Finger, 1970a). This scheme of cation distribution in anthophyllite has been confirmed by infrared and Mössbauer resonance spectroscopy (Bancroft et al., 1966; Seifert and Virgo, 1974).

Primitive Cummingtonite and the $P2_1/m \rightarrow C2/m$ Phase Transition

As in anthophyllite, there are two crystallographically distinct double silicate chains in primitive cummingtonite to allow for the denser packing around the small M4 cation (Fig. 10). In a manganese-poor primitive cummingtonite studied by Ghose and Wan (1982) the coordination of the M1, M2, and M3 sites are regular octahedral with average M-O distances of 2.084, 2.079, and 2.074 Å, respectively. The M4 site is highly distorted with essentially $(4 + 1)$ coordination; four oxygen atoms are at an average distance of 2.070 ± 0.074 Å with a fifth one at 2.415 Å; two further oxygens at 2.868 and 2.889 Å may be considered as a part of the coordination sphere; this situation is very similar to that found in anthophyllite. This fivefold coordination of the M4 cation is distinctly different from that found in C-centered cummingtonite, where the M4 coordination is sixfold $(4 + 2)$ (Table 9). In both anthophyllite and primitive cummingtonite, the $(4 + 1)$ coordination of M4 is accomplished

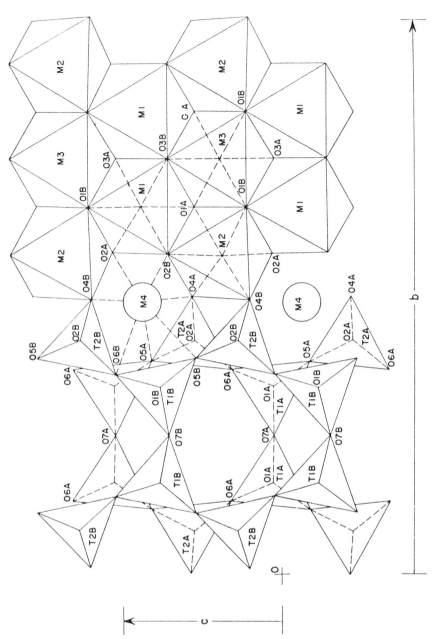

Fig. 9. Projection of the anthophyllite crystal structure down a^* (after Finger, 1970a).

Table 9. M-O bond lengths in anthophyllite, primitive and C-centered cummingtonites.

	Anthophyllite[a]	Primitive Cummingtonite[b]	Cummingtonite[c]	Grunerite[d]
M1 Site Occupancy	Fe: 0.040(3) Mg: 0.960(3)	Fe: 0.002(4) Mg: 0.998	Fe: 0.16 Mg: 0.84	Fe: 0.848(8) Mg: 0.152
M1-O1A	2.062(3)	2.041(5)	2.076	2.082(5)
-O1B	2.053(4)	2.061(5)	2.076	2.082(5)
-O2A	2.112(3)	2.105(5)	2.115	2.160(4)
-O2B	2.133(3)	2.144(5)	2.115	2.160(4)
-O3A (OH)	2.082(3)	2.087(5)	2.103	2.122(4)
-O3B (OH)	2.063(3)	2.065(5)	2.103	2.122(4)
Mean	2.084	2.084	2.098	2.121
M2 Site Occupancy	Fe: 0.027(3) Mg: 0.973	Fe: 0.00 Mg: 1.00	Fe: 0.05 Mg: 0.95	Fe: 0.773(7) Mg: 0.227(7)
M2-O1A	2.138(3)	2.142(5)	2.128	2.161(5)
-O1B	2.121(3)	2.121(5)	2.128	2.161(5)
-O2A	2.067(3)	2.073(5)	2.093	2.128(4)
-O2B	2.082(3)	2.088(5)	2.093	2.128(4)
-O4A	2.010(3)	2.008(5)	2.029	2.078(4)
-O4B	2.034(3)	2.042(7)	2.029	2.078(4)
Mean	2.076	2.079	2.083	2.121
M3 Site Occupancy	Fe: 0.034(4) Mg: 0.966(4)	Fe: 0.028 Mg: 0.972	Fe: 0.16 Mg: 0.84	Fe: 0.888(12) Mg: 0.112(12)
M3-O1A (×2)	2.075(3)	2.083(5)	2.102	2.118(5)
-O1B (×2)	2.079(3)	2.083(5)	2.102	2.118(5)
-O3A (OH)	2.055(3)	2.075(5)	2.070	2.103(6)
-O3B (OH)	2.059(3)	2.038(5)	2.070	2.103(6)
Mean	2.070	2.074	2.091	2.113
M4 Site Occupancy	Fe: 0.651 Mg: 0.349	Fe: 0.432 Mg: 0.568	Fe: 0.87 Mg: 0.13	Fe: 0.985(8) Mg: 0.015(8)
M4-O2A	2.156(3)	2.133(5)	2.176	2.135(5)
-O2B	2.128(3)	2.113(5)	2.176	2.135(5)
-O4A	2.044(3)	2.039(5)	2.041	1.988(5)
-O4B	1.996(3)	1.996(5)	2.041	1.988(5)
-O5A	3.481(3)	3.469(4)	3.147	3.298(5)
-O5B	2.865(3)	2.889(4)	3.147	3.298(5)
-O6A	2.387(3)	2.415(5)	2.699	2.757(5)
-O6B	2.867(3)	2.868(5)	2.699	2.757(5)
Average of 6 (Mean)	(2.263)	(2.261)	2.305	2.293
8	(2.491)	(2.490)	2.516	2.544
Average of 5	2.142	2.139		
Average of 7	2.349	2.350		

[a] Finger (1970a).
[b] Ghose and Wan (1982).
[c] Ghose (1961).
[d] Finger (1969).

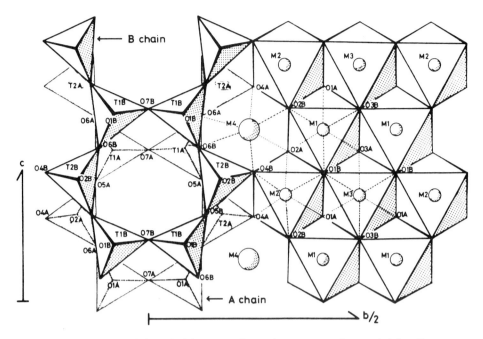

Fig. 10. Projection of the primitive cummingtonite structure down a^* (after Sueno *et al.* 1969).

by the further distortion of the B-chain to an O5-O6-O5 angle of 157° in anthophyllite and 159° in primitive cummingtonite as opposed to 172° in grunerite.

The primitive cummingtonite transforms rapidly to the C-centered cummingtonite at relatively low temperatures (Prewitt, Papike, and Ross, 1970). The transition temperature of a manganoan primitive cummingtonite with the composition $(Ca_{0.36}Na_{0.06}Mn_{0.96}Mg_{0.57})Mg_5Si_8O_{22}(OH)_2$ from Talcville, New York, is 100°C. The crystal structures of this cummingtonite below and above the transition temperature have been determined by Papike *et al.* (1969) and Sueno *et al.* (1972) respectively. The principal structural change due to the transition is the straightening out of the two tetrahedral double chains with kinking angles (O5-O6-O5) of 178.4° and 166.2° for the A- and B-chains respectively at room temperature. Above the transition, the two chains become crystallographically equivalent with a kinking angle of 173° at 275°C. This change is effected principally by changes in the coordination of the M4 cation as a function of temperature (Table 10), namely, the room temperature coordination $(4 + 1 + 1)$ becomes more symmetric $(4 + 2)$ above the transition. Because of the presence of considerable amounts of Ca and Mn in the M4 site, the M4 coordination is not as pronouncedly fivefold as it is in the nearly manganese-free primitive cummingtonite with the composition $(Ca_{0.07}Mn_{0.02}Fe_{0.97}Mg_{6.00})Si_8O_{22}(OH)_2$ investigated by Ghose and Wan (1982). Unlike the $P2_1/c \rightarrow C2/c$ transition in pigeonite, the $P2_1/m \rightarrow C2/m$ transition in cummingtonite is completely displacive and does not involve breakage

Table 10. M4-O bond lengths in primitive ($P2_1/m$) and C-centered Mn-rich cummingtonite below and above the transition temperature (100°C).

M4	Primitive Cummingtonite[a] (25 °C)	C-Centered Cummingtonite[b] (275 °C)
Site Occupancy	Ca 0.36 Na 0.06 Mn 0.96	Mg 0.57
M4-O2A	2.195 (6)	2.215 (4)
-O2B	2.208 (6)	2.215 (4)
-O4A	2.139 (8)	2.109 (5)
-O4B	2.074 (8)	2.109 (5)
-O6A	2.511 (8)	2.584 (5)
-O6B	2.650 (8)	2.584 (5)
-O5A	3.209 (6)	3.090 (5)
-O5B	2.932 (6)	3.090 (5)
Mean of 6	2.296 (3)	2.303

[a] Papike et al. (1969).
[b] Sueno et al. (1972).

and reformation of any M-O bonds. Because of low transition temperatures, this transition should have no effect on the cation distribution in cummingtonites, which presumably was quenched in at higher temperatures.

The C-Centered Cummingtonite

As noted earlier, the C-centered cummingtonites are richer in iron than anthophyllites and primitive cummingtonites. The M1, M2, and M3 octahedra are nearly regular, which increase in size approximately linearly as a function of the increasing iron content; on the other hand, the coordination of the M4 site changes in a very nonlinear fashion (Table 9). With the increasing iron content, the M4 cation is pulled closer to the four nearest oxygen atoms (O2 and O4), whereas the two farthest oxygen atoms (O6) are pushed further away (Fig. 11). Note that in spite of the increasing iron content, the M4-O4A and M4-O4B distances are nearly constant throughout the ferromagnesian amphiboles. Furthermore, in grunerite the M4-O4 distance is 1.988 Å, where the M4 site is nearly completely occupied by iron (Finger, 1969). Note that as in pyroxenes, the O4 oxygen atoms are slightly underbonded, being bonded to one silicon and two divalent cations each. The M4-O4 bonds are much shorter than the sum of the ionic radii of Fe^{2+} and O^{2-} (2.18 Å) and have considerable covalent character (Ghose, 1961). The slightly smaller Mössbauer isomer shift of Fe^{2+} at the M4 site as opposed to Fe^{2+} at the M1, M2, and M3 sites confirm the higher degree of covalent bonding of Fe^{2+} at the M4 site (Hafner and Ghose, 1971). As in pyroxenes, these covalent bonds are believed to be stablizing Fe^{2+} at this site (Ghose, 1961). As in anthophyllite and

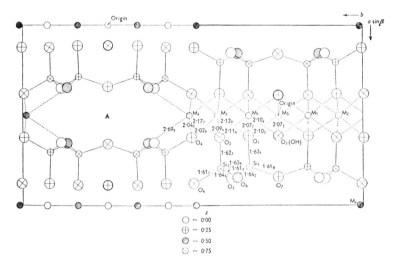

Fig. 11. The crystal structure of a *C*-centered cummingtonite projected down the *c* axis (after Ghose, 1961).

primitive cummingtonite, the M1 and M3 sites have similar (Mg, Fe^{2+}) content, whereas the M2 site strongly prefers Mg and the M4 site Fe^{2+}. This scheme of Mg–Fe distribution, first determined by X-ray diffraction (Ghose, 1961; Finger, 1969), have been subsequently confirmed by infra-red and Mössbauer spectroscopy (Bancroft *et al.*, 1967; Hafner and Ghose, 1981).

Crystal Chemistry of the New Pyriboles

Thompson (1970, 1978) considered the amphibole structure to consist of two modules, pyroxene (P) and trioctahedral mica or talc (M). The amphibole structure can then be considered (MP) and models can be constructed for structures intermediate between ferromagnesian amphibole and talc such as (MMP). Veblen and Burnham (1978b) describe the structures of new biopyriboles (MMP) and (MPMMP), two orthorhombic varieties of which have been named jimthompsonite and chesterite, respectively. The monoclinic forms of jimthompsonite (clinojimthompsonite) and chesterite exist as well.

Jimthompsonite and Chesterite

These minerals occur as fine lamellar intergrowths parallel to (010) in anthophyllite or cummingtonite. The structure of jimthompsonite, which crystallizes in the orthorhombic space group *Pbca*, is characterised by triple silicate chains, two of which sandwich between them Mg, Fe^{2+} cations in five distinct octahedral M sites (Figs. 12a, b, c). The M5 site, which occurs on the exterior of the octahedral band, is the most distorted one, where most of the

Fe^{2+} ions are concentrated. The M1, M2, M3, and M4 sites are nearly regular octahedral and are mostly occupied by Mg^{2+}. There are two crystallographically distinct triple silicate chains, the B-chain being more distorted than the A-chain. The structure of clinojimthompsonite (space group $C2/c$) consists of I-beams containing five distinct octahedral M-sites and symmetrically equivalent triple silicate chains (Figs. 12d, e). As in jimthompsonite, most of the Fe^{2+} ions are concentrated in the outer M5 site.

Chesterite crystallizes in the orthorhombic space group $A2_1ma$. Its crystal structure is characterized by an alternation of double and triple silicate chains, two each of which sandwich between them octahedral cations giving rise to two different types of I-beams (Figs. 13a, b, c). Of the two topologically distinct I-beams, one is similar to the I-beam of anthophyllite with A- and B-double silicate chains, and the other is similar to the I-beam of jimthompsonite with A- and B-triple silicate chains. The Fe^{2+} ions are

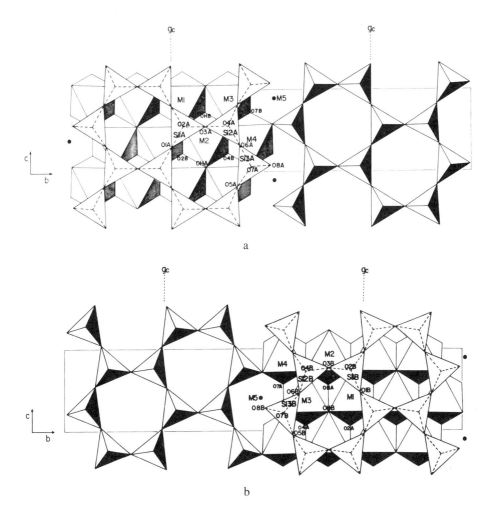

Fig. 12. (a)–(b) The crystal structure of jimthompsonite.

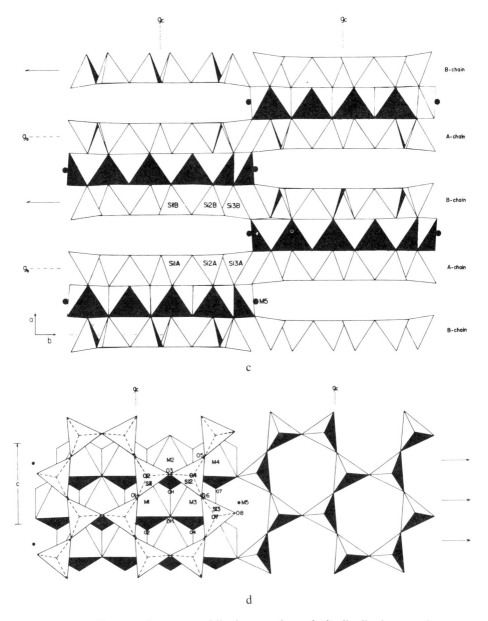

Fig. 12. (c) The crystal structure of jimthompsonite and (d) clinojimthompsonite.

concentrated in the two outer distorted M-sites, MD4 and MT5 within the anthophyllite and jimthompsonite type I-beams respectively. Note that in orthorhombic pyriboles jimthompsonite and chesterite, the outer distorted M-sites all have effectively fivefold $(4 + 1)$ coordination as in anthophyllite and primitive magnesian cummingtonite (Table 11). The coordination of the M5 site in clinojimthompsonite is sixfold $(4 + 2)$ and is comparable to that of the M4 site in C-centered cummingtonite. Furthermore, within each of these

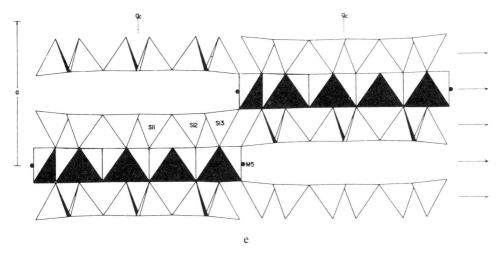

e

Fig. 12. (e) The crystal structure of clinojimthompsonite. (Figure after Veblen and Burnham, 1978b.)

outer M-polyhedra where the Fe^{2+} ions are concentrated, there are two sets of unusually short M-O bonds involving oxygen atoms, which are slightly charge deficient. Considerable covalency of the Fe^{2+}–O bonds at these sites is to be expected (cf. cummingtonite, Ghose (1961)).

The Crystal Chemistry of Olivines $(Mg, Fe)_2SiO_4$ and the Humite Group of Minerals

Olivines

The crystal structure of olivine (space group *Pbnm*) is based on hexagonal close-packed array of oxygen ions, in which one-eighth of the tetrahedral interstices are filled with silicon and one-half of the octahedral interstices by Fe^{2+}, Mg. There are two crystallographically distinct octahedral sites M1 and M2 with site multiplicities 4, 4 and point symmetries $\bar{1}$ and *m*, respectively. The M1 and M2 octahedra share edges and form a serrated band parallel to the *a*-axis (Fig. 14). Similar serrated edge-sharing octahedral chains also exist in humite minerals. These octahedral bands are cross-linked to each other by isolated $[SiO_4]$ tetrahedra. The M1 octahedron shares two of its edges with two other M1 octahedra, two M2 octahedra, and two $[SiO_4]$ tetrahedra. In contrast, the M2 octahedron shares two edges with two M1 octahedra and one with the silicate tetrahedron. Because of cation–cation repulsion, the shared edges are shorter than the nonshared edges. As a result, the M1 octahedron which shares six of its edges is more distorted than the M2 octahedron, which shares only three. Furthermore, the M1 octahedron is distinctly smaller than

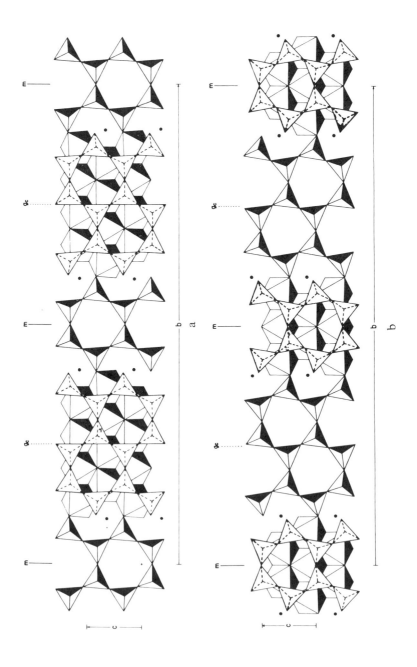

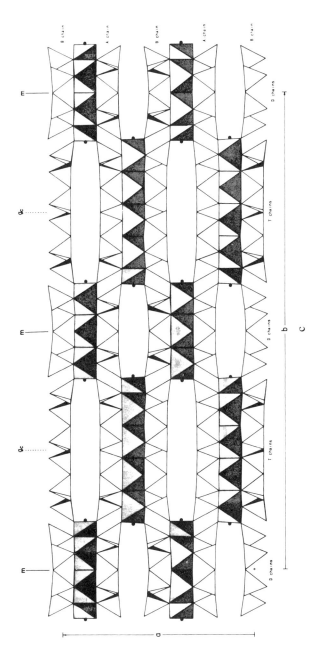

Fig. 13. (a)–(c) The crystal structure of chesterite (after Veblen and Burnham, 1978).

Table 11. M-O bond lengths within the distorted outer iron-rich M-sites in jimthompsonite, clinojimthompsonite, and chesterite.

M5 Site Occupancy	Jimthompsonite[a] Fe: 0.803(7), Mg: 0.197	Clinojimthompsonite[a] Fe: 0.845(18), Mg: 0.155
M5-O7A	2.167(5)	2.162(9)
-O7B	2.154(5)	2.162(9)
-O8A	2.042(5)	2.032(9)
-O8B	2.002(5)	2.032(9)
-O6A	2.443(5)	2.659(9)
-O6B	2.800(5)	2.659(9)
Mean	2.268	2.284

Chesterite[a]

MT5 Site Occupancy	Fe: 0.805(13), Mg: 0.195	MD4 Site Occupancy	Fe: 0.173(13); Mg: 0.187
MT5-OT7A	2.15(2)	MD4-OD2A	2.19(2)
-OT7B	2.20(2)	-OD2B	2.13(2)
-OD4A	2.04(2)	-OT8A	2.02(2)
-OD4B	1.99(2)	-OT8B	2.44(2)
-OD6A	2.42(2)	-OT6A	2.44(2)
-OD6B	2.82(2)	-OT6B	2.81(2)
Mean	2.27	Mean	2.27

[a] Veblen and Burnham (1978b).

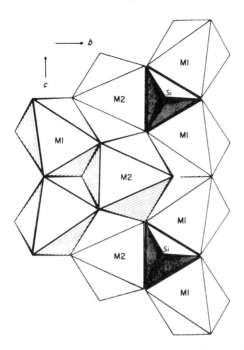

Fig. 14. Polyhedral diagram of olivine (after Papike and Cameron, 1976).

Table 12. Octahedral M-O bond distances (Å) in olivines.

M1 Site Occupancy	Forsterite Mg: 1.00	Hortonolite Fe: 0.50 Mg: 0.50	Fayalite Fe^{2+}: 1.00
($\times 2$) M1-O1	2.085(1)	2.101(3)	2.121(5)
($\times 2$) M1-O2	2.069(1)	2.101(3)	2.131(4)
($\times 2$) M1-O3	2.132(1)	2.181(3)	2.219(3)
Mean	2.095	2.128	2.157
M2 Site Occupancy	Mg: 1.00	Fe: 0.50 Mg: 0.50	Fe^{2+}: 1.00
M2-O1	2.183	2.205(4)	2.221(5)
M2-O2	2.051	2.081(4)	2.103(3)
($\times 2$) M2-O3	2.216	2.272(3)	2.296(3)
($\times 2$) M2-O3″	2.067	2.058(3)	2.080(3)
Mean	2.133	2.158	2.179

Smyth and Hazen (1973); pure Mg_2SiO_4.
Brown (1980); $(Mg_{0.49}Fe_{0.49}Ca_{0.01}Mn_{0.01})_2SiO_4$.
Smyth (1975); pure Fe_2SiO_4.

the M2 octahedron. The average M1-O distances in forsterite and fayalite are 2.095 and 2.133 Å and the average M2-O distances 2.157 and 2.179 Å respectively (Table 12). Numerous refinements of olivine crystal structures (see Brown (1980)) indicate that the average M1-O and M2-O bond distances are linear functions of the average size of the cations occupying these sites.

On the basis of the ionic size difference between Fe^{2+} and Mg^{2+}, Ghose (1962) predicted that Fe^{2+} will be preferred at the M2 site. Initial attempts to detect $Mg-Fe^{2+}$ ordering in olivines by X-ray diffraction techniques were unsuccessful (Birle et al., 1968). Subsequently, slight Fe^{2+} enrichment in the smaller M1 site in terrestrial and lunar olivines was determined by single crystal X-ray diffraction and Mössbauer resonance spectroscopy (Finger, 1970b; Finger and Virgo, 1971). Since then, a number of lunar and terrestrial volcanic olivines investigated by X-ray diffraction showed a slight preference of Fe^{2+} for the M1 site (Brown and Prewitt, 1973; Wenk and Raymond, 1973; Ghose et al., 1976). On the other hand, for some terrestrial Mg-rich olivines X-ray diffraction and Mössbauer resonance studies indicated a slight preference of Fe^{2+} for the M2 site (Wenk and Raymond, 1973; Shinno et al., 1974; Ghose et al. 1976).

Using Mössbauer resonance spectroscopy, Shinno (1974) showed that olivines synthesized at high temperatures ($\sim 1400°C$) indicate considerable ordering of Fe^{2+} at the M1 site. Prolonged heat treatment of these samples at lower temperatures down to 600°C resulted in an increase in the cation disorder (K_D at 1150, 950, and 800°C were 3.16, 1.85, and 1.32, respectively). This result is contrary to our expectations from the point of view of entropy, i.e., the crystal should be more disordered with increasing T. Will and Nover (1979) and Nover and Will (1981) using X-ray diffraction have shown that the degree of $Mg-Fe^{2+}$ order in olivine depends on the oxygen partial pressure under which they have been equilibrated. Under low oxygen pressure (10^{-16} and 10^{-21} bar) K_D increased from 1.09 and 1.06 to 1.20, indicating a higher degree of Fe^{2+} order in M1. Under high oxygen partial pressures, K_D decreased to 0.80, indicating a higher preference of Fe^{2+} for M2. However, since volcanic olivines crystallized under high temperatures and high oxygen fugacities show higher degree of order (Fe^{2+} preferring the M1 site) than the metamorphic olivines, the temperature effect on the degree of Mg-Fe order must be more significant than the effect of partial oxygen pressure in natural olivines.

In contrast to Fe^{2+}, other divalent cations, such as Co^{2+}, Ni^{2+}, and Zn^{2+} show strong preference for the M1 site, and Mn^{2+} for the M2 site with respect to Mg^{2+} (Ghose and Wan, 1974; Rajamani et al., 1975; Ghose and Weidner, 1974; Ghose et al., 1975; Francis and Ribbe, 1980). In the model we propose (Ghose and Wan, 1974; Ghose et al., 1975; Ghose et al., 1976), there are two competing factors which determine the ordering of the divalent transition metal ions in olivine with respect to Mg^{2+}: (a) the larger of the two ions prefers the larger octahedral M2 site and (b) the transition metal ions (including Fe^{2+}) prefer the smaller M1 site, which allows a greater degree of covalent bonding with the oxygen ions. For Fe^{2+}, these two competing factors nearly cancel each other, resulting in slight or no $Mg-Fe^{2+}$ order. The higher degree of covalency for Fe^{2+} at the M1 site is indicated by the smaller isomer shift, as determined by Mössbauer resonance spectroscopy (Finger and Virgo, 1971; Shinno et al., 1974). If one considers the dynamic Jahn–Teller effect and the different distortions of the M1 and M2 octahedra, the difference in crystal field stabilization energies of Fe^{2+} at the two different octahedral sites in olivine is nearly zero (Walsh et al., 1974). Hence, crystal field effects play at best a minor role in terms of Fe^{2+}–Mg ordering in olivines.

The Humite Series

The humite series of minerals are essentially hydroxylated magnesium ortho-silicate minerals, which along with olivine belong to a polysomatic series with the composition $Mg_{(2n+1)}(OH,F)_2(SiO_4)_n$, where $n = 1$ to 4 (Table 13). Each of these humite minerals have two unit cell dimensions a and b, which are the same as in olivine. A serrated chain of edge-sharing octahedra running parallel to the c-axis is a common structural unit in all these minerals (Fig. 15). The oxygen and (OH) or F ions are hexagonally close packed with half the octahedral sites occupied. As in olivine, the isolated silicate tetrahedron shares three edges with octahedra. The numbers and types of octahedral sites in humite minerals are shown in Table 14.

Chondrodite, $Mg_5(F,OH)_2(SiO_4)_2$, has one-tenth of its tetrahedral sites filled. The $M2_5$ octahedron with the coordination $MO_5(OH,F)$ is more distorted than the M2 site in olivine. The M1 site, which is the only site with no OH or F ligand, seems to prefer Fe^{2+} over Mg (Gibbs et al., 1970). *Humite*, $Mg_7(F,OH)_2(SiO_4)_3$, has 3/28 of the tetrahedral sites filled and there are two different types of tetrahedra. In a humite with $0.35Fe^{2+}$, from site occupancy refinement by single crystal X-ray diffraction Ribbe and Gibbs (1971) showed that Fe^{2+} prefers the more distorted octahedral sites and those with more polarizable ligands, i.e., $M(2)O_6$ $0.12Fe^{2+}$, $M(1)O_6$ $0.09Fe^{2+}$, $M(2)O_5(F,OH)$

Table 13. Unit cell and space groups of olivine and humite series of minerals.

Name	Chemical Composition	a (Å)	b (Å)	c (Å)	α (°)	Space Group
Forsterite[a]	Mg_2SiO_4	4.756(1)	10.207(1)	5.980(1)	—	$Pbnm$
Fayalite[b]	Fe_2SiO_4	4.818(2)	10.471(3)	6.086(2)	—	$Pbnm$
Norbergite[c]	$Mg_2SiO_4 \cdot MgF_2$	4.707	10.265	8.724	—	$Pbnm$
Chondrodite[c]	$2Mg_2SiO_4 \cdot MgF_2$	4.725	10.249	7.788	109.2	$P2_1/b$
Humite[d]	$3Mg_2SiO_4 \cdot MgF_2$	4.735	10.243	20.72	—	$Pbnm$
Clinohumite[c]	$4Mg_2SiO_4 \cdot MgF_2$	4.740	10.226	13.582	100.9	$P2_1/b$
Hydroxyl-chondrodite[e]	$2Mg_2SiO_4 \cdot Mg(OH)_2$	4.752	10.350	7.914	108.7	$P2_1/b$
Hydroxyl-clinohumite[f]	$4Mg_2SiO_4 \cdot Mg(OH)_2$	4.747	10.284	13.695	100.6	$P2_1/b$

[a] Smyth and Hazen (1973).
[b] Smyth (1975).
[c] Duffy (1977).
[d] Van Valkenburg (1961).
[e] Yamamoto and Akimoto (1974).
[f] Yamamoto and Akimoto (1977).

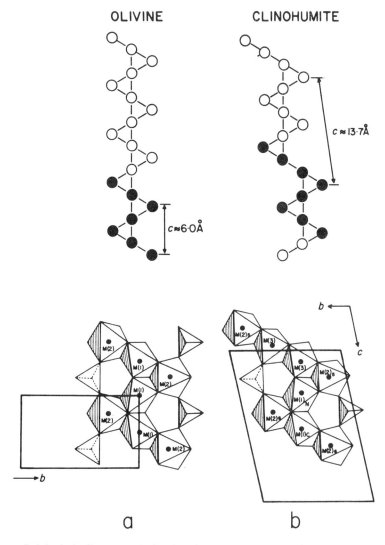

Fig. 15. Polyhedral diagram of the humite series of minerals (after Papike and Cameron, 1976).

$0.03Fe^{2+}$ and $M(3)O_4(OH,F)_2$ $0.01Fe^{2+}$. *Clinohumite,* $Mg_9(OH,F)_2(SiO_4)_4$ contains a considerable part of the olivine structure. The X-ray site occupancy refinement of a Ti-poor clinohumite indicated the same trend of Mg–Fe distribution as in humite (Robinson *et al.*, 1973). A similar pattern has been found in a titanoclinohumite, where all the Ti have been restricted to the M3 site (Kockman and Rucklidge, 1973). Because of the rarity of these minerals and the small Fe^{2+} content, no systematic studies of $Mg–Fe^{2+}$ distribution as a function of composition and temperature have been undertaken either by X-ray diffraction or spectroscopic techniques.

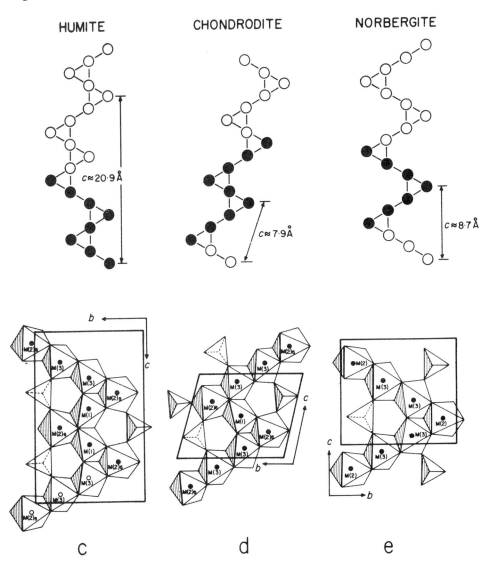

Fig. 15. (*continued*)

Techniques for the Determination of Cation Distribution in Ferromagnesian Silicates

Single Crystal X-ray Diffraction

The X-ray scattering factors, f for Mg ($Z = 12$) and Fe ($Z = 26$) are quite different, f_{Fe} being more than twice that of f_{Mg}. The effective scattering factor for the cations in a given site will be strongly dependent on the Fe/(Fe + Mg)

Table 14. Multiplicities and point symmetries of octahedral sites in olivine and humite series of minerals.

Name	Space Group	Site	Multiplicity	Point
Forsterite	*Pbnm*	$M1_6$	4	$\bar{1}$
		$M2_6$	4	m
Norbergite	*Pbnm*	$M2_4$	4	m
		$M3_4$	8	1
Chondrodite	$P2_1/b$	$M1_6$	2	$\bar{1}$
		$M2_5$	4	1
Humite	*Pbnm*	$M1_6$	8	1
		$M2_6$	4	m
		$M2_5$	8	1
		$M3_4$	8	1
Clinohumite	$P2_1/b$	$M1_6$	2	$\bar{1}$
		$M1_6$	4	1
		$M2_5$	4	1
		$M2_6$	4	1
		$M3_4$	4	1

Note: Subscripts of the M sites indicate the number of oxygen ligands.

ratio. Ghose (1961, 1965a) took advantage of this fact during the least-squares refinement of the cummingtonite and orthopyroxene structures by manually adjusting the scattering factors for random cation sites until the isotropic temperature factors, B, refined to nearly equal values. The isotropic temperature factor, B (Woolfson, 1970), represents an adjustment to the scattering factor

$$f_{\text{eff}} = f_0 e^{-B(\sin\theta/\lambda)^2},$$

where f_0 is the scattering factor of the atom at rest, θ is the Bragg scattering angle, and λ is the wavelength of the X-ray. The adjustments in B during the least-squares refinement will indicate a change in the scattering factor, f_0. However, it is to be expected that the B parameter for Mg and Fe atoms with different atomic weights will be different. In spite of this fact, Ghose's (1961, 1965b) conclusions regarding the strong site preference of Fe^{2+} for the M4 site in cummingtonite and the M2 site in hypersthene have been subsequently confirmed by X-ray diffraction, Mössbauer resonance, and optical and infrared spectroscopy.

During the least-squares refinement it is now possible to adjust the site occupancy factors independently of the temperature factors. The effective scattering factor can be written as

$$f_{\text{eff}} = \left[nf_{\text{Mg}} + (1 - n)f_{\text{Fe}} \right] e^{-B(\sin\theta/\lambda)^2},$$

where f_{Mg} and f_{Fe} are the scattering factors for Mg and Fe, respectively, n is the atomic fraction of Mg in the cation site, and B is the overall temperature factor (Burnham et al., 1971). This method is applicable to strictly binary (Fe, Mg) silicates such as hypersthene and cummingtonite. For pigeonite, where Ca is the third species which has to be considered, the above technique can be modified to restrict all the available Ca in the M2 site and distribute the available Mg and Fe between the rest of the available M2 and M1 sites. This procedure involves the assumption that Ca is completely ordered in the M2 site. Finger (1969) developed a least-squares refinement technique, where the site occupancy in one site is dependent on the site occupancy in the other; this technique allows the adjustment in several site occupancies simultaneously with the constraint that the total amounts of the atomic species have to agree with those determined by the chemical analysis. This constraint allows the site occupancies to be precisely determined not only for a two-site case, such as olivine, hypersthene, or pigeonite, but also the four-site case, such as anthophyllite and cummingtonite. Under favorable circumstances, where the chemical composition is essentially binary, site occupancy refinement of single crystal X-ray diffraction data yield precision of the order of ± 0.01 atom at a given site.

^{57}Fe Mössbauer Resonance Spectroscopy

The technique of Mössbauer resonance involves recoilless resonance absorption of gamma rays (Wertheim, 1964; Bancroft, 1976; Gütlich et al., 1978). In the case of ^{57}Fe (natural abundance, 2%), the resonant absorption corresponds to the nuclear transition of the iron nucleus from the ground state (nuclear spin, $I = \frac{1}{2}$) to the first excited state ($I = \frac{3}{2}$). ^{57}Co acts as the source of the 14.4 gamma rays. The ^{57}Fe Mössbauer resonance spectra of ferromagnesian silicates at temperatures down to 77 K are characterized by electric monopole and quadrupole interactions which give rise to the isomer shift and the quadrupole splitting respectively.

Isomer Shift

The electric monopole interaction involves the electrostatic coulomb interaction between the nuclear charge distribution in the ground and excited states and the electronic charge density distribution at the nucleus. The electronic charge distribution at the nucleus is due to the s-electrons which penetrate the nucleus and spend a fraction of their time in the nucleus. As a result of this interaction, the nuclear energy levels of the ground state and the excited states are shifted to different extents (Fig. 16a). In a Mössbauer experiment, one observes only the difference of the electrostatic shift between the source and the absorber. The isomer shift is measured by the shift of the center of gravity of the absorption line positions from zero Doppler velocity. It measures essentially the s-electron density at the nucleus and hence is strongly dependent, first, on the valence state and, second, on the chemical environment of the atom, i.e., the interaction of the atom with its neighbors, which is a

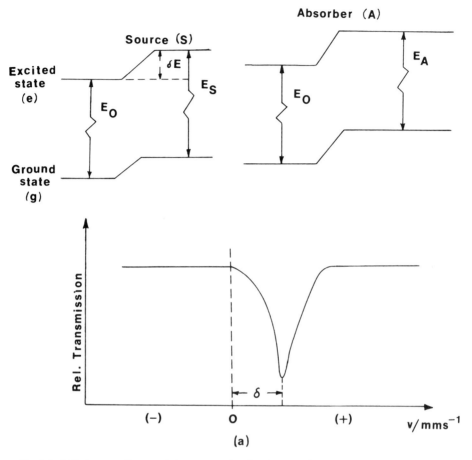

Fig. 16. Mössbauer effect and the energy levels of a transition metal. (a) Isomer shift.

characteristic of the structural site. The observed isomer shifts are usually reported with respect to ^{57}Fe in 99.999% pure iron. The isomer shift provides a measure of the covalency of the Fe^{2+}–O bonds at a given site, the degree of covalency being inversely proportional to the isomer shift.

Quadrupole Splitting

The electric quadrupole interaction originates from a coupling of the electric quadrupole moment, eQ of the ^{57}Fe nucleus in its excited state and the electric field gradient at the nuclear site. For Fe^{2+}, the electric field gradient is primarily caused by the anisotropic electron density distribution in the valence shell of ^{57}Fe (valence contribution), and secondarily by the electric charges on the neighboring ions which surround ^{57}Fe in a noncubic structural site (lattice contribution). The sixth d-electron in Fe^{2+} is distributed in one or the other of the five d-levels, causing a strong anisotropy in the electric field gradient and

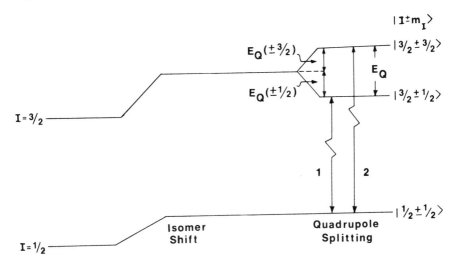

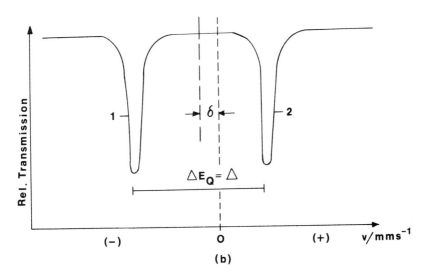

Fig. 16. (b) Quadrupole splitting.

hence a large quadrupole splitting. Furthermore, the occupancy of any particular d-level by this electron is strongly temperature dependent. Hence, for
Fe^{2+} the quadrupole splitting is very sensitive to temperature. The electrostatic interaction of the quadrupole moment of the nucleus with the electric
field gradient at the nuclear site gives rise to a splitting of the $(2I + 1)$-fold
degenerate energy levels of a nuclear state with $I = \frac{3}{2}$ to $I = \pm \frac{3}{2}$ and $I = \pm \frac{1}{2}$.
The nuclear transitions from the nuclear ground state $I = \pm \frac{1}{2}$ (which is

unsplit because eQ = 0) to these two excited levels give rise to two resonance absorption lines, the separation (ΔE_Q) of which is called the quadrupole splitting (Fig. 16b). Since the quadrupole moment of ^{57}Fe is constant in iron compounds, the quadrupole splitting essentially gives a measure of the electric field gradient at the nuclear site.

Mössbauer Resonance Spectra of Ferromagnesian Silicates

In orthopyroxenes, where Fe^{2+} occurs in two different octahedral sites, M1 and M2, two quadrupole split doublets are expected (Fig. 17). The inner doublet is due to Fe^{2+} at the distorted octahedral site, and the outer doublet is

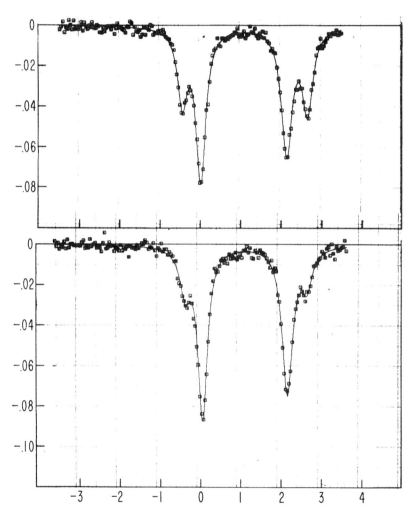

Fig. 17. Mössbauer resonance spectra of orthopyroxenes at 77°K (after Virgo and Hafner, 1969). Lower spectrum unheated orthopyroxene, upper spectrum heated (1000°C) orthopyroxene.

due to Fe^{2+} in the more regular octahedral site, M1. At lower temperature, the quadrupole splitting for Fe^{2+} in M2 does not change appreciably. The weak temperature dependence of the M2 doublet reflects a large splitting of the lowest crystalline field states (t_{2g}) caused by the very strong distortion of the M2 site.

The Mg–Fe^{2+} distribution in pigeonites can be determined from Mössbauer spectroscopy, based on the assumption that all of the Ca atoms occupy the M2 site. Comparison of site occupancies in low-Ca clinopyroxenes determined by Mössbauer spectroscopy and X-ray diffraction are closely comparable. However, the determination of site occupancies in high calcium clinopyroxenes by Mössbauer spectroscopy yields anomalous results. This anomaly can be clarified from the postulate that four distinct quadrupole-split doublets result from four possible next-nearest-neighbor configurations for iron cations in M1, arising from random distribution of Ca and Fe in M2. Some of these M1 doublets overlap with the M2 doublets at room temperature and also partially at liquid nitrogen temperature (Dowty, Ross, and Cuttita, 1972; Dowty and Lindsley, 1973).

In cummingtonites, the structural characteristics of the three nearly regular octahedral sites M1, M2, and M3 are nearly the same. Hence, their electric field gradients are nearly the same, in spite of the difference in the chemical environments of the M2 from the M1 and M3 sites. As a result, the quadrupole split doublets due to Fe^{2+} at M1, M2, and M3 overlap so strongly that they cannot be distinguished even at 77 K. Hence, the Mössbauer spectra of cummingtonites are characterized by two apparent doublets, the inner doublet being from Fe^{2+} at the distorted octahedral M4 site and the outer one from Fe^{2+} at the M1, M2, and M3 sites (Fig. 18). As in orthopyroxene, the splitting of the outer doublet increases considerably at lower temperatures. The configuration of the octahedral sites in anthophyllites are very similar to the magnesium-rich primitive cummingtonites and their Mössbauer spectra are very similar.

In olivine the nearly regular octahedral sites M1 and M2 are not very different from each other. This fact is reflected by the strong overlap of the two quadrupole split doublets from Fe^{2+} in the M1 and M2 sites (Fig. 19). Only at high temperatures in fayalite the quadrupole splittings are substantially different (Eibschutz and Ganiel, 1967). Because of this difficulty, the site assignments of the quadrupole doublets has been ambiguous, accentuated by the fact that Fe^{2+} does not show considerable site preference in olivine (Virgo and Hafner, 1972; Malysheva et al., 1969). However, from a comparison of the observed slight site preference of Fe^{2+} for the M1 site by X-ray diffraction and Mössbauer resonance, the site assignment of the spectra has been made correctly (Finger and Virgo, 1971); the inner doublet is from Fe^{2+} at M1. This assignment has been corroborated from the Mössbauer spectra of Fe, Mn olivines, where Mn^{2+} is known to have a strong site preference for the M2 site (Shinno, 1974).

The precision attainable in the site occupancy determinations by the Mössbauer resonance technique strongly depends on the resolution of the

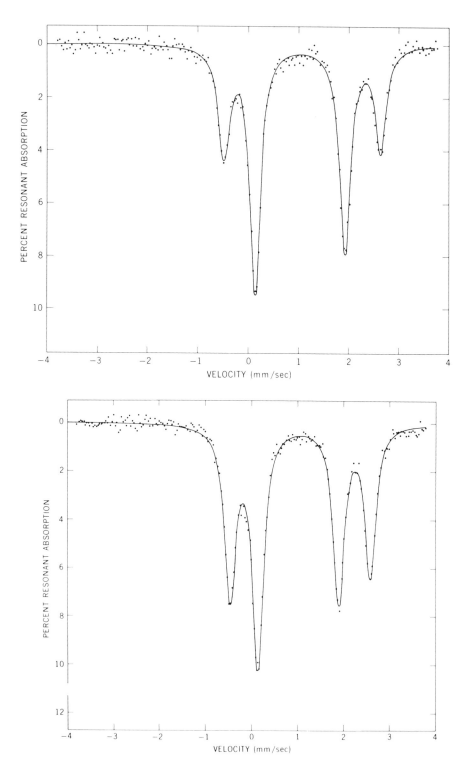

Fig. 18. Mössbauer resonance spectra of cummingtonite at 77°K. Upper spectrum unheated sample, lower spectrum heated (700°) sample (after Ghose and Weidner, 1972).

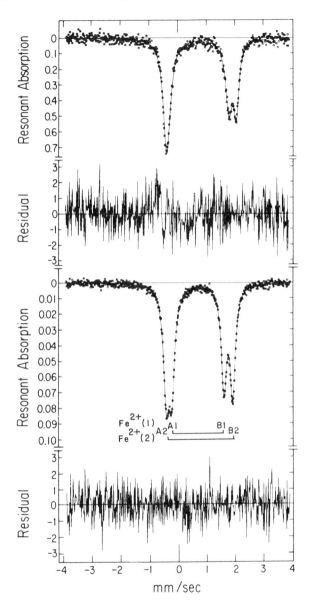

Fig. 19. Mössbauer resonance spectra of olivines (after Virgo and Hafner, 1972). Site assignment after Finger and Virgo (1971).

partially overlapping quadrupole split doublets since the site occupancies are determined from the area ratios of the Mössbauer peaks. For intermediate to high iron orthopyroxenes and cummingtonites the two doublets are clearly resolved at 77 K and the precision is high, being comparable to that obtainable from the single crystal X-ray diffraction technique (Burnham *et al.*, 1971; Hafner and Ghose, 1971). On the other hand, for magnesium-rich orthopyroxenes, cummingtonites, and anthophyllites, the outer doublets amount to

Table 15. Comparison of the site occupancies in orthopyroxene and grunerite determined by single crystal X-ray diffraction and Mössbauer techniques.

Specimen	Total Fe/(Fe + Mg)	Method	Site Occupancy Fe/(Fe + Mg)	
Orthopyroxene (XYZ)	0.850	X-ray[a]	M1 0.743(3)	M2 0.957(3)
		Mössbauer[b] (77 K)	M1 0.760(2)	M2 0.940(2)
Grunerite (Klein 1B)	0.886	X-ray[c]	M1, M2, M3 0.826(4)	M4 0.985(9)
		Mössbauer[b] (77 K)	M1, M2, M3 0.839(8)	M4 0.965(8)

[a] Burnham *et al.* (1971).
[b] Hafner and Ghose (1971).
[c] Finger (1969).

small shoulders on the strong inner doublets and are not very clearly resolved even at 77 K, and the precision is considerably lower. Because of the strong overlap of the two doublets, the precision of site occupancy determination in olivine by the Mössbauer technique is low. Another point which should be kept in mind is that the recoilless fraction of the gamma rays for ^{57}Fe at various crystallographic sites may not be exactly equal, although they are assumed to be equal for the purposes of site occupancy determinations in ferromagnesian silicates. Comparisons between site occupancies determined by Mössbauer and X-ray techniques on orthopyroxene and grunerite are shown in Table 15. Although the agreement is satisfactory, further comparisons would be highly desirable to evaluate the accuracy of the site occupancies determined by the Mössbauer technique, since this technique is much cheaper and faster and is applicable to powders rather than single crystals required for the site occupancy determination by X-ray diffraction.

Nuclear Magnetic Resonance Spectroscopy

The ^{57}Fe nucleus (natural abundance, 2%) in the ground state has a nuclear spin, $I = \frac{1}{2}$. In such a case, a single magnetic resonance absorption line results from Fe^{2+} in one crystallographic site. In orthopyroxene, where the Fe^{2+} ions at M1 and M2 sites have different internal magnetic fields, the resonance frequencies are different. When the intensities of the resonance absorption lines are corrected for the frequency difference, their area ratios (assuming that the lines have Lorentzian line shapes) would yield the Fe^{2+} distribution between the M1 and M2 sites, as in Mössbauer resonance spectroscopy. Khristoforov *et al.* (1974) have used this technique to determine the Fe^{2+} site distribution in orthopyroxenes based on NMR spectra taken at 77 K. Although an accuracy of ± 0.01 atomic units is claimed, it is difficult to judge the accuracy because no spectra have been published.

Electronic Absorption Spectroscopy

Orthopyroxenes

Because of the strong site preference of Fe^{2+} for the distorted octahedral M2 site in orthopyroxenes, the electronic absorption spectra in the visible region for magnesium-rich orthopyroxenes is almost exclusively due to Fe^{2+} in the M2 site (Bancroft and Burns, 1967). Furthermore, because of the high degree of distortion of this site, the Fe^{2+} absorption bands in M2 are usually intense and dominate the spectrum, a fact which has been utilized to determine mineral composition on planetary surfaces from remotely sensed optical absorption spectra (Adams, 1974). The intense bands at 10,500 cm^{-1}, 5000 cm^{-1}, and 2350 cm^{-1} polarized in α, β, and γ, respectively, are due to $^{5}A_{1} \rightarrow {}^{5}A_{1}$, $^{5}A_{1} \rightarrow {}^{5}B_{1}$, and $^{5}A_{1} \rightarrow {}^{5}B_{2}$ transitions from Fe^{2+} in the M2 site (Fig. 20) (Goldman and Rossman, 1977). Only in heated intermediate orthopyroxenes two components of the spin allowed transition of Fe^{2+} in the M1 site at about 13,000 cm^{-1} and 8500 cm^{-1} in γ can be clearly resolved (Rossman, 1979). Goldman and Rossman (1979) have correlated linearly the intensities of the α, β, and γ components of the 10,500–11,000 and 4900–5400 cm^{-1} bands with the Fe^{2+} concentration of 10 mole/l of Fe^{2+} at the M2 site, up to 66% occupancy. For Fe^{2+} (M2) concentrations higher than this value, the intensities of both bands in β and the 5000-cm^{-1} band in α fall close to the linear trend. From the molar absorptivity (ξ) values determined for all bands in the linear regions, the M2 (Fe^{2+}) concentration can be determined from any band using the equation

$$Fe^{2+} (M2) \, (\text{mole}/l) = Abs/(T, \xi),$$

where Abs is the optical absorbance, $\log(I_0/I)$, T is the thickness in cm, and ξ is the molar absorptivity for the band. The technique is promising for magnesium-rich and intermediate orthopyroxenes. Hopefully a similar technique can be developed for magnesium-rich cummingtonites and anthophyllites.

Infrared Spectroscopy

Anthophyllites and Cummingtonites

We recall that the (OH) ion is simultaneously bonded to one M3 and two symmetry related M1 ions in the amphibole structure. The O–H stretching vibration frequency is dependent on the occupancy of the M1 and M3 sites. For pure end members, such as grunerite and tremolite, where the M1 and M3 sites are occupied by FeFeFe or MgMgMg, a single fundamental peak

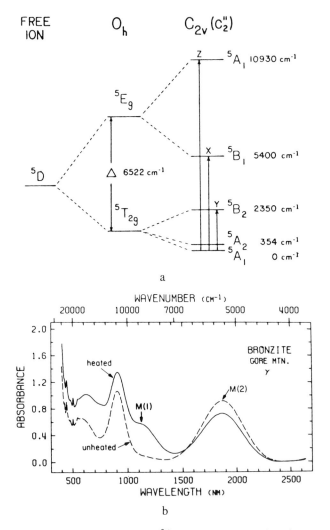

Fig. 20. (a) Energy level diagram for Fe^{2+} in the M2 site of orthopyroxene (after Goldman and Rossman, 1977). (b) Optical absorption spectra of heated and unheated intermediate orthopyroxene (after Rossman, 1979).

maximum 3600 cm^{-1} is observed. However, in intermediate members where the M1 and M3 sites are occupied by both Fe and Mg, this peak is split into four whose frequencies depend on the four possible near-neighbor configurations: FeFeFe, FeFeMg, FeMgMg, and MgMgMg (Burns and Strens, 1966). Figure 21 shows the splitting of the fundamental O–H stretching vibration frequencies in cummingtonites. Based on the observed area ratios of these peaks and the known chemical composition, Fe^{2+}–Mg partitioning between two sets of sites (M1, M3) and (M2, M4) can be determined. In case the M1 and M3 sites contain other cations such as Al^{3+}, Fe^{3+}, etc., the possible cation configurations become numerous and this method is no longer applica-

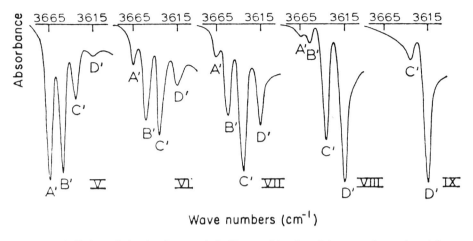

Fig. 21. Splitting of the fundamental O–H stretching band in cummingtonites (after Burns and Strens, 1966).

Table 16. Comparison of site occupancies determined by single crystal X-ray diffraction and infra-red spectroscopy in a grunerite.

Specimen	Total Fe/(Fe + Mg)	Method	(M1, M3) Fe/(Fe + Mg)	(M2, M4) Fe/(Fe + Mg)
Klein, 1B	0.888	X-rays[a]	0.87(1)	0.88(1)
		Infra-red[b]	0.85	0.92

[a] Finger (1969).
[b] Burns and Strens (1966).

ble. The accuracy and precision of the site occupancy determination from ir spectra is difficult to evaluate. The comparison of the site occupancies in a grunerite determined by single crystal X-ray diffraction and ir techniques is shown in Table 16. The agreement between the two sets of results is not satisfactory. Further comparisons of precise site occupancy determinations by these two techniques are necessary before any reliance can be put on the accuracy of the ir results.

Conclusion

In ferromagnesian silicates the degree of ordering of Fe^{2+} and Mg^{2+} ions in different crystallographic sites depends on the difference in crystal-chemical environment of these sites as well as temperature and pressure. In pyroxenes, amphiboles, and other pyriboles, the Fe^{2+}–Mg^{2+} ordering is quite pronounced even at fairly high temperatures. As a result, these minerals serve as very useful indicators of the cooling history of rocks in which they occur. On the other hand, in olivines and humite group of minerals, the degree of Fe^{2+} –Mg^{2+} ordering is much less pronounced, and its use as indicators of rock cooling history is marginal. However, in both cases a knowledge of the intra-crystalline Fe^{2+} –Mg^{2+} distribution is a prerequisite for the evaluation of the thermodynamic mixing properties of the Fe and Mg end-members.

Acknowledgments

This paper would never have been written, except for the friendly but insistent demand by the editor, Professor S. K. Saxena. A critical review of this paper by Professor J. Ganguly has resulted in considerable improvements. It was written while I was a Visiting Professor at the Laboratoire de Cristallographie, C.N.R.S. Grenoble, France. I am indebted to Professor E. F. Bertaut and Dr. M. Marezio for their kind hospitality. This research has been partially supported by the NSF Grant EAR 7904886.

References

Adams, J. B. (1974) Visible and near infrared diffuse reflectance spectra of pyroxenes as applied to remote sensing of solid objects in the solar system, *J. Geophys. Res.* **79**, 4829–4836.

Bancroft, G. M., Burns, R. G. and Maddock, A. G. (1967) Determination of cation distribution in the cummingtonite-grunerite series by Mössbauer spectra, *Amer. Mineral.* **52**, 1009–1026.

Bancroft, G. M., Maddock, A. G., Burns, R. G., and Strens, R. G. J. (1966) Cation distribution in anthophyllite from Mössbauer and infra-red spectroscopy, *Nature* **212**, 913–915.

Bancroft, G. M., and Burns, R. G. (1967) Interpretation of the electronic spectra of pyroxenes, *Amer. Mineral.* **52**, 1278–1287.

Bancroft, G. M. (1974) *Mössbauer Spectroscopy: An Introduction for Inorganic Chemists and Geochemists.* Academic Press, New York.

Birle, J. D., Gibbs, G. V., Moore, P. B., and Smith, J. V. (1968) Crystal structures of natural olivines, *Amer. Mineral.* **53**, 807–824.

Brown, G. E. (1980) Olivines and the silicate spinels. *Reviews in Mineralogy, 5, Orthosilicates.* Mineral. Soc. America, 275–381.

Brown, G. E., Prewitt, C. T., Papike, J. J., and Sueno, S. (1972) A comparison of the structures of low and high pigeonite, *J. Geophys. Res.* **77**, 5778–5789.

Brown, G. E. and Prewitt, C. T. (1973) High temperature crystal chemistry of hortonolite. *Amer. Mineral.* **58**, 577–587

Brown, M. G. and Gay, P. (1957) Observations on pigeonite. *Acta Crystallogr.* **10**, 440–441.

Brown, W. L., Morimoto, N., and Smith, J. V. (1961) A structural explanation of the polymorphism and transitions of $MgSiO_3$. *J. Geol.* **69**, 609–616.

Burnham, C. W. (1966) Ferrosilite, *Carnegie Inst. Washington Yearbook* **65**, 285–290.

Burnham, C. W., Ohashi, Y., Hafner, S. S., and Virgo, D. (1971) Cation distribution and atomic thermal vibrations in an iron rich orthopyroxene. *Amer. Mineral.* **56**, 850.

Burns, R. G. (1970) *Mineralogical Applications of Crystal Field Theory*. Cambridge University Press, Cambridge.

Burns, R. G. and Strens, R. J. G. (1966) Infrared study of the hydroxyl bands in clinoamphiboles. *Science* **153**, 890–892.

Buseck, P. R., and Iijima, S. (1975) High resolution electron microscopy of enstatite. II. Geological application. *Amer. Mineral.* **60**, 771–784.

Cameron, M. and Papike, J. J. (1981) Structural and chemical variations in pyroxenes. *Amer. Mineral.* **66**, 1–50.

Dowty, E., Ross, M., and Cuttita, F. (1972) Fe^{2+}, Mg distribution in Apollo 12021 clinopyroxenes: Evidence for bias in Mössbauer measurements, and relation of ordering to exsolution, Proceedings 3rd Lunar Science Conf., *Geochim. Cosmochim. Acta* **1**, 481–492.

Dowty, E., and Lindsley, D. H. (1973) Mössbauer study of synthetic hedenbergite–ferrosilite pyroxenes, *Amer. Mineral.* **58**, 850–868.

Duffy, C. J. (1977) Phase equilibria in the system $MgO–MgF_2–SiO_2–H_2O$, Ph.D. Dissertation, University of British Columbia, Vancouver, B.C.

Eibschutz, M. and Ganiel, U. (1967) Mössbauer studies of Fe^{2+} in paramagnetic fayalite (Fe_2SiO_4). *Solid State Commun.* **5**, 267–270.

Evans, B. J., Ghose, S., and Hafner, S. (1967) Hyperfine splitting of Fe^{57} and Mg–Fe order–disorder in orthopyroxenes ($MgSiO_3$ solid solution). *J. Geol.* **75**, 306–322.

Evans, B. W., Ghose, S., Rice, J. M., and Trommsdorff, V. (1974) Cummingtonite-anthophyllite phase transformation in metamorphosed ultramafic rocks, Ticino, Switzerland. *Trans. Am. Geophys. Union* **55**, 469.

Finger, L. W. (1969) The crystal structure and cation distribution of a grunerite. *Mineral. Soc. Am. Spec. Pap. No.* 2, 95–100.

Finger, L. W. (1969) Determination of cation distribution by least squares refinement of single crystal x-ray data. *Carnegie Inst. Washington Year Book* **67**, 216–217.

Finger, L. W. (1970a) Refinement of the crystal structure of an anthophyllite. Carnegie Inst. Washington *Year Book* **68**, 283–288.

Finger, L. W. (1970b) Fe/Mg ordering in olivines. *Carnegie Inst. Washington Year Book*, **69**, 302–305.

Finger, L. W. and Virgo, D. (1971) Confirmation of Fe/Mg ordering in olivines. *Carnegie Inst. Washington Year Book* **70**, 221–225.

Forsyth, J. B. (1980) The Chemical Interpretation of Magnetization Density Distributions in *Electron and Magnetization Densities in Molecules and Crystals*, edited by P. Becker, pp. 791–821. Plenum, New York.

Francis, C. A. and Ribbe, P. H. (1980) The forsterite–tephroite series. I. Crystal structure refinements. *Amer. Mineral.* **65**, 1263–1269.

Ghose, S. (1960) Fe–Mg ordering in some ferromagnesian minerals (abstract). Program and Abstracts. American Crystallographic Assoc. Washington Meeting, p. 19.

Ghose, S. (1961) The crystal structure of a cummingtonite. *Acta Crystallogr.* **14**, 622–627.

Ghose, S. (1962) The nature of $Mg^{2+}-Fe^{2+}$ distribution in some ferromagnesian silicate minerals, *Amer. Mineral.* **47**, 388–394.

Ghose, S. (1965a) $Mg^{2+}-Fe^{2+}$ order in an orthopyroxene, $Mg_{0.93}Fe_{1.07}Si_2O_6$. *Z. Kristallogr.* **122**, 81–99.

Ghose, S. (1965b) A scheme of cation distribution in the amphiboles. *Mineral. Mag.* **35**, 46–54.

Ghose, S. (1970) Book review: "Mineralogical Application of Crystal Field Theory," *Trans. Amer. Geophys. Union* **51**, 613.

Ghose, S., McCallum, I. S., and Tidy, E. (1973) Luna 20 pyroxenes: exsolution and phase transformation as indicators of petrologic history. *Geochim. Cosmochim. Acta* **37**, 831–839.

Ghose, S. and Wan, C. (1974) Strong site preference of Co^{2+} in olivine, $Co_{1.10}Mg_{0.90}SiO_4$. *Contrib. Mineral. Petrol.* **47**, 131–140.

Ghose, S., Wan, C., and McCallum, I. S. (1976) $Fe^{2+}-Mg^{2+}$ order in an olivine from the lunar anorthosite 67075 and the significance of cation order in lunar and terrestrial olivines. *Indiana J. Earth Sci.* **3**, 1–8.

Ghose, S., Wan, C., Okamura, F. P., Ohashi, H., and Weidner, J. R. (1975) Site preference and crystal chemistry of transition metal ions in pyroxenes and olivines (abstract). *Acta Crystallogr.* **31A**, 576.

Ghose, S. and Wan, C. (1982) Crystal chemistry of magnesium rich primitive cummingtonites (In press).

Ghose, S. and Weidner, J. R. (1972) $Mg^{2+}-Fe^{2+}$ order–disorder in cummingtonite $(Mg, Fe)_7Si_8-O_{22}(OH)_2$: A new geothermometer. *Earth Planet. Sci. Letts.* **16**, 346–354.

Ghose, S. and Weidner, J. R. (1974) Site preference of transition metal ions in olivine (abstr.) *Geol. Soc. Amer.* **6**, 751.

Gibbs, G. V. (1969) Crystal structure of protoamphibole. *Mineral. Soc. Amer. Spec. Pap. No.* 2, 101–109.

Gibbs, G. V., and Ribbe, P. H. (1969) The crystal structures of the humite minerals. I. Norbergite, *Amer. Mineral.* **54**, 376–390.

Gibbs, G. V., Ribbe, P. H., and Anderson, C. W. (1970) The crystal structures of the humite minerals. II. Chondrodite, *Amer. Mineral.* **55**, 1182–1194.

Goldman, D. S., and Rossman, G. R. (1977) The spectra of iron in orthopyroxene revisited: the splitting of the ground state, *Amer. Mineral.* **62**, 151–157.

Goldman, D. S., and Rossman, G. R. (1979) Determination of quantitative cation distribution in orthopyroxenes from electronic absorption spectra, *Phys. Chem. Minerals.* **4**, 43–53.

Gütlich, P., Link, R., and Trautwein, A. (1978) *Mössbauer Spectroscopy and Transition Metal Chemistry.* Springer-Verlag, Berlin.

Hafner, S. S., and Ghose, S. (1971) Iron and magnesium distribution in cummingtonites, *Z. Kristallogr.* **133**, 301–326.

Iijima, S., and Buseck, P. R. (1975) High resolution electron microscopy of enstatite. I. Twinning, polymorphism and polytypism, *Amer. Mineral.* **60**, 758–770.

Jones, N. W., Ribbe, P. H., and Gibbs, G. V. (1969) Crystal chemistry of the humite minerals, *Amer. Mineral.* **54**, 391–411.

Khristoforov, K. K., Nikitina, L. P., Krizhansky, L. M., and Yekimov, S. P. (1974) Kinetics of disordering of distribution of Fe^{2+} in orthopyroxene structures, *Dokl. Akad. Nauk SSSR* **214**, 909–912.

Kockman, V., and Rucklidge, J. (1973) The crystal structure of a titaniferous clinohumite, *Can. Mineral.* **12**, 39–45.

Malysheva, T. V., Kurash, V. V., and Ermakov, A. N. (1969) Study of the isomorphic replacement of magnesium and iron (II) in olivines by Mössbauer resonance spectroscopy, *Geokhimiya* **11**, 1405–1408.

Morimoto, N. (1956) The existence of monoclinic pyroxenes with the space group $C_{2h}^5–P2_1/c$, *Japan. Acad.* **32**, 750–752.

Morimoto, N., Appleman, D. E., and Evans, H. T., Jr. (1960) The crystal structure of clinoenstatite and pigeonite, *Z. Kristallogr.* **114**, 120–147.

Mueller, R. F. (1973) System $CaO–MgO–FeO–SiO_2–C–H_2–O_2$; some considerations from nature and experiment. *Amer. J. Sci.* **273**, 152–170.

Nover, G., and Will, G. (1981) Structure refinements of seven natural olivine crystals and the influence of the oxygen partial pressure on the cation distribution. *Z. Kristallogr.* **155**, 27–45.

Ohashi, Y., and Finger, L. W. (1976) The effects of Ca substitution on the structure of clinoenstatite. *Carnegie Inst. Washington Year Book* **75**, 743–746.

O'Nions, R. K. and Smith, D. G. W. (1973) Bonding in silicates: an assessment of bonding in orthopyroxene. *Geochim. Cosmochim. Acta* **37**, 249–257.

Papike, J. J., and Cameron, M. M. (1976) Crystal chemistry of silicate minerals of geophysical interest. *Rev. Geophys. Space Phys.* **14**, 37–80.

Papike, J. J., Prewitt, C. T., Sueno, S., and Cameron, M. (1973) Pyroxenes: comparisons of real and ideal structural topologies. *Z. Kristallogr.* **138**, 254–273.

Papike, J. J., and Ross, M. (1970) Gedrites: crystal structures and intracrystalline cation distributions *Amer. Mineral.* **55**, 1945–1972.

Papike, J. J., Ross, M., and Clark, J. R. (1969) Crystal chemical characterization of clinoamphiboles based on five new structure refinements. *Mineral. Soc. Amer. Spec. Pap. No.* 2, 117–136.

Pauling, L. (1960) *The Nature of the Chemical Bond.* Cornell University Press, Ithaca, New York, 3rd. Ed.

Prewitt, C. T., Papike, J. J., and Ross, M. (1970) Cummingtonite: a reversible nonquenchable transition from $P2_1/m$ to $C2/m$ symmetry. *Earth Planet Sci. Letters,* **8**, 448–450.

Prewitt, C. T., Brown, G. E., and Papike, J. J. (1971) Apollo 12 clinopyroxenes: high temperature x-ray diffraction studies. *Proc. 2nd Lunar Sci. Conf.* **1**, 59–68.

Rabbit, J. C. (1948) A new study of the anthophyllite series. *Amer. Mineral.* **33**, 263–275.

Rajamani, V., Brown, G. E., and Prewitt, C. T. (1975) Cation ordering in Ni–Mg olivine. *Amer. Mineral.* **60**, 292–299.

Ribbe, P. H., and Gibbs, G. V. (1971) Crystal structures of the humite minerals. III. Mg/Fe ordering in humite and its relation to other ferromagnesian silicates. *Amer. Mineral.* **56**, 1155–1173.

Rice, J. M., Evans, B. W., and Trommsdorff, V. (1974) Widespread occurrence of magnesiocummingtonite in ultramafic schists, Cima di Gagnone, Ticino, Switzerland, *Contrib. Mineral. Petrol.* **43**, 245–251.

Robinson, K., Gibbs, G. V., and Ribbe, P. H. (1973) The crystal structures of the humite minerals. IV. Clinohumite and titanoclinohumite, *Amer. Mineral.* **58**, 43–49.

Ross, M., Papike, J. J., and Shaw, H. R. (1969) Exsolution textures in amphiboles as indicators of sub-solidus thermal histories. *Mineral Soc. Amer. Spec. Pap.* **2**, 275–299.

Rossman, G. R. (1979) Structural information from quantitative infra-red spectra of minerals. *Trans. Amer. Crystallographic Assoc.* **15**, 77–91.

Sadanaga, R., Okamura, F. P., and Takeda, H. (1969) X-ray study of the phase transformation of enstatite. *Mineral. J. (Japan)* **6**, 110–130.

Saxena, S. K., and Ghose, S. (1971) Mg^{2+}–Fe^{2+} order–disorder and the thermodynamics of the orthopyroxene crystalline solution. *Amer. Mineral.* **56**, 532–559.

Seifert, F. and Virgo, D. (1974) Temperature dependence of intra-cyrstalline Fe^{2+}–Mg distribution in a natural anthophyllite. Carnegie Inst. Wash. *Year Book*, **73**, 405–411.

Shannon, R. D. and Prewitt, C. T. (1970) Revised values of effective ionic radii. *Acta Crystallogr.* **B26**, 1046–1048.

Shinno, I. (1974) Mössbauer studies of olivine—the relation between Fe^{2+} site occupancy number T_{Mi} and interplanar distance d_{130}. *Mem. Geol. Soc. Japan* **1**, 11–17.

Shinno, I., Hyashi, M., and Kuroda, Y. (1974) Mössbauer studies of natural olivines, *Mineral. J. Japan* **7**, 344–358.

Smyth, J. R. (1971) Protoenstatite: a crystal structure refinement at 1100°C. *Z. Kristallogr.* **134**, 262–274.

Smyth, J. R. (1973) An orthopyroxene structure up to 850°C. *Amer. Mineral.* **58**, 636–648.

Smyth, J. R. (1974) The high temperature crystal chemistry of clinohypersthene. *Amer. Mineral.* **59**, 1069–1082.

Smyth, J. R. (1975) High temperature crystal chemistry of fayalite. *Amer. Mineral.* **60**, 1092–1097.

Smyth, J. R. and Burnham, C. W. (1972) The crystal structures of high and low clinohypersthene. *Earth Planet Sci. Letts.* **14**, 183–189.

Smyth, J. R., and Hazen, R. M. (1973) The crystal structures of forsterite and hortonolite at several temperatures up to 900°C. *Amer. Mineral.* **58**, 588–593.

Smyth, J. R., and Ito, J. (1977) The synthesis and crystal structure of a magnesium–lithium–scandium protopyroxene, *Amer. Mineral.* **62**, 1252–1257.

Stephenson, D. A., Sclar, C. B., and Smith, J. V. (1966) Unit cell volumes of synthetic orthoenstatite and low clinoenstatite, *Mineral. Mag.* **35**, 838–846.

Sueno, S., Cameron, M., and Prewitt, C. T. (1976) Orthoferrosilite: high temperature crystal chemistry, *Amer. Mineral.* **61**, 38–53.

Sueno, S., Papike, J. J., Prewitt, C. T., and Brown, G. E. (1972) Crystal structure of high cummingtonite, *J. Geophys. Res.* **77**, 5767–5777.

Tofield, B. C. (1975) The study of covalency by magnetic neutron scattering, in *Structure and Bonding*, Vol. 21, pp. 1–87. Springer-Verlag, New York, 1975.

Tofield, B. C. (1976) Covalency effects in magnetic interactions, in *Application of the Mössbauer effect*. J. de Physique C-6-539-570.

Thompson, J. B., Jr. (1970) Geometrical possibilities for amphibole structures: model biopyriboles, *Amer. Mineral.* **55**, 292–293.

Thompson, J. B., Jr. (1978). Biopyriboles and polysomatic series, *Amer. Mineral.* **63**, 239–249.

Van Valkenburg, A. (1961) Synthesis of the humites, $Mg_2SiO_4 \cdot Mg (F, OH)_2$, *J. Res. Nat. Bur. Standards, A. Phys. Chem.* **65**, 415–428.

Varret, F. (1976) Crystal field effects on high-spin ferrous ion. *J. de Physique.* **37**, C6-437–C6-456.

Veblen, D. R., and Burnham, C. W. (1978a) New biopyriboles from Chester, Vermont: I. Descriptive mineralogy, *Amer. Mineral.* **63**, 1000–1009.

Veblen, D. R., and Burnham, C. W. (1978b) New biopyriboles from Chester, Vermont. II. The crystal chemistry of jimthompsonite, clinojimthompsonite, and chesterite, and the amphibole–mica reaction, *Amer. Mineral.* **63**, 1053–1073.

Veblen, D. R., Buseck, P. R., and Burnham, C. W. (1977) Asbestiform chain silicates: new minerals and structural groups, *Science* **198**, 359–365.

Virgo, D., and Hafner, S. (1969) Fe^{2+}, Mg order–disorder in heated orthopyroxenes, *Mineral. Soc. Amer. Spec. Pap. No.* 2, 67–81.

Virgo, D., and Hafner, S. S. (1972) Temperature-dependent Mg, Fe distribution in a lunar olivine, *Earth. Planet. Sci. Lett.* **14**, 305–312.

Viswanathan, K., and Ghose, S. (1965) The effect of Mg^{2+} –Fe^{2+} substitution on the cell dimension of cummingtonites, *Amer. Mineral.* **50**, 1106–1112.

Walsh, D., Donnay, G., and Donnay, J. D. H. (1974) Jahn–Teller effects in ferromagnesian minerals: pyroxenes and olivines, *Bull. Soc. fr. Mineral. Cristallogr.* **97**, 170–183.

Wenk, H. R., and Raymond, K. N. (1973) Four new structure refinements of olivine, *Z. Kristallogr.* **137**, 86–105.

Wertheim, G. K. (1964) *Mössbauer Effect: Principles and Applications.* Academic Press, New York.

Will, G. and Nover, K. (1979) Influence of oxygen partial pressure on the Mg/Fe distribution in olivines. *Phys. Chem. Minerals* **4**, 199–208.

Woolfson, M. M. (1970) *An introduction to X-ray crystallography.* Cambridge Univ. Press, 380 pp.

Yamamoto, K. and Akimoto, S. (1974) High pressure and high temperature investigations in the system $MgO–SiO_2–H_2O$. *J. Solid State Chem.* **9**, 187–195.

Yamamoto, K., and Akimoto, S. (1977) The system $MgO–H_2O–SiO_2$ at high pressures and temperatures–stability field for hydroxyl-chondrodite, hydroxyl-clinohumite and 10 Å-phase. *Amer. J. Sci.* **277**, 288–312.

II. Thermodynamics, Kinetics, and Geological Applications (J. Ganguly)

Theory of Intracrystalline Distribution

On the basis of Boltzman distribution law in statistical mechanics, Bragg and Williams (1934) formulated a theory of statistical equilibrium governing the distribution of a pair of atoms i and j over two nonequivalent sites α and β in a binary alloy as function of temperature (see Nix and Shockley, 1938, for a critical review). The disordering, which increases with increasing temperature, involves the interchange of an i atom in α site with a j atom in β site, culminating, at some high temperature, in the completely random distribution of i and j between α and β.[1] A modification of the B and W theory by Dienes (1954) considers the order–disorder process to be describable by a chemical exhange reaction, as follows

$$i(\alpha) + j(\beta) \rightleftarrows j(\alpha) + i(\beta), \tag{a}$$

which is similar to that governing the distribution of the species i and j between two coexisting phases A and B. In the latter case, the expression for the equilibrium fractionation of i and j can be derived from the condition of the equality of chemical potentials of the reactants and products at equilibrium and is as follows:

$$K = \exp\left(-\frac{\Delta G^{\circ}}{RT}\right) = \left[\frac{X_i^B X_j^A}{X_i^A X_j^B}\right]\left[\frac{\gamma_i^B \gamma_j^A}{\gamma_i^A \gamma_j^B}\right]. \tag{1}$$

Here ΔG° is the standard free-energy change of the heterogeneous exchange reaction

$$i(A) + j(B) \rightleftarrows j(A) + i(B), \tag{b}$$

and X_i^A and γ_i^A are respectively the mole-fraction and activity coefficient of the i-component in the phase A, and so on. For the sake of brevity, we shall designate the terms within the first and second square brackets as K_D and K_{γ}, respectively. In the case of homogeneous or intracrystalline exchange, however, the concept of chemical potentials of the species in the crystallographic sites faces formal problems which have been discussed by Mueller *et al.* (1970) and Grover and Orville (1970). The thermodynamic approach to derive the equilibrium intracrystalline distribution at a constant P and T will be to minimize the Gibbs free energy of the entire crystal with respect to an

---

[1] Besides this ordering between nonequivalent lattice sites, which is known as *long-range order*, there is also *short-range* or *local order*, which depends on how, on the average, each atom is surrounded in the immediate neighborhood (Bethe, 1935; see Nix and Shockley, 1938); the state of maximum local order is defined to be the one in which an atom is surrounded completely by unlike neighbors.

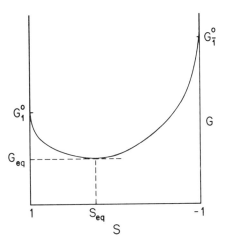

Fig. 1. Hypothetical variation of the molar Gibbs free energy of a disordered crystal $(i, j)^{\alpha}_{0.5}(i, j)^{\beta}_{0.5}F$ of fixed chemical composition as a function of a long-range ordering parameter $S = x^{\alpha}_A - x^{\beta}_A$. The equilibrium site composition corresponds to the minimum of G versus S curve. The end members corresponding of $S = 1$ and $S = -1$ are $i^{\alpha}j^{\beta}F$ (completely ordered) and $j^{\alpha}i^{\beta}F$ (antiordered), respectively. The ΔG° for the exchange reaction $i(\alpha) + j(\beta) \rightleftharpoons j(\alpha) + i(\beta)$ is given by $(G^{\circ}_{\bar{1}} - G^{\circ}_1)$.

appropriate ordering or site occupancy function, as illustrated in Fig. 1. This method was pioneered by Nix and Shockley (1938) (also see Thompson, 1969, and Navrotsky, 1971). Dienes (1954), on the other hand, derived the condition of equilibrium intracrystalline distribution by considering the dynamic balance between the rates of forward and backward exchange reactions. We shall adopt here the approach of Dienes, as modified and expanded by Mueller (1967), since it not only leads to the condition of equilibrium intracrystalline distribution, but can be further extended to analyse the kinetics of the order–disorder process, which is of considerable importance in the understanding of the cooling history of rocks. Dienes assumed that homogeneous reactions of the type (a) are second-order chemical reactions (first order with respect to the concentration in one site)[2] so that, from conventional rate theory, one can write

$$- \frac{dC^{\alpha}_i}{dt} = \vec{K} C^{\alpha}_i C^{\beta}_j - \overleftarrow{K} C^{\alpha}_j C^{\beta}_i, \qquad (2)$$

where $\vec{K}$ and $\overleftarrow{K}$ are the specific rate constants for the forward (or disordering) and backward (or ordering) reactions, respectively, and the C's refer to the number of ions of the indicated species in the specified sublattice per unit volume (cm^3) of the crystal. In general, these rate constants are functions not only of P and T, but also of composition. Expression (2) can be transformed into a rate equation describing the change of the atomic fraction of an ion

---

[2] The reduction of the rate theory to the Bragg and Williams equation at equilibrium condition constitutes *a posteriori* justification for this assumption.

within a lattice site, as follows (Mueller, 1967). Let C_α and C_β be the total number of α and β sites, respectively, per unit volume of the crystal, $C_O = C_\alpha + C_\beta$, $p = C_\alpha/C_O$, and $q = C_\beta/C_O$. Thus,

$$C_i^\alpha = C_\alpha X_i^\alpha = pC_O X_i^\alpha,$$
$$C_i^\beta = qC_O X_i^\beta,$$

(3)

and similarly for C_j^α and C_j^β, where $X_i^\alpha = (n_i/\sum n_i)^\alpha$, and so on. The unit cell volume of a silicate varies very little with the change of its site occupancies at constant bulk composition. Thus, C_O can be essentially treated as a constant. Combining Eqs. (2) and (3), we then obtain

$$-\frac{dX_i^\alpha}{dt} = qC_O\left[\vec{K}X_i^\alpha X_j^\beta - \overleftarrow{K}X_j^\alpha X_i^\beta\right].$$

(4)

Let $\Delta\overline{U}$ be the potential energy difference (which is essentially equal to the enthalpy difference, ΔH) between the ordered and disordered states, and let $E(O)$ be the minimum potential energy barrier, or the activation energy, in going from a disordered to an ordered state (Fig. 2). We then have, from the theory of absolute reaction rate (see Darken and Gurry, 1953, for a lucid discussion)

$$\vec{K} = \vec{\nu}\exp\left(-\frac{E(O) + \Delta\overline{U}}{RT}\right),$$
$$\overleftarrow{K} = \overleftarrow{\nu}\exp\left(-\frac{E(O)}{RT}\right),$$

(5)

where the preexponential terms, which are known as frequency factors,

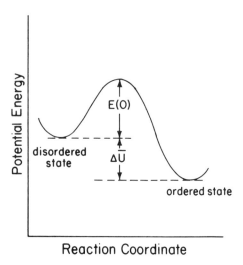

Fig. 2. Schematic illustration of potential energy variation in an order–disorder process. Reaction coordinate refers to the path of ion exchange between lattice sites along which the potential energy barrier is minimum. (Note that the potential energy of the ordered state must be lower than that of the disordered state.)

include the entropies of activation. At equilibrium, $dX_i^\alpha / dt = 0$, so that Eq. (4) reduces to

$$\frac{X_j^\alpha X_i^\beta}{X_i^\alpha X_j^\beta} = \frac{\vec{K}}{\vec{K}} = \frac{\vec{v}}{\vec{v}} \exp\left(-\frac{\Delta \bar{U}}{RT}\right). \tag{6}$$

The ratio of the frequency factors essentially equals $\exp(\Delta S / R)$, where ΔS is the entropy difference between the ordered and disordered states arising from a difference of their vibrational spectrums. ΔS is usually a very small quantity (see later), and indeed Bragg and Williams formulated their equilibrium theory on the basis of only ΔU. An important aspect of the Bragg and Williams theory is the recognition of a cooperative relation between the disordering energy, that is, the energy difference between the ordered and disordered states, and the extent of disorder. As the crystal becomes disordered, the geometric and, hence, energetic differences between the two sites decrease, which, in turn, facilitates further disordering. This is known as "convergent" disordering, which is most readily seen in the metal alloys, in which the geometric distinction between the two sites depends solely on their compositions. Bragg and Williams (1934) assumed that $\Delta \bar{U} = S \Delta \bar{U}^\circ$, where S is a long-range ordering parameter defined such that $1 \leqslant S \leqslant 0$, the limits of unity and zero corresponding to the states of complete order and complete disorder, respectively, and ΔU° is the energy required to exchange one mole of $i(\alpha)$ with one mole of $j(\beta)$ when the crystal is perfectly ordered. Similarly one can write (Dienes, 1954) that $\Delta S = S \Delta S^\circ$. The consequent variation of S as a function T for an alloy of composition $A_{0.5} B_{0.5}$ is illustrated in the Fig. 3. It

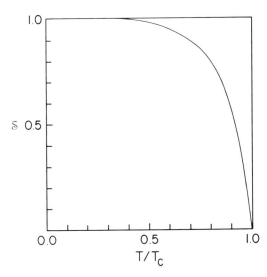

Fig. 3. Disordering of an *AB* alloy according to the Bragg and Williams convergent disordering model. $S = 1$ and $S = 0$ correspond to the states of maximum order and maximum disorder, respectively. T_c is the temperature of maximum disorder (after Swalin 1972).

shows that in the Bragg and Williams type convergent disordering, most of the disordering takes place over a rather restricted temperature interval near the temperature of complete disordering.

As discussed by Mueller (1962), an important distinction between the disordering processes in metal and silicate systems lies in the extent of involvement of the cooperative phenomenon. In an alloy, all the lattice sites participate in the order–disorder process, and the geometric distinction between the sites is a result of their compositional difference. As opposed to this, only a small fraction of the total lattice sites participate in the order–disorder process in a silicate, and the geometries of these sites are determined primarily by the nonparticipating rigid framework, rather than their site occupancies. For example, in orthopyroxene there are two geometrically distinct octahedral sites, M1 and M2, between which the divalent cations are distributed, and these sites remain distinct from one another even at the end member compositions, $MgSiO_3$ and $FeSiO_3$, that is, even when the sites have identical compositions. Thus, disordering of the cations should have comparatively smaller effect on the geometric and energetic distinction between the sites. At any rate, we can, in general, write

$$\vec{K} = \phi_f \vec{K}^\circ = \phi_f \left(\vec{\nu}^\circ \exp\left(-\left(E(O)^\circ + \Delta\overline{U}^\circ \right)/RT \right) \right), \tag{7.1}$$

$$\overleftarrow{K} = \phi_b \overleftarrow{K}^\circ = \phi_b \left(\vec{\nu}^\circ \exp\left(- E(O)^\circ/RT \right) \right), \tag{7.2}$$

where $\vec{K}^0$ and $\overleftarrow{K}^0$ are the specific rate constants between some reference states of order or disorder, and are, hence functions of only P and T, and ϕ_f and ϕ_b are appropriate functions that account for the compositional dependences of $\vec{K}$ and $\overleftarrow{K}$, respectively.[3] Thus, Eq. (6) can be expressed in a general form

$$\left[\frac{X_j^\alpha X_i^\beta}{X_i^\alpha X_j^\beta} \right] \phi = \exp\left(- \frac{\Delta\overline{U}_a^0 - T\Delta S_a^\circ}{RT} \right)$$

$$= \exp\left(- \frac{\Delta G_a^\circ}{RT} \right) \tag{8.1}$$

or

$$\left[\frac{\left(1 - \overline{X}_i^\alpha\right)\overline{X}_i^\beta}{\overline{X}_i^\alpha\left(1 - \overline{X}_i^\beta\right)} \right] \phi = \exp\left(- \frac{\Delta G_a^\circ}{RT} \right), \tag{8.2}$$

where $\overline{X}_i^\alpha = n_i/(n_i + n_j)^\alpha$, etc., and $\phi = \phi_b/\phi_f$. As mentioned earlier, ΔS_a^0 is a relatively small quantity so that ΔG_a^0 should respond weakly to temperature changes. It can be shown (Thompson, 1969), that ΔG^0 is the difference between the free energies of the crystal in the completely ordered (all i in α) and antiordered (all i in β) states (Fig. 1).

---

[3] Such resolution of specific rate constant into a constant and a compositional-dependent term is a standard practice in chemical kinetics (see, e.g., Darken and Gurry, 1953).

Expression (8) is of the same functional form as that governing heterogeneous distribution, Eq. (1), as if the sublattices α and β are two distinct phases. Thus, to model the compositional dependence of the intracrystalline-distribution coefficient, k_D (the term within the square bracket in (8)), ϕ may be treated as a function similar to the K_γ function in Expression (1),

$$\phi = \phi_j^\alpha \phi_i^\beta / \phi_i^\alpha \phi_j^\beta, \tag{9}$$

in which ϕ_i^α is expressed in terms of the compositions of the sublattice α in the same way as a macroscopic activity coefficient. This approach was adopted by Saxena and Ghose (1971), Hafner and Ghose (1971), and Ganguly (1976). Thus, if a term like ϕ_i^α (which has been called "partial activity coefficient" by Saxena and Ghose, 1971) in a binary solution is expressed by a "simple mixture" type function of X_j^α, $RT \ln \phi_i^\alpha = W^\alpha (1 - X_i^\alpha)^2$, and so on, where W is a constant at constant P and T, then Eq. (8) reduces to

$$RT \ln k_D = RT \ln K_a - RT \ln \phi$$
$$= -\Delta G_a^0(T) + W^\alpha(1 - 2X_i^\alpha) - W^\beta(1 - 2X_i^\beta). \tag{10}$$

Expression (9) implies that $\phi_b = C\phi_j^\alpha \phi_i^\beta$ and $\phi_f = C\phi_i^\alpha \phi_j^\beta$, where C is a proportionality constant. It can be shown from transition state theory (Glasstone et al., 1941, p. 403) that C is the reciprocal of the activity coefficient of the activated complex. Since the intracrystalline exchange produces very little volume change of a silicate or oxide solid solution, it can be easily seen by considering the pressure derivative of expression (10) that k_D for such solid solutions should be insensitive to changes of pressure.

Mueller (1967, 1969) obtained an integrated expression, as follows, relating the change of site occupancy with time under isothermal (and isobaric) condition.

If $\overline{X}_i = n_i / n_i + n_j$ for the whole crystal, then

$$\overline{X}_i = p\overline{X}_i^\alpha + q\overline{X}_i^\beta. \tag{11}$$

Noting from Eq. (6) that $k_D = \vec{K}/\overset{\leftarrow}{K}$, we obtain from Eqs. (4) and (8.2)

$$-\frac{d\overline{X}_i^\alpha}{dt} = C_0\vec{K}\left[a\left(\overline{X}_i^\alpha\right)^2 + b\overline{X}_i^\alpha + c \right] \tag{12}$$

or

$$-C_0\vec{K}\int_{t_0}^{t}dt = \int_{\overline{X}_i^\alpha(t_0)}^{\overline{X}_i^\alpha(t)}\left[\frac{1}{\left(a\left(\overline{X}_i^\alpha\right)^2 + b\overline{X}_i^\alpha + c\right)} \right]d\overline{X}_i^\alpha, \tag{13.1}$$

where

$$a = p\left[1 - \left(k_D^{-1}\right)\right],$$
$$b = q - \overline{X}_i + \left(k_D^{-1}\right)\left(\overline{X}_i + p\right), \tag{13.2}$$

and

$$c = -k_D^{-1}\overline{X}_i.$$

The integration of Eq. (13.1) depends on whether b^2 is greater than, equal to, or less than $4ac$. Since $k_D > 0$, it can be easily shown from Eq. (13.2) that $b^2 > 4ac$. Thus, the integration of expression (13) at fixed T (and P), and bulk composition of the crystal yields

$$-C_0 \vec{K} \Delta t = \frac{1}{\left(b^2 - 4ac\right)^{1/2}} \left| \ln \frac{\left(2a\bar{X}_i^\alpha + b\right) - \left(b^2 - 4ac\right)^{1/2}}{\left(2a\bar{X}_i^\alpha + b\right) + \left(b^2 - 4ac\right)^{1/2}} \right|_{\bar{X}_i^\alpha(t_0)}^{\bar{X}_i^\alpha(t)} \quad (14.1)$$

when

$$\left(2a\bar{X}_i^\alpha + b\right) > \left(b^2 - 4ac\right)^{1/2},$$

or

$$-C_0 \vec{K} \Delta t = \frac{1}{\left(b^2 - 4ac\right)^{1/2}} \left| \ln \frac{\left(b^2 - 4ac\right)^{1/2} - \left(2a\bar{X}_i^\alpha + b\right)}{\left(2a\bar{X}_i^\alpha + b\right) + \left(b^2 - 4ac\right)^{1/2}} \right|_{\bar{X}_i^\alpha(t_0)}^{\bar{X}_i^\alpha(t)} \quad (14.2)$$

when

$$\left(2a\bar{X}_i^\alpha + b\right) < \left(b^2 - 4ac\right)^{1/2}.$$

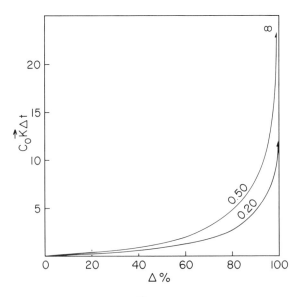

Fig. 4. The behavior of the function $(C_0\vec{K}\Delta t)$ for isothermal ordering of orthopyroxene at 500°C as function of the percentage ordering $\Delta\%$ (Eq. (15)) and bulk composition of the crystal. The numbers on the curve denote bulk $Fe/(Fe + Mg)$ ratio. The figure is modified from Kishina (1978), which is based on the equilibrium site occupancy data of Saxena and Ghose (1971). Note that the curve (0.20) approaches the 100% Δ value asymptotically and does not touch the axis.

$(2a\bar{X}_i^\alpha + b)$ will be greater or less than $(b^2 - 4ac)^{1/2}$ depending on whether the change of $\bar{X}_i^\alpha$ with time represents an ordering or disordering process, respectively. At the equilibrium value of $\bar{X}_i^\alpha$ at given T, P, and composition of the crystal, the numerator of the logarithmic term in Eq. (14) is zero, and, therefore, $C_0 \vec{K} \Delta t$ has an infinite discontinuity. The rate constant $\vec{K}(T)$ can be evaluated from (14) by subjecting the crystal at T for a fixed length of time (but less than that required to achieve a constant value within the resolution of measurement), and measuring the initial and final values of $\bar{X}_i^\alpha$, and by determining k_D at T. In practice, however, one should make several disequilibrium measurements at constant T, and derive a $\vec{K}$ that provides the best fit to these data according to expression (14) (see later). The behavior of the function $C_0 \vec{K} \Delta t$ is illustrated in Fig. 4 as a function of the bulk composition of the crystal and percentage of transformation from initial to equilibrium value of $\bar{X}_i^\alpha$ at T, which is defined as

$$\Delta\% = \frac{\bar{X}_i^\alpha(t) - \bar{X}_i^\alpha(t_0)}{\bar{X}_i^\alpha(\text{equilibrium}) - \bar{X}_i^\alpha(t_0)} . \tag{15}$$

It should be noted that at a fixed temperature, the time required to achieve a certain extent ($\Delta\%$) of ordering or disordering depends on the bulk composition of the crystal.

Relationship between Macroscopic and Site-Mixing Properties

A convenient approach to the problem of relating site and macroscopic activity coefficients in a disordered crystalline solution is provided by the concept of reciprocal ionic or salt solution, as developed by Flood, *et al.* (1954). The concept can be illustrated as follows. Let $A_{\nu_1} Y_{\nu_2} F$ be an end-member (standard state) component in a crystalline solution in which A and Y occupy distinctly separate sites, and assume further that the mixing in each site is random. According to the ionic solution model, which stems primarily from the work of Temkin (1945) (also see Bradley, 1962; Ramberg and DeVore, 1951; Kerrick and Darken, 1975), the activity of the component $A_{\nu_1} Y_{\nu_2} F$ in such a crystalline solution is expressed as[4]

$$^a A_{\nu_1} Y_{\nu_2} = (X_A)^{\nu_1} (X_Y)^{\nu_2} \gamma_{A_{\nu_1} Y_{\nu_2}}^{(\text{ionic})}, \tag{16}$$

where X_A and X_Y are respectively the site fractions of A and Y. It is important to recognize, however, that in a multisite solution of the type $(A, B, \ldots)_{\nu_1}$

---

[4] In general, the activity of a component in a solution can be expressed as $a_{A_{\nu_1} Y_{\nu_2}} = \tilde{X}\tilde{\gamma}$ where $\tilde{X}$ is a conveniently chosen function of the concentration of the component $A_{\nu_1} Y_{\nu_2}$, and $\tilde{\gamma}$ is the *corresponding* activity coefficient. The ionic solution model provides the simplest expression for the activity of a component in a solution of the type described here.

$(X, Y, \dots)_{\nu_2}F$, the chemical potentials of all the possible end-member components are not mutually independent. For example, there are four possible end-member components in the solution $(A, B)_{\nu_1}(X, Y)_{\nu_2}F$, but the chemical potentials of only three of these are independent since the four components are related by a reciprocal or metathetical equilibrium as follows:

$$A_{\nu_1}X_{\nu_2}F + B_{\nu_1}Y_{\nu_2}F \rightleftharpoons B_{\nu_1}X_{\nu_2}F + A_{\nu_1}Y_{\nu_2}F. \qquad \text{(c)}$$

The multisite solutions are, thus, called *reciprocal solutions*. In expressing the activity of a component of a multisite solution according to (16), we have implicitly chosen the maximum number of components to describe the properties of the solution. However, the fact that all of these are not mutually independent is reflected in the expression of the corresponding activity coefficient term, which can be shown to have the following form for the two-site solution involving four possible components with random mixing within each site (Flood *et al.*, 1954):

$$\gamma_{i_{\nu_1}j_{\nu_2}}^{(\text{ionic})} = (\gamma_i)^{\nu_1}(\gamma_j)^{\nu_2}\gamma_{i_{\nu_1}j_{\nu_2}}^{\text{reciprocal}}, \qquad (17.1)$$

$$\gamma_{i_{\nu_1}j_{\nu_2}}^{\text{reciprocal}} = \exp\left(\pm \frac{(1 - X_i)(1 - X_i)\Delta G_r^0}{RT} \right). \qquad (17.2)$$

Here γ_i and γ_j are the site-activity coefficients of i and j, respectively, and ΔG_r^0 is the standard free-energy change of the reciprocal reaction (c). The positive or negative sign for the exponential term depends on whether $i_{\nu_1}j_{\nu_2}$ is a reactant or a product component of the reciprocal reaction, as written. In this treatment it is assumed that the energetic interaction within one site is independent of the composition of the other sites.

Expression (17.2) can be generalized for two-site multicomponent solution as discussed by Wood and Nicholls (1978). The reciprocal solution theory can be also applied to *disordered* multisite solutions of the type $[(i, j)_{\nu_1}^{\alpha}(i, j)_{\nu_2}^{\beta}]F$ (e.g., $(\text{Fe}, \text{Mg})^{M2}(\text{Fe}, \text{Mg})^{M1}\text{Si}_2\text{O}_6$) in which mixings in both α and β sites involve the same pair of species i and j. We could consider i in the site α to be a distinct ion from i in β, and similarly j^{α} to be distinct from j^{β}. In this sense, the binary disordered solution $(i, j)_{\nu}F$, where ν is the total number of α and β sites, $\nu_1 + \nu_2$, can be thought to consist of four end-member components which are related by the reciprocal reaction

$$\left(i_{\nu_1}^{\alpha}i_{\nu_2}^{\beta}\right)F + \left(j_{\nu_1}^{\alpha}j_{\nu_2}^{\beta}\right)F \rightleftharpoons \left(i_{\nu_1}^{\alpha}j_{\nu_2}^{\beta}\right)F + \left(j_{\nu_1}^{\alpha}i_{\nu_2}^{\beta}\right)F. \qquad \text{(d)}$$

Assuming that the mixings in the α and β sites are random, and energetically independent on one another, we then have, according to (16) and (17)

$$a_{i,F} = \left[(X_i^{\alpha})^{\nu_1}(X_i^{\beta})^{\nu_2}\right]\gamma_{i,F}^{(\text{DS})}, \qquad (18.1)$$

$$\gamma_{i,F}^{(\text{DS})} = \left[(\gamma_i^{\alpha})^{\nu_1}(\gamma_i^{\beta})^{\nu_2}\right]\left[\exp(1 - X_i^{\alpha})(1 - X_i^{\beta})\Delta G_{\text{rs}/RT}^0\right], \qquad (18.2)$$

where ΔG_{rs}^0 is the standard free-energy change of the reciprocal (site) reaction (d). The superscript DS denotes "disordered site mixing model." Activity expression of the type (18.1) for disordered two-site binary solution was introduced, somewhat intuitively, by Mueller (1961), but he did not specify a more explicit form for the activity coefficient function. Expressions (18.1) and (18.2) were first derived by Thompson (1969, 1970) without recourse to the reciprocal salt formulation, the discussion of which is, however, beyond the scope of this review. The exponential term in (18.2), which arises from the reciprocal exchange–interaction between the two binary sites, can be generalized for the multicomponent two-site disordered solution according to the general procedure for the multicomponent two-site solution given by Wood and Nicholls (1979) and writing the possible reciprocal reactions following the conceptual basis of the reaction (d).

In a disordered crystal, there are energetic interactions of ions not only within the individual sites, but also between the sites. All these interactions must be reflected in the compositional dependence of k_D for the intracrystalline fractionation. Thus, the W^α and W^β terms in expression (10) should not only reflect the within-site interactions, but also the between-site interactions. If it is assumed that the departure from ideal interaction within the site α can also be expressed by a "simple mixture" type form $RT\ln\gamma_i^\alpha = w^\alpha(X_i^\alpha)^2$, where γ_i^α is the activity coefficient solely due to within-site interaction, then it can be shown (Sack, 1980) that

$$W^\alpha = w^\alpha - \tfrac{1}{2}\Delta G_{rs}^0 \qquad (19)$$

and similarly for W^β.

As emphasized by several workers (Mueller, 1962, 1967; Matsui and Banno, 1965; Grover, 1974), an "ideal" activity expression according to (18.1), $a_{i_\nu F} = (X_i^\alpha)^{\nu_1}(X_i^\beta)^{\nu_2}$, which is often called the "ideal two-site model," implies that the activity of a component $i_\nu F$ ($\nu > 1$) is less than its molecular fraction, $a_{i_\nu F} < X_{i_\nu F}$. As shown below, this behavior is the result of the stoichiometric relation (11). Equating the molecular activity expression

$$a_{i_\nu F} = x_{i_\nu F}\gamma_{i_\nu F}^{(M)} = X_i\gamma_{i_\nu F}^{(M)} \qquad (20)$$

with (18.1), and setting $\gamma^{(DS)} = 1$, we have

$$\gamma_{i_\nu F}^{(M)} = \frac{\left[(X_i^\alpha)^p(X_i^\beta)^q\right]^\nu}{pX_i^\alpha + qX_1^\beta} \qquad (21)$$

which is less than unity (for $\nu = 1$, it is <1 except in the case of complete disorder). Further, noting that $(X_i\gamma_{i_\nu F}^{(ionic)})^\nu = X_i\gamma_{i_\nu F}^{(M)}$, expression (21) yields

$$\gamma_{i_\nu F}^{(ionic)} = \frac{(X_i^\alpha)^p(X_i^\beta)^q}{pX_i^\alpha + qX_i^\beta} \qquad (22)$$

which is $\leqslant 1$, the equality holding in the limiting case of complete disorder, $X_i^\alpha = X_i^\beta = X_i$. Equations (21) or (22) provide important restrictions on the domain of applicability of the "two-site ideal model" for the disordered solid solution. Thus, for example, the observed exsolution in natural $C2/C$ clinopyroxene solid solution $(Ca, Mg, Fe)(Mg, Fe)Si_2O_6$, which implies $\gamma_{Mg_2Si_2O_6} > 1$, precludes a "two-site ideal" activity expression for $Mg_2Si_2O_6$ in $C2/C$ clinopyroxene solid solution.

When the pure state of a component is disordered, a normalization factor has to be introduced in the activity expression of the type (18.1), as pointed out by Powell (1976), in order that a_i equals X_i at the limiting composition of $X_i = 1$. For example, the activity of $CaAl_2SiO_6$(CaTs) component in a pyroxene solid solution should be expressed in terms of the site compositions as

$$a_{CaTs}^{Px} = 4(X_{Ca}^{M2} X_{Al}^{M1} X_{Al}^{T} X_{Si}^{T}) \gamma_{CaTs}^{(DS)} \tag{23}$$

since at $X_{CaTs} = 1$, $X_{Al}^{T} = X_{Si}^{T} = \frac{1}{2}$, and $X_{Ca}^{M2} = X_{Al}^{M1} = 1$.

Systematics and Thermodynamic Analysis of Intracrystalline Distribution of Fe^{2+} and Mg in Silicates

Pyroxenes

Virgo and Hafner (1969) have annealed several natural orthopyroxenes with $Fe^{2+}/(Fe^{2+} + Mg)$ ratio between 0.178 and 0.877 at 1000°C, and an ortho-

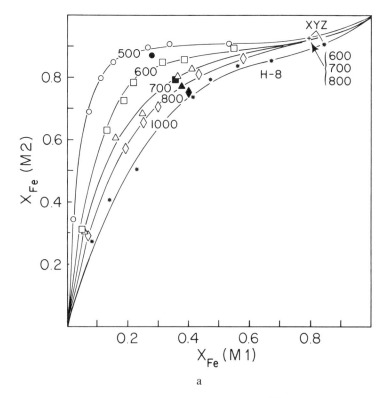

a

Fig. 5. (a) Experimentally determined distribution of Fe^{2+} and Mg between the M1 and M2 sites of orthopyroxene. $X_{Fe} = Fe^{2+}/(Fe^{2+} + Mg)$.

pyroxene (sample No. 3209) with $Fe^{2+}/(Fe^{2+}+Mg)$ ratio of 0.574 at 500, 600, 700, 800, and 1000°C and determined the intracrystalline distribution of Fe^{2+} in the quenched samples by Mössbauer spectroscopy. A more systematic study of the Fe^{2+}–Mg fractionation in orthopyroxene was later undertaken by Saxena and Ghose (1971). The results of these two groups of workers, all of which are based only on disordering experiments, are illustrated in Fig. 5(a). It is clear that Virgo and Hafner's data on sample No. 3209 between 500 and 800°C are in serious disagreement with the results of Saxena and Ghose. Recently, Besancon (1981) has studied the disordering kinetics of two ortho-pyroxene samples with $Fe^{2+}/(Fe^{2+}+Mg)$ ratio of 0.13 (HC) and 0.51 (TZ) between 600 and 800°C. The last sample had a small amount of Fe^{3+} which was resolvable in Mössbauer spectroscopy, but microprobe analysis and charge and site stoichiometry considerations suggested that Fe^{3+} was less than 2% of total iron (Besancon, 1981). The inferred equilibrium ordering data for

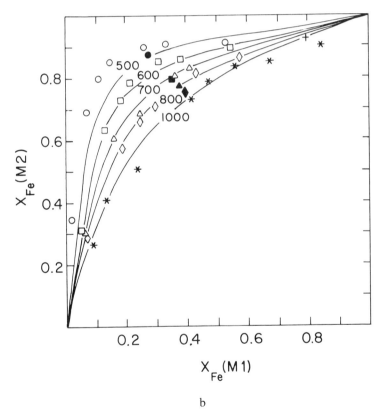

b

Fig. 5. (b) Calculated distribution isotherms for orthopyroxene on the basis of revised estimates of K and site-interaction parameters. The isotherms are compared with the experimental distribution data. Star: 1000°C (data from Virgo and Hafner, 1969), diamond: 800°C, triangle: 600°C, circle: 500°C. Open symbols: data from Saxena and Ghose (1971); filled symbols: data for sample No. 3209 ($X_{Fe^{2+}(total)} = 0.574$) of Virgo and Hafner (1969). All data have been obtained by disordering experiments and Mössbauer spectroscopy.

this sample is in good agreement with those of Saxena and Ghose. We, therefore, suspect some systematic error in Virgo and Hafner's data on sample No. 3209. Saxena and Ghose (1971) have analyzed their isothermal distribution data using a "simple mixture" type site-mixing formulation (see Eq. (10)) and found the interaction parameters $W(M1)$ and $W(M2)$ to be sensitive to temperature. The values of both interaction parameters derived from the data at 600, 700, and 800°C defined linear trends versus $1/T$ ($W(M1) = 3525(10^3/T) - 1667$, $W(M2) = 2548(10^3/T) - 1261$), but those at 500°C differed considerably from these trends. These discrepancies were most likely due to disequilibrium and/or low degree of accuracy in the determination of the very small M1 site occupancy of Fe^{2+} at 500°C, especially in the Mg-rich samples, and the sensitivity of k_D to errors in the site occupancy at very small concentrations of Fe(M1) (Saxena and Ghose, 1971). For any given sample, as illustrated in Fig. 6 (see Table 1 for regression data), the k_D at 500°C is in serious disagreement with the $\ln k_D$ versus $1/T$ trend defined by the data between 600 and 800°C. Thus, as already suggested by Saxena (1973), the 500°C data of Saxena and Ghose should be rejected. The equilibrium constant for the exchange reaction

$$Fe(M2) + Mg(M1) \rightleftharpoons Mg(M2) + Fe(M1) \tag{e}$$

at 600, 700, and 800°C, derived by these workers from their distribution data, along with that at 1000°C derived from the data of Virgo and Hafner using

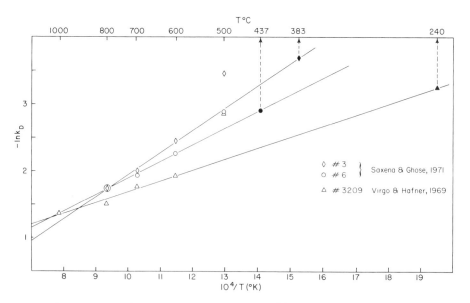

Fig. 6. $\ln k_D$ versus $1/T$ for the intracrystalline Fe^{2+} –Mg distribution for the orthopyroxene samples No. 3 (diamond) and No. 6 (circle) of Saxena and Ghose (1971) and No. 3209 (triangle) of Virgo and Hafner (1969). Note that the data at 500°C for all samples sharply disagree with the linear trends defined by the higher-temperature data. Filled symbols are plots of $\ln k_D$ of the natural samples, as determined by Virgo and Hafner (1969), on their respective $\ln k_D$ versus $1/T$ trends (ignoring the 500°C data).

Table 1. Regressed constants for $\ln K_D$ versus $1/T$ data shown in Figs. 6, 11, and 13. $\ln K_D = A/T + B$.

Figure No.	Sample	A	B	Source of Data
6	OPx 3	− 3348.71(270)	1.399	Saxena and Ghose (1971)
	OPx 6	− 2500.21(210)	0.604	Saxena and Ghose (1971)
	OPx 3209	− 1644.26(205)	− 0.049	Virgo and Hafner (1969)
11	CPx nodule	− 2370.55	0.206	McCallister (pers. comm.)
	Kakanui angite	− 3208.90	0.705	McCallister (pers. comm.)
13	Anthophyllite/AG23	− 3339.20	0.837	Seifert (1978)
	Anthophyllite/AG17	− 3443.88	1.189	Seifert (1978)
	Anthophyllite Rabbitt 30	− 2698.54	0.586	Seifert (1978)
	Anthophyllite/AG12	− 2255.66	0.111	Seifert (1972)
	Cummingtonite/DH7-482	− 1366.8	− 0.465	Ghose and Weidner (1972)
	Cummingtonite USNM	− 1836.5	− 0.502	Ghose and Weidner (1972)

expression (10), are illustrated in Fig. 7 in terms of a $\ln K$ versus $1/T$ plot. The data of Saxena and Ghose, which show a surprisingly small temperature dependence, are highly incompatible with that of Virgo and Hafner in that $\ln K$ should essentially be a linear function of $1/T$, considering that the order–disorder process should be associated with very little ΔCp effect.

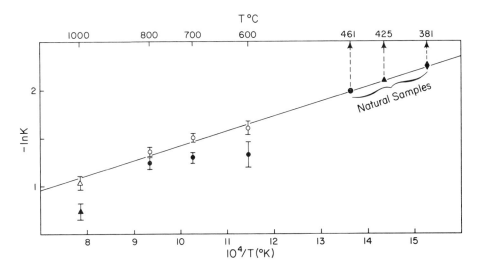

Fig. 7. $\ln K$ versus $1/T$ for intracrystalline Fe^{2+}–Mg exchange equilibrium in orthopyroxene. Small circles and triangle with uncertainty (σ) bars are data from Saxena and Ghose (1971) and Virgo and Hafner (1969), respectively; filled symbols: $\ln K$ derived from all isothermal data shown in Fig. 5(a); open symbols: $\ln K$ derived by ignoring the two most iron-rich samples XYZ and H-8. The natural samples have the same symbols as in Fig. 6.

However, the two sets of data are found to become mutually compatible if only two samples are ignored, sample XYZ and H-8, which are the two most Fe^{2+}-rich samples. (In these works, as well as in other studies of Fe^{2+}–Mg order–disorder by Mössbauer spectroscopy, the site occupancy of Mg has been determined by assuming that the sites are essentially binary mixtures of Fe^{2+} and Mg, so that $X_{Mg(M1)} \approx 1 - X_{Fe^{2+}(M1)}$. However, at very low concentrations of Mg, this procedure can introduce significant error in Mg site occupancy (also see Hafner and Ghose, 1971). The results, which are illustrated in Fig. 7, yield

$$\ln K \simeq 0.1435 - 1561.81/T. \tag{24}$$

The associated values of the interaction parameters are illustrated in Fig. 8. It is evident that while the data of Saxena and Ghose (1971) still seem to suggest linear dependence of the W–s on $1/T$, the entire set of data is suggestive of nearly constant values of the interaction parameters:

$$W(M1) \approx 1524 \text{ cal/mole},$$
$$W(M2) \approx 1080. \tag{25}$$

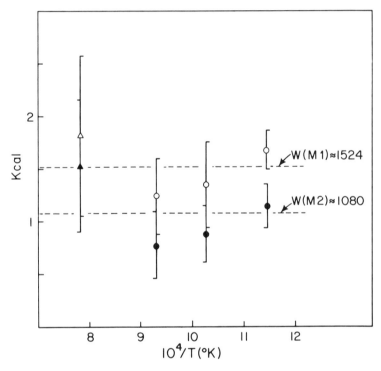

Fig. 8. Site interaction parameters for orthopyroxenes, $W(M1)$ and $W(M2)$, derived from the intracrystalline Fe^{2+}–Mg distribution data of Saxena and Ghose (1971) (circles) and Virgo and Hafner (1969) (triangles), neglecting samples XYZ and H-8 (Fig. 5). Open symbols: $W(M1)$, closed symbols: $W(M2)$, vertical bars: uncertainty (σ).

Using these W values and Eq. (10), the $\ln K$ values calculated for sample TZ of Besancon (1981), on the basis of the site occupancy data at 600, 700, 750, and 800°C, agree within ~ 0.01 with those predicted by expression (24). However, the $\ln K$ values calculated in the same way for Besancon's magnesian-rich sample, HC, sharply disagree with this expression. This discrepancy may be due to the problem associated with the determination of the M1 site occupancy of Fe^{2+} at very small concentrations, as discussed earlier, and the small amount of anthophyllite impurity in the sample.

I have calculated the isothermal Fe^{2+}–Mg distributions in orthopyroxene, and $\ln k_D$ as a function of $Fe^{2+}/(Fe^{2+}+Mg)$ ratio, on the basis of the above estimates of $\ln K$ and site-interaction parameters (Eqs. (24) and (25)). The site-occupancy calculations have been done numerically by using the program REGSOL2 of Saxena (1973), which solves Eq. (10) for X_i^α for any specified value of X_i^β and T (K_d in Eq. (10) being replaced by the atomic fraction ratio as defined by 8.2). These calculations are illustrated in Figs. 5(b) and 9. Snellenburg (1975) has performed computer simulation of the Fe^{2+}–Mg distribution in orthopyroxene, assuming that the exchange is limited to nearest-neighbor cations. As in Fig. 9, this work also shows a minimum in $\ln k_D$ versus X_{Fe}.

By considering the preliminary Fe^{2+}–Mg fractionation data, as determined by Mössbauer spectroscopy, in a pigeonite ($Wo_9En_{59}Fs_{32}$) and an augite ($Wo_{36}En_{44}Fs_{20}$) annealed at 1000°C, along with Virgo and Hafner's data on calcium-free orthopyroxenes, Hafner et al. (1971) have suggested that the ΔG for exchange reaction (e) increases linearly with the increasing $CaSiO_3(Wo)$ content of the pyroxenes. Saxena et al. (1974) have subsequently made a more controlled study on the cation distribution in synthetic low-calcium clino-

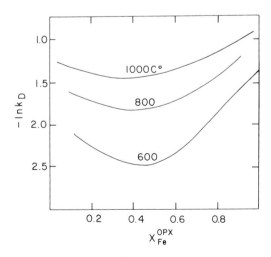

Fig. 9. Calculated variation of $\ln k_D(Fe^{2+}$–Mg) for orthopyroxene as a function of the bulk $Fe^{2+}/(Fe^{2+}+Mg)$ ratio on the basis of revised estimates of K and the site interaction parameters.

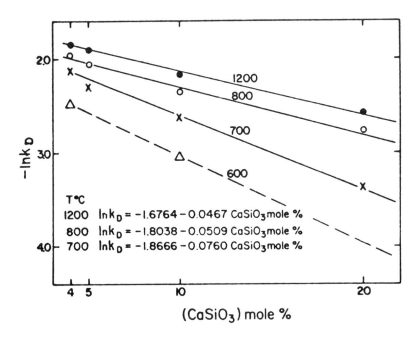

Fig. 10. Variation of $\ln k_D (Fe^{2+} - Mg)$ for clinopyroxene as a function of $CaSiO_3$ content (from Saxena *et al.* (1974)).

pyroxenes (pigeonites, $P2_1/C$, and a subcalcic augite, $C2/C$), with compositions in the ternary plane $FeSiO_3 - MgSiO_3 - CaSiO_3$, as a function of temperature and Ca content. All samples had $MgSiO_3/FeSiO_3 \approx 1$, while $CaSiO_3$ varied from 4 to 20%. At constant temperature, $\ln k_D$, as determined from Mössbauer data (with the restriction that all Ca is in the M2 site), was found to decrease linearly with increasing Wo content (Fig. 10), which is in agreement with the suggestion of Hafner *et al.* For pyroxenes with Wo = 8–9%, the site occupancies obtained from Mössbauer spectra are in good agreement with those determined from single-crystal X-ray intensities. (See Dowty and Lindsley (1973) for discussion about the problem of Fe(M2) estimation by Mössbauer spectra in Ca-rich samples.)

McCallister *et al.* (1976) have studied the intracrystalline $Fe^{2+} - Mg$ equilibria in two natural clinopyroxenes as a function of temperature, the results of which are illustrated in Figs. 11(a) and 11(b). One of these is a discrete nodule, showing exsolved pigeonite, from the Thaba Putsua Kimberlite pipe, Lesotho, while the other is a discrete nodule from Kakanui nephelinite breccia in New Zealand. Unlike the other studies discussed in this section, the site occupancies have been obtained from crystal-structure refinement. Neglecting the minor amounts of Cr, Ti, Mn, and Fe^{3+}, the structural formula for these samples are as follows. Thaba Putsua clinopyroxene (PHN 1600E4): $(Na_{0.118}Ca_{0.553}Mg_{1.059}Fe^{2+}_{0.138}Al_{0.119})Si_{1.987}O_6$; Kakanui augite: $(Na_{0.093}Ca_{0.634}Mg_{0.885}Fe^{2+}_{0.134}Al_{0.350})Si_{1.822}O_6$.

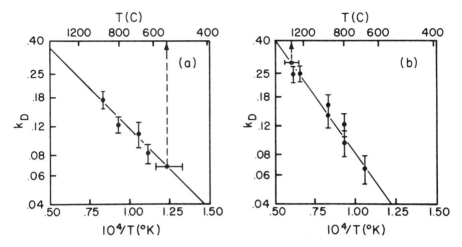

Fig. 11. $\ln k_D (Fe^{2+} - Mg)$ versus $1/T$ for two clinopyroxene samples according to McCallister *et al.* (1976). (a) A discrete nodule from Thaba Putsua Kimberlite pipe, Lesotho; (b) Kakanui Augite, New Zealand. The k_D of natural samples are shown as horizontal bars.

Amphiboles

Hafner and Ghose (1971) have studied the intracrystalline distribution in natural cummingtonites, mostly collected from the Precambrian metamorphosed iron formation in the Bloom Lake and neighboring area, Quebec. As discussed earlier (Ghose, pt. 1), there are four nonequivalent octahedral sites in cummingtonites, M1, M2, M3, and M4, over which Fe^{2+} and Mg are distributed. However, due to the close similarity of the geometry of the M1, M2, and M3 sites, it was not possible to determine the fractionation among these sites from Mössbauer spectra. The fractionation among these sites is expected to be small, especially considering that the electronic states of Fe^{2+} in these sites are almost identical, with slight preference of Mg for the M2 site (see also Papike and Ross, 1970; Kretz, 1978). Thus, following Mueller (1961), the M1, M2, and M3 sites were treated, as a first approximation, as a single group of sites. Hafner and Ghose also fitted the distribution data according to both ideal and simple mixture behavior of the sites. The observed fractionation of Fe^{2+} between M4 and (M1, M2, M3) sites from the Bloom Lake area, which should have been quenched from similar conditions, is illustrated in Fig. 12. I have refitted these data according to expression (10), which yields

$$\ln K = -1.8682 \ (\pm 0.7381), \tag{26.1}$$

$$\frac{W^{M1,2,3}}{RT} = 1.544 \ (\pm 0.381), \tag{26.2}$$

$$\frac{W^{M4}}{RT} = \approx 0. \tag{26.3}$$

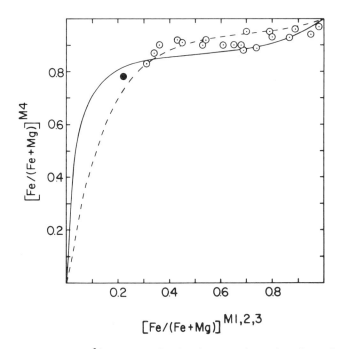

Fig. 12. Intracrystalline Fe^{2+}–Mg distribution in cummingtonites from the Precambrian metamorphosed Iron Formations, Bloom Lake area, Ontaria, Canada, according to Hafner and Ghose (1971). Solid line: fit of the data according to a "simple mixture type" site-mixing model; dashed line: distribution predicted by Mueller. The filled symbol represents the sample DH7-482 which has been investigated by Ghose and Weidner (1972) and discussed later in the section on "cooling history."

The distribution curve, calculated according to the above data, is shown in Fig. 12. Mueller (1961) predicted a similar distribution curve for the Bloom Lake samples on the basis of the observed fractionation of Fe^{2+} and Mg between coexisting cummingtonite and actinolite in this area, and the assumption that the (M1, M2, M3) group of sites in the two minerals, being geometrically similar, are also energetically equivalent, so that $X_{Mg}(Act) \approx X_{Mg(M1,2,3)}(Cum)$ (note that in actinolite, the M4 site is occupied almost completely by Ca). Mueller's predicted intracrystalline distribution for the Bloom Lake samples is also shown in Fig. 12 for comparison.

Ghose and Weidner (1972) have studied the Fe^{2+}–Mg fractionation in two natural cummingtonite samples, referred to as USNM 118125 and Mueller DH7-482 (from Bloom Lake), as a function of temperature between 600 and 800°C at 2 kbars H_2O pressure. Their results are illustrated in Fig. 13. From these data, we obtain $375 \pm 20°C$ as the apparent intracrystalline equilibration temperature, T_{ae}, of the Bloom Lake sample, DH7-482.[5] Consequently, from

[5]This is about 45°C lower than the T_{ae} calculated by Ghose and Weidner (1972). This difference is due to their treatment of the data in the form $\ln K_D = A/T$ instead of $\ln k_D = A/T + B$.

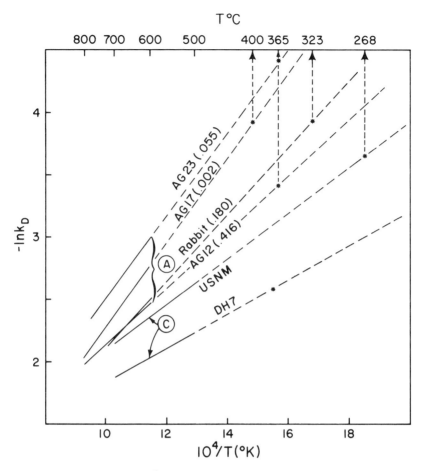

Fig. 13. $\ln k_D$ versus $1/T$ for Fe^{2+}–Mg distribution in anthophyllites (curves A) and cummingtonites (curves C). The solid segments of the lines denote the range of experimental calibrations. The k_D of the natural samples are shown by asterisks. The Al content of the anthophyllite samples, based on 46 negative charges, are shown within the parentheses. The cummingtonite samples USNM and DH7-482 have very similar $Fe^{2+}/(Fe^{2+} + Mg)$ ratios (0.364 and 0.378, respectively). The USNM sample probably has crystallized at low temperature close to that of apparent intracrystalline equilibration (Ghose and Weidner, 1972).

Eqs. (25), $\Delta G^0 = 2405 \pm 950$, $W(M1,2,3) = 1988 \pm 491$, and $W(M4) \approx 0$ cal for cummingtonites, which are similar to the corresponding values for ortho-pyroxene at 375°C ($\Delta G^0 = 2948$, $W(M1) \approx 1524$, $W(M2) \approx 1080$) except that the M4 site in cummingtonite seems to be somewhat more ideal than the M2 site of orthopyroxene.

   Seifert and Virgo (1974) and Seifert (1978) have studied Fe^{2+}–Mg fraction-ation in natural anthophyllites as a function of temperature by Mössbauer spectroscopy. The $Fe^{2+}/(Fe^{2+} + Mg)$ ratio of these samples varied from 0.75

to 0.87. The data for each sample are found to define a linear relation between $\ln k_D$ and $1/T$, and those for a few selected samples are illustrated in Fig. 13. Seifert (1975) noticed that the values of both constants in the relation $\ln k_D = A/T + B$ systematically decrease (or, in other words, $-\ln k_D$ versus $1/T$ plots rotate clockwise) with increasing Al content of the minerals (Fig. 13). He rationalized this observation on the basis of the fact that increasing Al content reduces the geometric dissimilarity, and hence the energetic difference, between the M4 and M1,2,3 sites. The qualitative dissimilarity between the Fe^{2+}–Mg fractionation in the two cummingtonite samples studied by Ghose and Weidner, which have very similar $Fe^{2+}/(Fe^{2+} + Mg)$ ratios (0.364 and 0.378), seem compatible with Seifert's idea, in that Al_2O_3 content of the USNM cummingtonite is 0.28, whereas that of the Mueller sample is 0.40. Thus, in cummingtonites the intracrystalline Fe^{2+}–Mg fractionation should depend not only on $Fe^{2+}/(Fe^{2+} + Mg)$ ratio, as discussed above, but also on the Al_2O_3 content (or Gedrite component) of the samples. By analogy with cummingtonites and orthopyroxenes, Fe^{2+}–Mg fractionation in anthophyllite is also expected to depend on the $Fe^{2+}/(Fe^{2+} + Mg)$ ratio. In the anthophyllite samples studied by Seifert, this ratio has a limited variation, from 0.75 to 0.87, but is positively correlated with Al ($Al = -0.3827 + 2.953(\pm 0.919)X_{Fe}$; cor. coeff.: 0.821). Consequently, it seems likely that Seifert's (1975) deduction about the effect of Al on Fe^{2+}–Mg fractionation in anthophyllite, as shown in Fig. 13, represents the combined effects of Al and $Fe^{2+}/(Fe^{2+} + Mg)$.

General Considerations

It is evident from the above discussions that the intracrystalline fractionation of Fe^{2+} and Mg in ferromagnesian silicates depends on their compositions, especially $Fe^{2+}/(Fe^{2+} + Mg)$ ratios. The compositional dependence of k_D can be adequately modeled by assuming a "simple mixture" type mixing of Fe^{2+} and Mg in the various sites (Eq. (10)). It is found that in both orthopyroxene and amphibole, for which there are adequate data for this type of modeling, the interaction parameters are positive, but the one for the relatively Fe^{2+}-rich site, which is larger and more distorted, is somewhat smaller than that for the Mg-rich ones. It should be recalled, however, that these interaction parameters are not independent of the cooperative influence among the various sites.

The Fe^{2+}–Mg fractionation data within the silicates, and the derived thermodynamic properties, as discussed above, corroborate Mueller's (1961) suggestion that geometrically similar sites in different silicates have similar energetic properties. This has also been the conclusion of Kretz (1978). The validity of this idea can be further tested by comparing the observed distributions of Fe^{2+} and Mg between coexisting silicates with those predicted on the basis of the assumption that Fe^{2+} and Mg do not fractionate between geometrically similar sites of the coexisting pairs. The observed and predicted distributions for three coexisting pairs, Ca–pyroxene–actinolite, cummingtonite–anthophyllite, and orthopyroxene–actinolite are illustrated in Fig. 14.

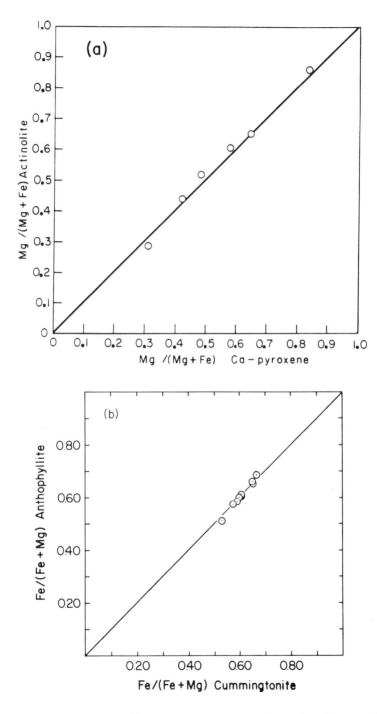

Fig. 14. The distribution of Fe^{2+} and Mg between coexisting minerals. (a) Actinolite and clinopyroxene, data from Mueller (1961). (b) Anthophyllite and cummingtonite, data from Stout (1972).

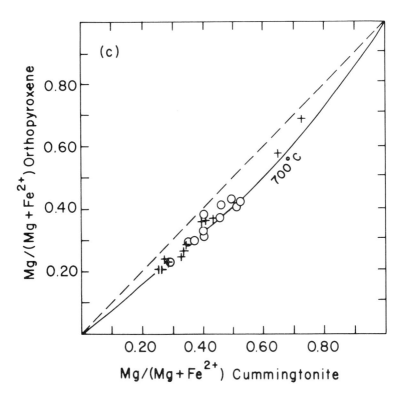

Fig. 14. (c) Orthopyroxene and cummingtonite. The solid lines are predicted distributions based on the assumption that Fe^{2+} and Mg do not fractionate between sites of similar geometries in the coexisting phases, data from Bonnichsen (1969) (cross) and Butler (1969) (circle). The Butler samples have crystallized at $\sim700°C$.

No fractionation of Fe^{2+} and Mg is expected in the first two pairs. In the third pair, Fe^{2+} and Mg should fractionate according to the relation derived below. From stoichiometric constraints (Eq. (11)), we have

$$\bar{X}_{Mg}^{OPx} = 0.5\bar{X}_{Mg}^{M2} + 0.5\bar{X}_{Mg}^{M1},$$

$$\bar{X}_{Mg}^{Cum} = 0.29\bar{X}_{Mg}^{M4} + 0.71\bar{X}_{Mg}^{M1,2,3}.$$

(27)

From the argument of the correspondence between the geometric and energetic similarities, $\bar{X}_{Mg}^{M2}(OPx) \approx \bar{X}_{Mg}^{M4}(Cum)$, $\bar{X}_{Mg}^{M1}(OPx) \approx \bar{X}_{Mg}^{M1,2,3}(Cum)$. Thus,

$$\bar{X}_{Mg}^{Cum} = \bar{X}_{Mg}^{OPx} + 0.21\left(\bar{X}_{Mg}^{M1} - \bar{X}_{Mg}^{M2}\right)^{OPx}.$$

(28)

Since in orthopyroxene $\bar{X}_{Mg}^{M1} > \bar{X}_{Mg}^{M2}$, it follows that $\bar{X}_{Mg}^{Cum} > \bar{X}_{Mg}^{OPx}$. The Butler samples, illustrated in Fig. 14(c), have crystallized $\sim700°C$ (see later). Using the intracrystalline Fe^{2+}–Mg fractionation data in orthopyroxene from

Saxena and Ghose (1971), I have calculated, according to the above expression, $\overline{X}_{Mg}^{Cum}$ versus $\overline{X}_{Mg}^{OPx}$ at 700°C. As illustrated in Fig. 14(c), the calculated distribution is in good agreement with the observational data.

Even though k_D is dependent on composition, it is found that for a given sample, it can be expressed as a function of T in the form of the "ideal expression" $\ln k_D = A + B/T$. This property can be understood in terms of expression (10), which can be recast as

$$\ln k_D = \frac{-\Delta H° + W^\alpha - W^\beta(1 - 4X_i) + \left[2X_i^\alpha(W^\beta - W^\alpha)\right]}{RT} - \frac{\Delta S°}{R}. \quad (29)$$

In this expression only the term within the square brackets depends on the site occupancy or ordering state of a crystal. But it can be easily seen from the data discussed above, e.g., those for orthopyroxenes, that the variation of this term even for a change of temperature of 400 to 1000°C is small compared to the sum of the other terms appearing as coefficient of $1/T$. Thus, for a given crystal, $\ln k_D$ effectively varies linearly as a function of $1/T$ if the interaction parameters are constants. It should also be noted from expression (29) that the intercept term relates simply to the change of the vibrational entropy between the ordered and antiordered states, and not to the "excess configurational entropy contribution," as recently suggested by some workers. Even though $\Delta S°$ should be very small, failure to take it into account in the expression of $\ln k_D$ as a function of T could lead to significant errors in the calculation of the apparent equilibrium temperature of ordering of natural minerals, as illustrated before for the cummingtonite sample DH7-482 (also see Seifert, 1975).

Kinetics of Fe^{2+}–Mg Order–Disorder in Silicates: Review and Analysis of Experimental Data

So far, the kinetics of intracrystalline Fe^{2+}–Mg distribution in only two ferromagnesian silicates, namely, orthopyroxene and anthophyllite, have been subjected to experimental investigation. In these works, the activation energies were determined according to Eq. (7.1), assuming the preexponential term to be constant. The pertinent experimental data have been summarized in Table 2 and Fig. 15. The compositions of the samples are given in Table 3.

Anthophyllite

Seifert and Virgo (1975a) have studied order–disorder kinetics in the Rabbitt anthophyllite, described in the previous section (Fig. 13). Rate experiments were performed in standard cold-seal bombs at 500 and 550°C, starting from untreated natural materials and material previously heated at 720°C for one

Table 2. Summary of the kinetic data for Fe^{2+}–Mg exchange in silicates. Disordering
rate constant $\vec{K} = K^* \exp(-Q/RT)$.

Sample	Experiments	C_0K^* (year^{-1})	Q (cal)	Source
Anthophyllite	Ordering	8.661×10^{17}	61,600	Seifert and Virgo (1975)
(Rabbitt 30)	Disordering	1.807×10^{17}	54,800	Seifert and Virgo (1975)
OPx 3209	Disordering	2.055×10^{10}	20,000	Virgo and Hafner (1969)
OPx TZ	Disordering	4.476×10^{11}	60,700	Besancon (1982)
OPx HC	Disordering	2.626×10^{11}	62,200	Besancon (1982)

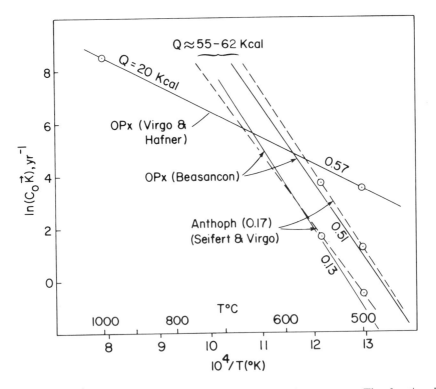

Fig. 15. $\ln(C_0\vec{K})$ versus $1/T$ for anthophyllite and orthopyroxene. The fractional
numbers denote $Fe^{2+}/(Fe^{2+} + Mg)$ ratios of the sample. The experimental measurements of Virgo and Hafner (1969) for orthopyroxene and of Siefert and Virgo (1974)
for anthophyllite are shown by circles. Besancon (1982) has determined $\vec{K}$ at 50°C
intervals for the orthopyroxene samples TZ ($X_{Fe^{2+}} = 0.51$) and HC ($X_{Fe^{2+}} = 0.13$) in
the range of 600–800°C and 650–800°C, respectively. His measured values of $\ln \vec{K}$
have a standard deviation of ~0.5.

Table 3. Composition of minerals subjected to kinetic studies (denoted by *) and cooling history calculations.

	At. Prop. Based on Six Oxygens Orthopyroxene				Oxide % Diopside	Anthophyllite	Cummingtonite
	3209*	TZ*	HC*	277	PHN1600E4	Rabbit 30*	DH7-482
Si	1.985	1.944	1.960		1.973	57.14	48.0
Al^{IV}	0.015	0.074	0.043		0.119	1.94	0.40
Al^{VI}	0.029						
Ca	0.040	0.028	0.021	0.021	0.553	0.64	2.2
Mg	0.819	0.936	1.702	1.084	1.059	26.82	18.5
Fe^{2+}	1.105[a]	0.992[a]	0.257[a]	0.859[a]	0.138	11.12	20.0[a]
Fe^{3+}					0.025	0.00	
Ti		0.002	0.002		0.009	Tr	< 0.01
Mn		0.034	0.006	0.037	0.004	0.11	1.35
Na		0.000	0.001		0.118	0.27	ND
Cr		0.000	0.011		0.008		
K						0.06	ND
F						0.00	ND
H_2O^+						2.06	ND
$Fe^{2+}/(Fe^{2+}+Mg)$	0.574	0.515	0.131	0.442	0.115	0.168	0.378

[a]Total iron as FeO.
Sources: OPx 3209: Virgo and Hafner (1969, microprobe); TZ and HC: Besancon (1982, microprobe); 277: Butler (1969, X-ray fluorescence for Fe^{2+}, Mn, Ca; $Mg = 2 - \sum X^{2+}$); diopside: McCallister *et al.* (1976, microprobe); anthophyllite: Rabbit (1948, wet chemistry); microprobe analysis gives a structural formula $Mg_{5.79}Fe^{2+}_{1.17}(Si_{7.81}Al_{0.18})O_{22}(OH)_2$, according to Seifert and Virgo (1975); cummingtonite: Mueller (1960, spectrograph); SiO_2 seems too low.

day. The oxygen fugacity was controlled by quartz–fayalite–magnetite buffer, and the site occupancies were determined by Mössbauer spectroscopy. The experimental data on the variation of $X_{Fe(4)}$ as a function of time at 550°C is shown in Fig. 16. These data can be fitted adequately according to Mueller's second-order kinetic model. Assuming again that M1, M2, and M3 sites can be treated as a single group, Seifert and Virgo analyzed the ordering and disordering data in terms of Eqs. (14.1) and (14.2), respectively, to obtain values of $C_0\vec{K}$ at 500 and 550°C. As shown in Table 2, the value of $C_0\vec{K}$ derived from the ordering experiments is considerably greater than that derived from the disordering experiments. Seifert and Virgo (1975) suggested that this might be due to the compositional dependence of $\vec{K}$. Figure 17 shows the values of $C_0\vec{K}$, derived from the data for each ordering and disordering experiment, as a function of $X_{Fe(4)}$. It is evident that the variation of $X_{Fe(4)}$ in the disordering experiments is nearly twice as much as the minimum difference of $X_{Fe(4)}$ between the ordering and disordering experiments; yet $\vec{K}$ is nearly constant for the disordering experiments. Therefore, the observed difference between the values of $\vec{K}$ derived from ordering and disordering data cannot be primarily due to the compositional dependence of $\vec{K}$. The reason for this difference is not clear and may lie in the physical properties of the samples and/or the amount of H_2O in the sample capsules. The physical state

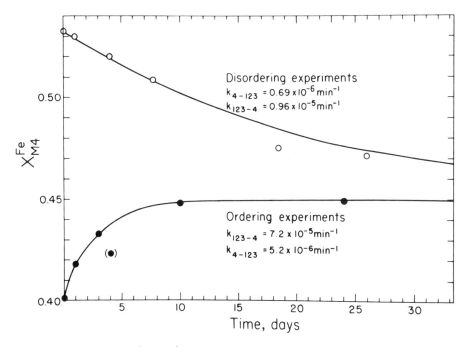

Fig. 16. Variation of $Fe^{2+}/(Fe^{2+}+Mg)$ in the M4 site of Rabbitt anthophyllite as a function of time at 550°C. The figure is taken from Seifert and Virgo (1975). The experimental data, which are shown as circles, have been fitted according to Muller's (1969) kinetic model. The bracketed ordering experiment has not been considered in the calculations. The starting materials for the disordering (open circles) and ordering experiments (closed circles) are respectively natural unheated anthophyllite and anthophyllite previously disordered at 720°C, 2 kbars for one day.

of the sample, such as plastic deformation, grain size, fluid inclusion, as well as the amount of H_2O have been found to affect the Al–Si order–disorder kinetics in feldspar (Yund and Tullis, 1980).

Orthopyroxene

The disordering kinetics in orthopyroxene has been studied by Virgo and Hafner (1969), Khristoforov *et al.* (1974), and Besancon (1981). The disordering rate constant, $\vec{K}$, derived by Virgo and Hafner and Besancon from their respective experimental data are illustrated in Fig. 15 as a function of $1/T$. In both works, the Fe^{2+} site occupancies were determined by Mössbauer spectroscopy. Khristoforov *et al.*, on the other hand, determined the site occupancies by NMR spectroscopy (see Ghose, pt. 1, for discussion). Khristoforov *et al.* rejected Mueller's model as being inadequate for their data, and proposed a first-order kinetic model of the form $Fe^{2+}(M1,t) = Fe^{2+}(M1,t_0)\exp(-\vec{K}t)$, with a first stage characterized by rapid disordering followed by a second

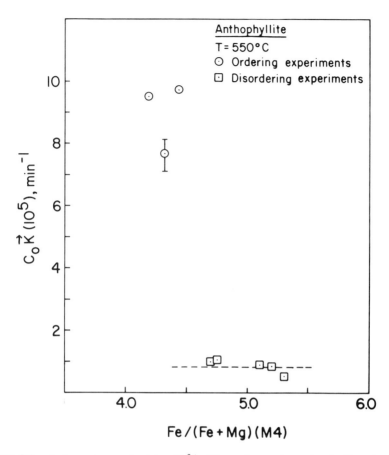

Fig. 17. Disordering rate constant for Fe^{2+}–Mg exchange in anthophyllite obtained from each ordering and disordering experiment of Seifert and Virgo (1975).

stage with much slower disordering. Besancon (1981a) has reanalyzed the data of Khristoforov *et al.* and has shown that their disordering data at only 500°C are good enough for kinetic analysis and these data can indeed be adequately fitted by Mueller's model to yield $C_0 \vec{K} = 2 \times 10^{-4}$ $min^{-1} = 1.051 \times 10^2$ $year^{-1}$. However, the problem still remains in accepting Khristoforov *et al.*'s data in that their most disordered sample ($X_{Fe^{2+}(M1)} \cong 0.32$) is apparently more disordered than permitted by the equilibrium site-occupancy constraint ($X_{Fe^{2+}(M1)} = 0.278$) according to Eqs. (24) and (25).

The activation energy for Fe^{2+}–Mg disordering in orthopyroxene derived by Besancon from his experimental data is in sharp disagreement with that of Virgo and Hafner, but is similar to Q (disordering) in anthophyllite (Fig. 15). In view of the geometric similarities of the octahedral sites in orthopyroxene and anthophyllite, one would, however, expect the energetics of Fe^{2+}–Mg exchange within these two silicates to be comparable. As shown earlier (Fig. 14), the assumption that geometrically similar sites in different silicates are energetically equivalent leads to prediction of Fe^{2+}–Mg fractionation between coexisting phases, notably in the anthophyllite–cummingtonite and cummingtonite–orthopyroxene pairs, that are in very good agreement with the observational data. Thus, Virgo and Hafner's activation energy of disordering in orthopyroxene seem to be too small. Besancon (1981a) has pointed out that their rate constant at 1000°C is based on inadequate data in that none of the disordered sample has Fe^{2+} site occupancy that deviates from the equilibrium value beyond one standard deviation. The rate constant at 500°C, on the other hand, is based on adequate disequilibrium experiments, as well as equilibrium site-occupancy data ($X_{Fe(M1)}$ = 0.278) which agree very well with that predicted (0.279) from Eqs. (24) and (25). Virgo and Hafner's K (500°C), therefore, seem reliable. This is, however, about eleven times the disordering rate constant according to Besancon's data on sample TZ, which has only a slightly different $Fe^{2+}/(Fe^{2+} + Mg)$ ratio (Table 3). Yund and Tullis (1980) have recently shown that evacuated atmosphere enhances the Al–Si disordering kinetics in felspars. It is, therefore, interesting to note that Virgo and Hafner conducted their experiments in evacuated (5×10^{-4} mm Hg) silica tubes, whereas Besancon's experiments were done in a controlled gas atmosphere.

As discussed earlier, Besancon's site occupancy determinations for the magnesium-rich sample HC might have been subject to some systematic error. Therefore, his kinetic data for this sample should be treated with caution. In summary, then, Virgo and Hafner's data at 500°C and Besancon's data for the sample TZ can be recommended as reliable. Using the disordering activation energy of 60.7 kcal, as determined by Besancon for sample TZ, Virgo and Hafner's $C_0\vec{K}$ (500°C) yields $C_0\vec{K}(T) = 7.283 \times 10^{12}\exp(-30{,}549/T)$ min^{-1}, as compared to $C_0\vec{K}(T) = 6.4761 \times 10^{11}\exp(-30{,}549/T)$ min^{-1} according to Besancon. These data apply to orthopyroxenes with $Fe^{2+}/(Fe^{2+} + Mg) \approx 0.50$–0.60.

Two-Step Mechanism of Order–Disorder

Virgo and Hafner (1970) have measured the intracrystalline Fe^{2+}–Mg fractionation in 30 orthopyroxene with a wide range of Fe^{2+}/Mg ratios from granulite facies rocks and noted that the distribution data fall close to the 500°C isotherm constructed on the basis of their experimental disordering data at 500°C for sample No. 3209 (Fig. 5a) and the assumption of "ideal" intracrystalline fractionation. Since such a high apparent equilibration temperature of ordering for the natural samples seems highly incompatible with the time scale of cooling of metamorphic rocks and the activation energy of ordering around 20 kcal, as deduced by Virgo and Hafner (1969), Mueller (1970) proposed a two-step mechanism for order–disorder kinetics in orthopyroxene. In this model, which is illustrated in Fig. 18, the order–disorder process is thought to be governed by two consecutive steps, one (H-step) characterized by low and the other (L-step) by high activation energies. In

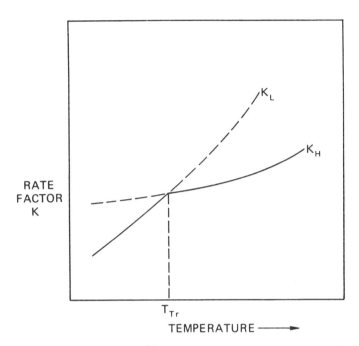

Fig. 18. Two-step mechanism for Fe^{2+} –Mg order–disorder kinetics in orthopyroxene suggested by Mueller (1970). At any T, the kinetics is governed by the smaller of the two rate constants.

consecutive processes, the overall rate is governed by the step for which the rate constant has the lowest value. Thus, the low-Q step would govern the high-temperature process, whereas the high-Q step would govern the low-temperature process, the K for both processes being equal around the inferred intracrystalline equilibration temperature of the granulite facies samples.

Saxena and Ghose (1971) have obtained distribution data at 500–800°C (Fig. 6, samples No. 3 and 6) in two of the granulite facies samples analyzed by Virgo and Hafner (1970). As illustrated in Fig. 6, the apparent equilibration temperatures of these samples are found to be 437 and 381°C when the k_D of the untreated natural samples are compared with the linear extrapolation of their respective $\ln k_D$ versus $1/T$ trends, ignoring the 500°C data for reasons discussed earlier. (This method will be referred henceforth as "distribution coefficient treatment.") Both of these samples are from the metamorphosed iron formation of the Gagnone area, Quebec, collected by Butler (1969; his samples No. 278 and 277, respectively). Similar apparent equilibration temperatures (461 and 381°C) are also obtained for these samples from the "equilibrium constant" treatment (Fig. 7). For sample No. 3209, the "distribution coefficient treatment" according to Virgo and Hafner's data yield equilibration temperature of 240°C, as compared to 461°C obtained from the "equilibrium constant" treatment. The first estimate is unlikely to be correct since, as discussed earlier, Virgo and Hafner's experimental distribution data above 500°C for this sample do not agree with those of Saxena and

Ghose (1971) and Besancon (1981a) for samples of similar compositions. The above equilibration temperatures of natural samples are, thus, up to $\sim 100°C$ lower than the earlier estimates of Virgo and Hafner. Further, as discussed earlier, the Q for Fe^{2+}–Mg order–disorder in orthopyroxene should be around 60 kcal instead of the earlier estimate of 20 kcal by Virgo and Hafner. Thus, the reasons for invoking a two-step consecutive mechanism for the order–disorder kinetics in orthopyroxene now seem much less compelling, and even less so when the inferred equilibration temperatures of natural anthophyllites and cummingtonites (Fig. 13), which are expected to have similar intracrystalline exchange kinetics as orthopyroxene, are taken into account. However, the possibility of a change of ordering mechanism at lower temperature cannot be excluded at this stage and needs to be tested by comparing the cooling rates of natural ferromagnesian silicates calculated on the basis of experimental data on order–disorder kinetics (see later) with those inferred from other methods.

Ordering in Natural Samples and Cooling History

General Considerations

Since the diffusive path lengths for intracrystalline exchange are extremely small, the state of ordering of a mineral in natural rock could undergo significant retrograde adjustments, especially if it cools slowly. Thus, the apparent equilibrium temperatures corresponding to the observed ordering state in natural minerals (Figs. 6, 7, 11, and 13) from plutonic and metamorphic rocks, in contrast to those from volcanic rocks (or from volcanic breccia), are much lower than the crystallization temperature of the host rocks, as can be inferred from heterogeneous partitioning data and other geological evidences. Although this observation severely limits the usefulness of intracrystalline partitioning data as geothermometers, these data can be utilized to constrain the cooling history of rocks.

As illustrated in Fig. 19, the ordering state of a crystal in a cooling system maintains equilibrium with temperature for a certain period of time, depending on the cooling rate and kinetics of ordering. As the system cools further, the kinetics of ordering, which decreases exponentially with T, becomes too sluggish for the crystal to maintain equilibrium with T. Under this condition, the kinetically controlled ordering path, K, diverges from the equilibrium ordering path, E, and *asymptotically* approaches a constant ordering state. Within the precision of our measurements, however, a constant ordering state is effectively established at some temperature T_Q, which we define as the quenching temperature. The apparent equilibrium temperature corresponding to this observed ordering state is often referred as the closure temperature, T_C (Dodson, 1976). Evidently, T_C depends on the cooling and ordering rates; for

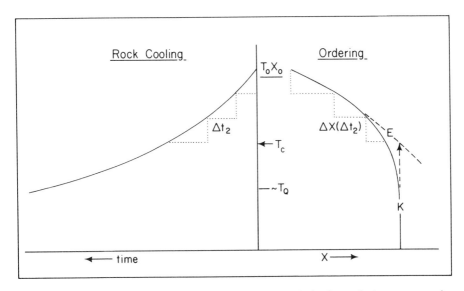

Fig. 19. Ordering in a continuously cooling system. X_0 is the ordering state at the temperature T_0. E is the equilibrium ordering path, whereas K is the actual ordering path which is controlled by the kinetics of intracrystalline exchange and cooling rate. T_C is the closure temperature or temperature of apparent equilibration corresponding to the quenched ordering state. T_Q is the temperature from which the observed ordering state has been effectively quenched.

example, for a given ordering rate, the faster the cooling rate, the higher the T_C. Thus, the cooling rate can be constrained on the basis of the quenched ordering state if the appropriate data for the order–disorder kinetics are known.

The cooling rate of minerals has been previously estimated (Seifert and Virgo, 1975; Besancon, 1981b) from ordering data through the use of temperature–log time–transformation or T–t–T plot. The method, which is discussed below, is based on the (usually unstated) assumption that the time required by a crystal to attain a certain amount of ordering under isothermal condition is the same as that in a continuously cooling system. Using the estimated value of X_i^α at the inferred crystallization temperature as the initial ordering state (Eq. (30)), the times required to achieve a series of values of X_i^α under isothermal condition are calculated for a set of T from Eq. (14.1). The data are contoured for various X_i^α values on a T–$\log t$ diagram to produce a kind of "kinetic phase diagram" as illustrated in Fig. 20. A series of cooling curves for the rock, based on an assumed cooling model, is then superimposed on the T–t–T plot. So long as the above assumption holds, a crystal cooling along a particular path passes through the ordering states given by the contours that the cooling curve intersects. Consequently, the final ordering state is given by the contour which appears as a tangent to the cooling curve since at a lower temperature the cooling curve intersects a contour representing a more disordered state. Therefore, the cooling curve that is finally chosen

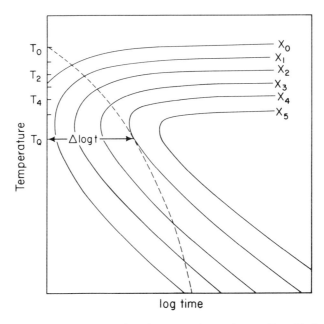

Fig. 20. Hypothetical temperature–log time–transformation or T–t–T plot of a mineral. $X_1, X_2, \ldots$ are site-occupancy numbers. Each contour goes through a minimum of $\log t$ and asymptotically approaches a constant T, which is the temperature of equilibrium ordering corresponding to the contour. For a given cooling path, such as shown by the dashed line, the quenched ordering state is given by the contour which appears as a tangent to the cooling path provided that the time to achieve a certain state of ordering under isothermal condition is the same as that in a continuously cooling system. Δt is the time of ordering from X_0 to X_4 isothermally at T_Q.

is the one that is tangent to a contour approximating the observed or quenched ordering state. It should, however, be clear that ultimately this "T–t–T method" is estimating Δt for isothermal ordering at $\sim T_Q$ from the initial ordering state to approximately the final observed state of order and finding the cooling rate based on this Δt and an assumed cooling model. However, the fact that a crystal orders continuously as it cools can be essentially and easily taken care of by considering that the ordering has taken place through a set of isothermal steps. In this method, which is illustrated in Fig. 19, an initial cooling curve is computed from an assumed cooling model, e.g., $1/T = 1/T_0 + \eta(\Delta t)$, and Δt for cooling from T_0 to T_C approximated from Eq. (14.1) for $\sim 99\%$ ordering at T_C. The cooling curve is, then, approximated by a series of isothermal steps. Using then the $\Delta t(T)$ permitted by the cooling curve, the final ordering state at each T is calculated from expression (31), given below. The process is repeated, varying the cooling curve (or cooling rate constant η), as necessary, until the calculated X versus T relation produces the observed quenched state.

The initial ordering state at the inferred crystallization temperature can be easily obtained by combining the stoichiometric constraint (11) with the expression of k_D, as defined by Eq. (8.2). Thus, requiring $\bar{X}_i^\alpha < 1.0$,

$$\bar{X}_i^\alpha(T_0) = \frac{-B + (B^2 - 4AC)^{1/2}}{2A}, \tag{30}$$

where

$$A = p(k_D^0 - 1),$$

$$B = k_D^0(q - \bar{X}_i) + \bar{X}_i + p,$$

$$C = -\bar{X}_i,$$

k_D^0 being the value of k_D at T_0.

Expression (14.1) can be rearranged to yield the following expression for $\bar{X}_i^\alpha(t)$ for isothermal ordering through Δt.

$$\bar{X}_i^\alpha(t) = \frac{(b^2 - 4ac)^{1/2}(1 + FD) - b(1 - FD)}{2a(1 - FD)}, \tag{31}$$

where

$$F = \exp\left[-C_0\vec{K}\Delta t(b^2 - 4ac)^{1/2}\right],$$

$$D = \frac{\left[2a\bar{X}_i^\alpha(t_0) + b\right] - (b^2 - 4ac)^{1/2}}{\left[2a\bar{X}_i^\alpha(t_0) + b\right] + (b^2 - 4ac)^{1/2}}.$$

The results of cooling history calculations for some natural samples are summarized in Table 4, and those of a cummingtonite sample are illustrated in Fig. 21. It has been assumed in these calculations that the rocks have cooled according to either $1/T = 1/T_0 + \eta t$ or $T = T_0\exp(-\eta t)$, which will be referred henceforth as the "asymptotic" and "exponential" cooling models. In both models, the rate of cooling decreases with decreasing T, and $t \to \infty$, as $T \to 0$. The calculated X versus T curves follow the equilibrium paths quite close to T_C or the apparent equilibrium temperatures for intracrystalline ordering. Variation of the crystallization temperatures up to the temperature (T^*) through which the kinetic and equilibrium curves are coincident has no effect on the calculated cooling rate. This implies that a cooling rate calculated on the basis of the quenched ordering state and $X(T_0) < X(T^*)$ is strictly applicable over the temperature interval T^* to T_Q.

The calculated cooling rates are found to be very sensitive to the uncertainties in the site-occupancy determinations. The reason for this lies in the high activation energies, $Q \approx 60$ kcal, of the order–disorder process and the nature of the function $C_0\vec{K}\Delta t$, as illustrated in Fig. 4. The high Q leads to values of T_Q close to the temperature of apparent equilibration T_c (Table 4). The observed site occupancy represents 100% ordering, as defined by expression (15), at T_c, and thus close to 100% ordering at T_Q. Since the function $C_0\vec{K}\Delta t$ is very sensitive to $\Delta\%$ near $\Delta\% = 100$, a small change in the site-occupancy number produces a very large change in $C_0\vec{K}\Delta t$. Consequently, the cooling

Table 4. Summary of cooling rate calculations of natural samples (see text for discussion about uncertainty of the time constant η). Asymptotic cooling model: $1/T = 1/T_0 + \eta t$; exponential cooling model: $T = T_0 \exp(-\eta t)$.

| Sample | Approximate Temperature (°C) of | | | X_{Fe} (M4 or M2) | | η (year^{-1}) | | Approximate Cooling Rate at T_C |
	$X'tn$	Apparent Equilibrium (T_c)	Quenching (T_Q)	Natural	Calculated at $\sim T_Q$	Asymptotic	Exponential	
Anthophyllite (Rabbitt 30)	625–750	295	245	0.533 (M4)	0.533	2.8156×10^{-10}		90°/10⁶ years
							0.1527×10^{-6}	85°/10⁶ years
Cummingtonite (DH7-482)	525	380	350	0.78 ± 0.01 (M4)	0.780	2.003×10^{-7}		5.5°/10² years
							2.0547×10^{-4}	13°/10² years
Orthopyroxene (Butler 277)	700	383	325	0.797 (M2)	0.7965	6.9570×10^{-7}		30°/10² years
Clinopyroxene (PHN1600E4)	1375	530	510	0.323 ± 0.013 (M2)	0.3236	1.0070×10^{-2}		18°/day
							8.325	18°/day

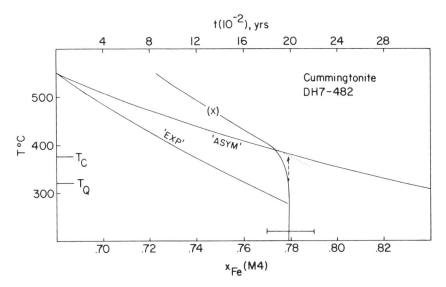

Fig. 21. The calculated cooling rate and correspondng evolution of $X_{Fe(M4)}$ versus T in a cummingtonite sample (DH7-482, Mueller, 1960) from the Precambrian Iron Formation, Bloom Lake, Ontario (Canada). $X_{Fe(M4)}$ determined by Mössbauer spectroscopy (Ghose and Weidner, 1972) in the natural sample is 0.78 ± 0.01. 'ASYM' and 'EXP' refer to cooling paths according to the 'asymptotic' and 'exponential' models (see text).

rates obtained from the ordering data cannot usually be relied upon, when all the uncertainties are considered, as anything better than "order of magnitude" numbers. The situation could be somewhat improved by considering multiple samples from the same rock and determining site occupancies by both Mössbauer and single-crystal techniques to minimize any systematic bias. The difference between the cooling rates obtained from the "asymptotic" and "exponential" models seems to be small compared to their uncertainties arising from those in the site-occupancy data. The cooling rates have been calculated for the following samples.

Rabbitt Anthophyllite

This is the sample investigated by Seifert and Virgo (1974), as discussed earlier, to obtain both kinetic and equilibrium intracrystalline distribution data as a function of temperature. The sample is from the Precambrian Dillion complex, Montana, and is associated, within an ultramafic body, with actinolite, serpentine, enstatite, clinohumite, spinel, etc. (Rabbitt, 1948). The quenched $X_{Fe(M4)}$, as measured by these workers, is 0.5327. Seifert and Virgo (1975) have also calculated the possible cooling rate for this mineral using the $T–t–T$ plots along with a linear cooling model and expressing the distribution coefficient in the form $\ln k_D = A/T$, where A is the average $\Delta G°$ over the experimental temperature range of 550–720°C. They concluded 100°C/10⁶ years as the maximum possible cooling rate for this sample. The cooling rate of this sample has been recalculated using the stepwise cooling algorithm and expression of $\ln k_D$ in the form of $A/T + B$ on the basis of Seifert and Virgo's data. (The recalculated expression, $\ln k_D = 0.36035 - 2.46158(10^3/T)$, yields $T_C = 302°C$, as compared to Seifert and Virgo's estimate of $T_C \approx 270°C$.) Comparison of the mineral assemblage and anthophyllite composition with the phase diagram of Mueller (1973) suggests crystallization temperature ~ 625–750°C. A cooling rate of ~ 75–100°C/10⁶ years, depending on whether the kinetic data from disordering or ordering experiments are used, yields $X_{Fe(M4)} = 0.533$. Seifert and Virgo (1974) did not report any uncertainty in their measured site-occupancy values. However, an error of 1% in the value of $X_{Fe(M4)}$ (see Ghose, pt. 1, for a discussion about precision in Mössbauer analysis) is found to affect the cooling rate by a factor of ~ 100. Seifert (1978) has recalculated the equilibrium site occupancies in this crystal (Fig. 13), but did not reduce the kinetic data using his revised scheme. His revised site-occupancy data and Seifert and Virgo's kinetic data yield about a 100–300 times faster cooling rate than that above. This fast cooling rate is probably an artifact of using two methods of site-occupancy calculations in the reduction of equilibrium and kinetic data. Seifert and Virgo (1975) did not give suffi-cient details of their kinetic data to enable us to reduce these data using Seifert's revised scheme of site-occupancy calculations.

Cummingtonite and Orthopyroxene from Iron Formation, Quebec

Although there are no data on the kinetics of Fe^{2+}–Mg exchange in cum-mingtonite, the cooling rate of the sample "Mueller DH7-482" (Fig. 13) from the Precambrian metamorphosed Iron Formation, Bloom Lake Area, Quebec, has been calculated on the basis of Seifert and Virgo's kinetic data for anthophyllite. A crystallization temperature $\sim 525°C$ has been suggested by Mueller (1973) for the Bloom Lake rocks. As discussed earlier, the energetics of the (Fe, Mg) sites in cummingtonite and anthophyllite are expected to be similar. It is assumed that the compositional difference between the Mueller cummingtonite and Rabbit anthophyllite (Table 3) does not have very signifi-cant effect on the Fe^{2+}–Mg order–disorder kinetics. The error introduced by this assumption is probably within the limits of that due to the uncertainties in site-occupancy determinations. Taking into account the quoted uncertainty in the determination of $X_{Fe(M4)}$ of ± 0.01 in the natural sample and the range of values of the kinetic data for anthophyllite, we obtain (Fig. 21) cooling rate $\sim 10(^\times_\div 10)°/10^2$ years around $T_C \approx 380°C$. (Most of the uncertainty ($\sim$two-thirds) in the inferred cooling rate arises from that in the site-occupancy data.) This is a surprisingly fast cooling rate for a metamorphic rock. However, similar fast cooling rate, $\sim 35°/10^2$ years, is also obtained for an orthopy-roxene sample (Butler sample No. 277 which is the same as Saxena and Ghose, 1971, sample No. 3; see Fig. 6), collected by Butler (1969) from the Iron Formations in the Gagnon Region, which is about 70 miles southwest of Bloom Lake but belongs to the same Labrador trough. The compositions of coexisting cummingtonite and orthopyroxene suggest a crystallization temper-ature, according to Mueller's (1973) phase diagram, $\sim 700°C$. The cooling rate calculations are based on Besancon's kinetic data for the orthopyroxene sample TZ (Fig. 15), which is of similar composition as the Butler sample (Table 3). Because of the sensitivity of the calculated cooling rates to errors in the site-occupancy data and uncertainty in the linear extrapolation of $\ln \vec{K}$ versus $1/T$ to low temperature, the inferred unusually fast cooling rate of the Iron Formation samples around $400°C$ should be considered as an interesting possibility rather than definitive. More work should be done, especially on the kinetics of intracrystalline exchange in cummingtonite and site-occupancy determinations of the Mueller sample DH7-482, to define the cooling rate. Also, the bulk compositions of both the cummingtonite and orthopyroxene samples need to be redetermined by microprobe, especially since the site-occupancy calculations from Mössbauer spectra are sensitive to errors in the bulk $Fe^{2+}/(Fe^{2+}+Mg)$ ratio.

Clinopyroxene from Lesotho Kimberlite Pipe

The cooling rate has been calculated for the discrete diopside nodule (PHN1600E4) from Lesotho kimberlite pipe described earlier (Fig. 11(a) and Table 3). The calculations are based on Besancon's kinetic data for $Fe^{2+}-Mg$ exchange in orthopyroxene, sample TZ (Fig. 15). It is assumed, as a first approximation, that the $Fe^{2+}-Mg$ exchange kinetics in the diopside $(Fe^{2+}/(Fe^{2+}+Mg)=0.323)$ is approximately the same as that in the ortho-pyroxene $(Fe^{2+}/(Fe^{2+}+Mg)=0.51)$ except for the difference arising from the differences in the numbers of Fe^{2+} and Mg centered M1 and M2 sites per unit volume in the two crystals. In other words, it is assumed that the essential difference between the $Fe^{2+}-Mg$ exchange kinetics in the two crystals lies in the values of C_0, p, and q. Even though it is a crude assumption, especially considering that the coordination number of M2 sites is 8 in diopside and 6 in orthopyroxene, it is believed that the error introduced by this assumption in the calculated cooling rate lies within the limits of the uncertainty due to that in the site-occupancy numbers. For the "PHN" diopside, we obtain from the data of McCallister et al. (1976), which are corrected for the exsolved pigeonite, $C_0 = 0.016 \times 10^{24}/cm^3 = 0.864 C_0(OPx)$, $p = 0.28$, $q = 0.72$. Nixon and Boyd (1973) estimated crystallization temperature of $\sim 1375°C$ for this discrete nodule. With these conditions, we obtain a cooling rate of $\sim 18°/day$ around $T_c \approx 530°C$. McCallister et al. quote an uncertainty of ± 0.004 in their site-occupancy numbers of the natural sample, which introduces an uncertainty of a factor of ~ 18 in the cooling rate around T_c. Although this is too large an uncertainty in the cooling rate, it is still interesting to speculate about the transport velocity of the kimberlite sample on the basis of the mean cooling rate and compare it with that inferred by other method.

On the basis of viscous partial thermoremanence acquired by accidental inclusions and wall rock samples close to the contact at the time of kimberlite emplacement and heat flow model including convection of the emplaced material, McFadden (1977) suggested emplacement temperature around 300°C for four Cretaceous kimberlite pipes in the Kimberly region. He further estimated that the inclusions were transported on the average of at least 70 m/hour. Assuming that our estimated mean cooling time constant of 0.2759×10^{-4} day^{-1} for the "asymptotic cooling model" holds to the temperature of ~ 300°C, we obtain ~ 22 days for the cooling of the discrete nodule from 600 to 300°C. It is reasonable to assume that the temperature–depth trajectory of the kimberlite was not steeper than the prevailing geothermal gradient in the temperature range 300–600°C. The geothermal gradient during the Cretaceous period in Lesotho and surrounding region was probably approximately the same as Clark and Ringwood's shield geotherm below ~ 1000°C (Boyd, 1973; Lane and Ganguly, 1980). Thus, the kimberlites were probably transported through at least 35 km as they cooled from 600 to 300°C. This implies a transport velocity of at least 66 m/hour, which is in excellent agreement with MacFadden's estimate. The close agreement is indeed surprising considering the large uncertainties associated with both calculations.

Acknowledgments

Thanks are due to Drs. S. Ghose, R. F. Mueller and S. K. Saxena for reading an early draft of the manuscript and suggesting improvements. Part of the work on the thermodynamic analysis of Fe–Mg distribution in orthopyroxene was carried out while the author was a guest of Dr. and Mrs. Saxena in New York. Their generous hospitality is gratefully acknowledged. Thanks are also due to Dr. Besancon for making available his manuscript before publication, and to Dr. McCallister for providing data on Fe–Mg distribution in clino-pyroxenes, PHN 1600E4 and Kakanui augite.

References

Besancon, J. R. (1981a) Rate of cation ordering in orthopyroxenes, *Amer. Mineral.* **66**, 965–973.

Besancon, J. R. (1981b) Cooling rate of orthopyroxene-bearing rocks estimated from magnesium-iron intersite distribution, *Trans. Amer. Geophys. Union EOS* **62**, 437.

Bonnischen, B. (1969) Metamorphic pyroxenes and amphiboles in the Biwabik Iron Formation, Dunka River Area, Minnesota. *Mineral. Soc. Amer.* Spec. Pap. **2**, 217–239.

Boyd, F. R. (1973) A pyroxene geotherm, *Geochim. Cosmochim. Acta* **37**, 2533–2546.

Bradley, R. S. (1962) Thermodynamic calculations of phase equilibria involving fused salts. Part 1. General theory and application to equilibria involving calcium carbonate at high pressure, *Amer. J. Sci.* **260**, 374–382.

Bragg, W. L., and Williams, E. J. (1934) The effect of thermal agitation on atomic arrangement in alloys, *Proc. Roy. Soc. London* **151A**, 540–566.

Butler, P. (1969) Mineral compositions and equilibria in the metamorphosed iron formation of the Gagnon Region, Quebec, Canada, *J. Petrology* **10**, 56–101.

Darken, L. S., and Gurry, R. W. (1953) *Physical Chemistry of Metals.* McGraw-Hill, New York.

Dienes, G. J. (1955) Kinetics of order–disorder transformation, *Acta Metall.* **3**, 549–557.

Dodson, M. H. (1976) Kinetic processes and thermal history of slowly cooling solids, *Nature* **259**, 551–553.

Flood, H., Førland, T., and Grjotheim, K. (1954) Uber den Zusammenhang zwischen konzentration and aktivitäten in geschmolzenen salzmischungen, *Z. Anorg. Allgem. Chem.* **276**, 290–315.

Ganguly, J. (1976) Energetics of natural garnet solid solutions. II. Mixing of the calciumsilicate end-members, *Contrib. Mineral. Petrol.* **55**, 81–90.

Ghose, S. (1982) Order–disorder in ferromagnesian silicates. I. Crystal-chemistry, in *Advances in Physical Geochemistry*, 2, edited by S. K. Saxena, Springer-Verlag, Berlin/Heidelberg/New York.

Ghose, S., and Weidner, J. R. (1972) $Mg^{2+}-Fe^{2+}$ order–disorder in cummingtonite, $(Mg, Fe)_7Si_8O_{22}(OH)_2$: A new geothermometer, *Earth Planet. Sci. Lett.* **16**, 346–354.

Glasstone, S. K., Laidler, J., and Eyring, H. (1941) *The Theory of Rate Processes.* McGraw-Hill, New York.

Grover, J. (1974) On calculating activity coefficients and other excess functions from the intracrystalline exchange properties of a double-site phase, *Geochim. Cosmochim. Acta* **38**, 1527–1548.

Grover, J., and Orville, P. (1970) Partitioning of cations in coexisting single- and multi-site phases: A reply with incidental corrections, *Geochim. Cosmochim. Acta* **34**, 1361–1364.

Hafner, S. S., and Ghose, S. (1971) Iron and magnesium distribution in cummingtonites $(Fe, Mg)_7Si_8O_{22}(OH)_2$, *Z. Kristallogr.* **133**, 301–326.

Hafner, S. S., Virgo, D., and Warburton, D. (1971) Cation distribution and cooling history of clinopyroxenes from Oceanus Procellarum, *Proc. Second Lunar Sci. Conf.* **1**, 91–108.

Kerrick, D. M., and Darken, L. S. (1975) Statistical thermodynamic models of ideal oxide and silicate solid solutions, with application to plagioclase, *Geochim. Cosmochim. Acta* **39**, 1431–1442.

Khristoforov, K. K., Nikitina, L. P., Krizhanskiy, L. M., and Yekimov, S. P. (1974) Kinetics of disordering of distribution of Fe^{2+} in orthopyroxene structures, *Dokl.*

Akad. Nauk SSSR **214**, 909–912 (transl. *Dokl. Akad. Nauk SSSR* **214**, 165–167 (1975)).

Kishina, N. R. (1978) Decomposition structures and Fe^{2+} –Mg ordering in pyroxenes as indicators of subsolidus cooling rates, *Geochem. Int.* **15**, 65–78.

Kretz, R. (1978) Distribution of Mg, Fe^{2+} and Mn in some calcic pyroxene hornblende–biotite–garnet gneisses and amphibolites from the Grenville Province, *J. Geology* **86**, 599–619.

Lane, D. L., and Ganguly, J. (1980) Al_2O_3 solubility in orthopyroxene in the system $MgO–Al_2O_3–SiO_2$: A reevaluation and mantle geotherm, *J. Geophys. Res.* **85**, 6963–6972.

Matsui, Y., and Banno, S. (1965) Intracrystalline exchange equilibrium in silicate solid solutions, *Proc. Japan Acad.* **41**, 461–466.

McCallister, R. H., Finger, L. W., and Oshashi, Y. (1976) Intracrystalline Fe^{2+} –Mg equilibria in three natural Ca-rich clinopyroxenes, *Amer. Mineral.* **61**, 671–676.

McFadden, P. L. (1977) A paleomagnetic determination of emplacement temperature of some South African kimberlites, *Geophys. J. R. Astr. Soc.* **50**, 587–604.

Mueller, R. F. (1960) Compositional characteristics and equilibrium relations in mineral assemblages of a metamorphosed iron formation, *Amer. J. Sci.* **258**, 449–497.

Mueller, R. F. (1961) Analysis of relations among Mg, Fe and Mn in certain metamorphic minerals, *Geochim. Cosmochim. Acta* **25**, 267–296.

Mueller, R. F. (1962) Energetics of certain silicate solid solutions, *Geochim. Cosmochim. Acta* **26**, 581–598.

Mueller, R. F. (1967) Model for order–disorder kinetics in certain quasi-binary crystals of continuously variable composition, *J. Phys. Chem. Solids* **28**, 2239–2243.

Mueller, R. F. (1969) Kinetics and thermodynamics of intracrystalline distribution, *Mineral. Soc. Amer. Spec. Pap.* No. 2, 83–93.

Mueller, R. F. (1970) Two-step mechanism for order–disorder kinetics in silicates, *Amer. Mineral.* **55**, 1210–1218.

Mueller, R. F. (1973) System $CaO–MgO–FeO–SiO_2–C–H_2–O_2$: Some correlations from nature and experiment, *Amer. J. Sci.* **273**, 152–170.

Mueller, R. F., Ghose, S., and Saxena, S. K. (1970) Partitioning of cations between coexisting single- and multi-site phases: A discussion, *Geochim. Cosmochim. Acta* **34**, 1356–1360.

Navrotsky, A. (1971) The intracrystalline cation distribution and the thermodynamics of solid solution formation in the system $FeSiO_3–MgSiO_3$, *Amer. Mineral.* **56**, 201–211.

Nix, F. C., and Shockley, W. (1938) Order–disorder transformation in alloys, *Rev. Mod. Phys.* **10**, 1–67.

Powell, R. (1976) Activity–composition relationships for crystalline solutions, in *Thermodynamics in Geology*, edited by D. G. Fraser, Reidel, Dordrecht, The Netherlands.

Papike, J. J., and Ross, M. (1970) Gedrites: crystal structures and intracrystalline cation distributions, *Amer. Mineral.* **55**, 1945–1972.

Rabbitt, J. C. (1948) A new study of the anthophyllite series, *Amer. Mineral.* **33**, 263–323.

Ramberg, H., and DeVore, G. W. (1951) The distribution of Fe^{2+} and Mg^{2+} in coexisting olivines and pyroxenes, *J. Geology* **59**, 193–210.

Sack, R. O. (1980) Some constraints on the thermodynamic mixing properties of Fe–Mg orthopyroxenes and olivines, *Contrib. Mineral. Petrol.* **71**, 257–269.

Saxena, S. K. (1973) *Thermodynamics of Rock-Forming Crystalline Solutions*. Springer-Verlag, Berlin/Heidelberg/New York.

Saxena, S. K., and Ghose, S. (1971) Mg^{2+}–Fe^{2+} order–disorder and the thermodynamics of the orthopyroxene crystalline solution, *Amer. Mineral.* **56**, 532–559.

Saxena, S. K., Ghose, S., Turnock, A. C. (1974) Cation distribution in low-calcium pyroxenes: dependence on temperature and calcium content and the thermal history of lunar and terrestrial pigeonites, *Earth and Plant. Sci. Lett.* **21**, 194–200.

Seifert, F. (1975) Equilibrium Mg^{2+}–Fe^{2+} cation distribution in anthophyllite, *Amer. J. Sci.* **278**, 1323–1333.

Seifert, F., and Virgo, D. (1974) Temperature dependence of intracrystalline Fe^{2+}–Mg distribution in a natural anthophyllite, *Carnegie Inst. Washington Yearbook* **73**, 405–411.

Seifert, F., and Virgo, D. (1975) Kinetics of Fe^{2+}–Mg order–disorder reaction in anthophyllites: Quantitative cooling rates, *Science* **188**, 1107–1109.

Snellenburg, J. W. (1975) Computer simulation of the distribution of octahedral cations in orthopyroxenes, *Amer. Mineral.* **60**, 441–447.

Stout, J. H. (1972) Phase petrology and mineral chemistry of coexisting amphiboles from Telemark, Norway, *J. Petrology* **13**, 99–145.

Temkin, M. (1945) Mixtures of fused salts as ionic solutions, *Acta Physicochim. URSS* **20**, 411–420.

Thompson, J. B. (1969) Chemical reactions in crystals, *Amer. Mineral.* **54**, 341–375.

Virgo, D., and Hafner, S. S. (1969) Order–disorder in heated orthopyroxenes, *Mineral. Soc. Amer. Spec. Pap.* No. 2, 67–81.

Virgo, D., and Hafner, S. S. (1970) Fe^{2+}–Mg order–disorder in natural orthopyroxenes, *Amer. Mineral.* **55**, 210–223.

Wood, B. J., and Nicholls, J. (1978) The thermodynamic properties of reciprocal solid solutions, *Contrib. Mineral. Petrol.* **66**, 389–400.

Yund, R. A., and Tullis, J. (1980) The effect of water, pressure, and strain on Al/Si order–disorder kinetics in feldspar, *Contrib. Mineral. Petrol.* **72**, 297–302.

Chapter 2

Redistribution of Ca, Mg, and Fe during Pyroxene Exsolution; Potential Rate-of-Cooling Indicator

R. Kretz

Introduction

The exsolution of pyroxene minerals forms an interesting example of a naturally occurring solid-state chemical reaction. During the past few years, the orientation, structure, and composition of exsolution lamellae in several igneous and metamorphic pyroxene crystals have been investigated by optical microscopy, X-ray diffraction, electron probe analysis, transmission electron microscopy, and analytical electron microscopy. One might suppose, therefore, that the reaction is now fully understood, but this is not so, for the more closely the crystals are examined, the more complex and wondrous they are found to be. The present study is concerned principally with the diffusion of Ca, Mg, and Fe that evidently takes place while the reaction is in progress, an aspect of the reaction that has received relatively little attention in the past. A previously derived equation for the augite slope of the solvus surface will be used to estimate the temperature at which ionic mobility within an augite crystal from the Skaergaard complex declined and the reaction came to a close. This temperature estimate may possibly provide an indication of the rate of cooling.

Exsolution in Crystals of Augite

Several kinds of exsolution lamellae have been observed in augite. Jaffe *et al.* (1975), in their examination of augite from metamorphic rocks, found ortho-pyroxene lamellae parallel to (100) of host augite, and two sets of pigeonite lamellae, one of which lies nearly parallel to (001) of host augite (referred to as pigeonite "001") and the other nearly parallel to (100) of host augite (referred to as pigeonite "100"). Robinson *et al.* (1977) in their examination of augite from the Bushveld complex found as many as five sets, three of which are varieties of pigeonite "001." These studies have shown that the orientation of the lamellae was precisely controlled by the lattice dimensions of the host and lamellae phases at the time and temperature of exsolution, and that the best possible fit of the two lattices was achieved. The resulting planar lamallae–host boundaries, which lie parallel or nearly parallel to (001) or (100) of host augite, are referred to as optimal phase boundaries.

The chemical composition of pigeonite lamellae and augite host in a specimen from the Bushveld complex was determined by Boyd and Brown (1969), who found host and lamellae phases to be nearly identical to lamellae and host phases of the associated grains of inverted pigeonite. This important study led to the conclusion that equilibrium or near-equilibrium with regard to intercrystalline element distribution was maintained as the exsolution reaction progressed. Similar results were obtained by Nobugai *et al.* (1978), Coleman (1978), and Buchanan (1979). The results of Nobugai *et al.* (1978) were obtained from Skaergaard ferrogabbro 4430, for which bulk pyroxene analyses were obtained by Brown (1957), and these data will be used in the present study. Recently Nobugai and Hafner (1980) confirmed that the "001" lamellae analysed by Nobugai *et al.* (1978) possess the pigeonite structure ($P2_1/c$).

Equilibrium Distribution of Ca, Mg, and Fe between Augite and Associated Ca-Poor Pyroxene

Abundant evidence may be cited for the close attainment of equilibrium between primary augite and associated pigeonite or orthopyroxene during the crystallization of many igneous and metamorphic rocks. The evidence exists in the form of trends and restrictions in the Ca:Mg:Fe ratio of the associated minerals, as found for example by Brown (1957) in the Skaergaard Intrusion and by Davidson (1968) in the Quairading metamorphic rocks. A consideration of the equilibrium distribution of Ca, Mg, and Fe between associated pyroxene phases therefore forms a convenient beginning to an inquiry into the redistribution of these elements between host and lamellae phases during exsolution.

Recently, Kretz (1982) suggested that equilibrium with regard to associated pyroxene minerals may be viewed in relation to a transfer reaction, which governs the Ca content of the minerals, and an exchange reaction, which governs the relative Mg:Fe ratio of the minerals. Data from the Skaergaard Intrusion and the Quairading metamorphics were used to project the experimental Ca–Mg solvus of Lindsley and Dixon (1976) into the Fe-bearing quadrilateral volume, thereby obtaining an equation for a portion of the augite slope of the solvus surface. The Skaergaard and Quairading data where then used to propose a relation between the Mg-Fe distribution coefficient and temperature. The equations are as follows:
Transfer reaction:

$$< 1100°C: \quad T = 1000/0.054 + 0.608X^C - 0.304\ln(1 - 2[Ca]); \quad (1)$$

Exchange reaction:

$$T = 1130/(\ln K_D + 0.505),   \tag{2}$$

where T is temperature in K, X is atomic $Fe^{2+}/(Mg + Fe^{2+})$, [Ca] is atomic $Ca/(Ca + Mg + Fe^{2+})$ in augite, and K_D is the distribution coefficient, defined as

$$K_D = \frac{X^L}{1 - X^L} \frac{1 - X^C}{X^C}$$

with C and L denoting augite and Ca-poor pyroxene, respectively. K_D varies slightly with X^L; Eq. (2) as written applies where $X^L \approx 0.3$, as it does in the pyroxene considered here.

The Ca-poor pyroxene in the experimental results and in the Quairading rocks is orthopyroxene, but in the Skaergaard rocks is principally pigeonite (now inverted), and the assumption that the $Ca:Mg:Fe^{2+}$ ratio of augite is only slightly dependent on whether the associated phase is orthopyroxene or pigeonite is the major simplification used in deriving Eqs. (1) and (2).

Some isotherms for the augite slope of the solvus surface, as defined by Eq. (1) are plotted in Fig. 1. One of the specimens used in the construction of Eqs. (1) and (2) is Skaergaard ferrogabbro 4430, reported by Brown (1957, specimen 7), and it is from this specimen that Nobaguai et al. (1978) obtained analyses for host and lamellae phases. The bulk pyroxene analyses, as shown in Fig. 1, are taken as the starting compositions for the exsolution reactions that occurred in both augite and pigeonite crystals on cooling. The pre-exsolution distribution coefficient for the specimen is 1.41.

Equilibrium Distribution of Mg and Fe^{2+} between M1 and M2 Sites in Augite and Associated Ca-Poor Pyroxene

In the study referred to above, Kretz (1982) also estimated the $Ca:Mg:Fe^{2+}$ ratios on M1 and M2 sites of the Skaergaard and Quairading pyroxene crystals, at the deduced temperatures of crystallization. The calculations were based on the intracrystal distribution relations that were experimentally obtained by Virgo and Hafner (1969), Hafner et al. (1971) and McCallister et al. (1976). The main conclusion to emerge from this study was that the crystallization composition of the M1 sites of the associated augite and orthopyroxene or pigeonite were nearly identical, as was postulated by Mueller (1962), and, surprisingly, that in pairs of rocks (one Skaergaard and one Quairading) of

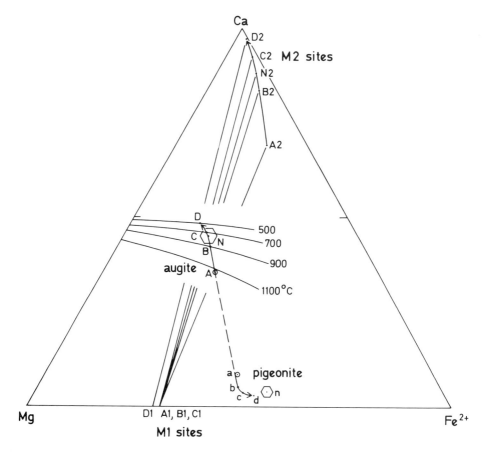

Fig. 1. Ca–Mg–Fe^{2+} composition triangle, showing grain composition of augite (A) and inverted pigeonite (a) in Skaergaard ferrogabbro 4430 (Brown, 1957), and of host augite (N) and lamellar pigeonite (n) of augite (A) (Nobugai *et al.* 1978). Also shown are 500, 700, 900, and 1100°C contours on the augite slope of the solvus (Eq. (1)), the calculated cooling path for augite A (A-B-C-D) (Eqs. (1) and (2)), and for pigeonite lamallae (a-b-c-d) (Eq. (2) and Ross and Huebner, 1979), the calculated M1 and M2 site composition for augite A (points A1 and A2) and the deduced site compositions for augite B, C, D and N.

nearly identical "total" pyroxene composition, all M1 sites were occupied by nearly identical Mg:Fe^{2+} ratios. This suggested that the transfer and exchange reactions that result from a change in temperature leave the M1 sites virtually unchanged in Mg and Fe^{2+} content, the required compositional changes taking place almost entirely in the M2 sites.

The calculated M1 and M2 site composition for the augite from Skaergaard 4430 at the estimated temperature of crystallization ($\sim$1100°C) is shown in Fig. 1 as points A1 and A2, respectively. The corresponding intracrystal distribution coefficient is 4.8.

Changes in Crystal and Site Composition Accompanying Exsolution within Augite from Skaergaard Ferrogabbro

Points A (augite) and a (pigeonite) in Fig. 1 presumably represent points on the pyroxene solvus at about 1100°C. On cooling, point A is expected to move down the augite slope of the solvus surface, following a path which may be located by use of Eqs. (1) and (2). This is accomplished, for example, by moving point A to the 900°C contour, where it is now referred to as point B, and then along the contour, causing a rotation of the B-b tie line, until (by iteration) the distribution coefficient for 900°C (Eq. (2)) is obtained, whence point B comes to rest. This assumes, of course, that equilibrium with the exsolving lamellae is maintained. Points C (700°C) and D (500°C) were similarly obtained, to define path A-B-C-D, to show the expected direction of change of composition of augite crystals in Skaergaard 4430 during cooling and exsolution, and a path along which, at some point, the composition of augite may have come to rest, depending on the temperature at which the exsolution reaction ceased to operate.

Note that with decreasing temperature, the augite host is expected to become richer in Ca (as determined by the transfer reaction) and richer in Mg relative to Fe (as determined by the exchange reaction). The calculated path is very slightly dependent on adopted changes in the Ca content of pigeonite lamellae, which were here approximated, using a phase diagram proposed by Ross and Huebner (1979, Fig. 1) as a guide. This results in path a ($\sim$1100°C)-b($\sim$900°C)-c($\sim$700°C)-d($\sim$500°C) for the pigeonite lamellae, as shown in Fig. 1. The corresponding change in the distribution coefficient is from 1.41 at the A-a tie line to 2.60 at the D-d tie line (not shown) as demanded by Eq. (2). Because the volume of augite host is much greater than the volume of pigeonite lamellae, small changes in path a-d will have only a slight effect on the position of proposed path A-D.

The anticipated change in M2 site occupancy of augite as the temperature decreases from $\sim$1100°C to $\sim$500°C may be estimated as follows. Beginning at calculated point A2, the path must approach the apex of the triangle. Assuming that Ca is confined to M2, the Ca content of M2 at 900, 700, and 500°C can be deduced from points B, C, and D, respectively. Assuming also that the Mg:Fe^{2+} ratio in M1 remains nearly fixed, as discussed above, Fe^{2+}/(Mg + Fe^{2+}) in M2 is estimated to increase in response to an increase in the intracrystal distribution coefficient, with decreasing temperature (McCallister et al., 1976). These considerations produce curved path A2-B2-C2-D2, as shown in Fig. 1. The magnitude of the curvature in path A-D seems to demand that the Mg:Fe^{2+} occupancy of the M1 sites does change slightly, from A1 to D1 on the base of the triangle.

These anticipated compositional changes may now be compared with the analyses of augite host and pigeonite lamellae in augite from Skaergaard 4430,

R. Kretz

Table 1. Atomic proportions (percent) of Ca, Mg, and Fe^{2+} in grains, host, and lamellae phases within Skaergaard ferrogabbro (4430). Data from Brown (1957) and Nobugai *et al.* (1978).

	Ca	Mg	Fe^{2+}
Augite grain	35.6	38.1	26.3
Augite host	45.0	35.0	20.0
Pigeonite lamellae	4.1	41.4	54.5
Pigeonite grain	8.7	46.1	45.2
Orthopyroxene host	2.8	48.5	48.5
Augite lamellae	43.7	37.6	18.6

as obtained by Nobugai *et al.* (1978). The reported Ca:Mg:Fe (total) ratios are first transformed to $Ca:Mg:Fe^{2+}$ ratios, assuming that Fe^{2+}/Fe (total) in these phases is the same as in the grains of augite and pigeonite, as bulk-analysed by Brown (1957), namely, 0.915 for augite and 0.977 for pigeonite. The transformed ionic ratios, as listed in Table 1, appear in Fig. 1 as two hexagons, the size of which reflects approximately the uncertainty in the analyses.

The composition of the augite host (point N, Fig. 1) lies directly upon the predicted path, with a Ca content corresponding to 750°C (Eq. (1)) but the pigeonite lamellae, plotted as n in Fig. 1, fall at some distance from the predicted path a-d. If the first-formed small lamellae of pigeonite did indeed have a $Ca:Mg:Fe^{2+}$ ratio identical to that of the pigeonite grains exterior to the augite crystal, as proposed by Nobugai *et al.* (1978), then the path followed by the lamellae extended from a to n. Tie line n-N (not shown) does not, however, pass through point A, as required. The position of n, lying well to the right of point a, is the cause for the very low host–lamellae distribution coefficient, as calculated by Nobugai *et al.* (1978), corresponding to a temperature of 540°C (Eq. (2)).

Some of the disagreement between observation and expectation, as described above, may result from analytical uncertainty, as is indicated by the fact that tie line N-n (host–lamellae analyses) does not pass through point A (bulk analysis). Some of the disagreement may arise from metastability, in the sense that pigeonite lamellae did not invert to orthopyroxene, as did the pigeonite grains that are associated with augite, and are evidently metastable, as was discussed by Jaffe *et al.* (1975). Equations (1) and (2) at temperatures below about 1000°C were built largely on experimental and natural data for augite-orthopyroxene, and the distribution of Ca, Mg, and Fe between augite and metastable pigeonite may not be adequately expressed by the equations.

These considerations, taken together with the finding that augite host and augite lamellae in the same rock differ slightly in composition, as do orthopyroxene host and pigeonite lamellae, as reported by Nobugai *et al.* (1978) and Coleman (1978) indicate that the transfer and exchange reactions that take place during cooling are not the exact reverse of those that would take place between discrete crystals of pyroxene during heating.

Despite the difficulties that remain in the interpretation of the data from Skaergaard 4430, one may conclude that exsolution of pigeonite in the augite crystals continued to about 750°C, whence the lamellae ceased growing and Ca became essentially immobile, and that the Fe–Mg exchange reaction continued to operate to a lower temperature (very approximately 540°C), beyond which no further change in phase composition occurred. Within the augite host, these changes are reflected by a change in the site occupancy of M2 from that represented by point A2 (Fig. 1) to that represented approximately by point N2.

Exsolution within Crystals of Pigeonite from Skaergaard Ferrogabbro

The crystallization of pigeonite in Skaergaard ferrogabbro 4430, and its inversion to orthopyroxene and exsolution of augite were described by Brown (1957). The host and lamellae phases in these grains, which coexist with the augite grains described above, were analysed by Nobugai et al. (1978), and these analyses were transformed to $Ca:Mg:Fe^{2+}$ ratios by the same procedure that was used for the associated augite. Application of Eqs. (1) and (2) gave temperatures of 860 and 590°C, respectively, which may be compared with 750 and 540° for the augite. If the difference between the values 860 and 750°C are real, exsolution in augite continued to a slightly lower temperature than did exsolution in the neighboring crystals of orthopyroxene.

Redistribution of Si, Al, Ti, Mn, and Na during Exsolution

Boyd and Brown (1969) have shown that exsolution in pyroxene minerals from the Bushveld complex brought about a redistribution, not only of Ca, Mg, and Fe, but also of Al, Ti, Mn, Na, and even Cr, which is present in trace amounts (150 ppm). Mn, together with Fe and Mg, was extracted from the augite host, while the remaining elements, together with Ca, became more concentrated in the host. Evidently no net migration of Si occurred. In order to determine the number of ions that were displaced, the initial augite composition, prior to exsolution of pigeonite, was estimated assuming that the host:lamellae volume ratio is 80:20, as is indicated by a profile shown by Boyd and Brown (1969, Fig. 2). The calculated net change in augite composition is indicated in Table 2, which shows that relatively few atoms are involved, less than 1 of each kind per 600 oxygen ions. It is nevertheless noteworthy that

Table 2. Change in composition of a Bushveld augite upon exsolution of lamellae of Ca-poor pyroxene. Data from Boyd and Brown (1968).

| | Atoms/600 oxygen ions | | |
	p = Original augite	h = Host augite (after exsolution)	Change $(h - p)$
Si	196.0	195.0	—
Ti	0.9	1.0	+ 0.1
Al	3.5	4.0	+ 0.5
Al	2.6	2.8	+ 0.2
Fe	43.4	34.0	− 9.4
Mn	1.2	1.0	− 0.2
Ca	69.2	85.7	+ 16.5
Mg	82.0	74.8	− 7.2
Na	1.7	2.0	+ 0.3

ions such as Al, Ti, and Cr were sufficiently mobile within the augite crystals for equilibrium or near-equilibrium to be maintained in the distribution of these ions between the augite host and the newly formed pigeonite lamellae.

Mechanism of Exsolution

The mechanism of exsolution as a chemical reaction was reviewed by Burke (1965) and Raghaven and Cohen (1975), who regarded the reaction as a cooperation of three principal processes: nucleation, crystal growth, and diffusion. Applications to minerals were considered by Yund and McCallister (1970). Following is a brief discussion of the mechanism of the exsolution of pigeonite from augite, with emphasis on the diffusion of Ca, Mg, and Fe in augite.

The first step in the exsolution of pigeonite from homogeneous augite is the creation of small domains of pigeonite, which may form by nucleation or by spinodal decomposition. Copley *et al.* (1974) in their examination of augite from pitchstone found evidence for the heterogeneous nucleation of pigeonite at grain boundaries, and possibly at imperfections, and suggested that heterogeneous nucleation is the dominant mechanism on rapid cooling, and spinodal decomposition on slow cooling. McCallister and Yund (1977) evidently reproduced both processes experimentally. Nobugai and Morimoto (1979) considered that pigeonite lamellae in Skaergaard augite formed by heterogeneous nucleation, and in the discussion to follow, this mechanism will be assumed to dominate.

If pigeonite lamellae have indeed formed by nucleation, then their preferred orientation, as described by Jaffe *et al.* (1975) and Robinson *et al.* (1977)

would present a clear demonstration of the highly selective nature of the nucleation process. According to the classical theory of nucleation, pigeonite embryo of diverse size, shape, and orientation may be visualized, but only those that possess the preferred lamellar shape, with optimal phase boundaries of lowest possible interfacial and strain energy, will survive to become nuclei.

Important evidence concerning the mechanism of crystal growth was presented by Copley et al. (1975), who observed growth ledges on the "surfaces" of pigeonite lamellae, similar to those previously discovered by Champness and Lorimer (1973) on augite lamellae in orthopyroxene. Thus the thickening or growth of the lamellae evidently occurred by the spreading of growth layers on the pigeonite–augite interfaces. This mechanism was adopted by Nobugai and Morimoto (1979) for the Skaergaard crystals, even though growth ledges were not detected.

The presence of two or more sets of exsolution lamellae in single crystals of augite raises some interesting questions concerning discontinuities in nucleation and growth. Robinson et al. (1977) found four sets of pigeonite lamellae in augite from Bushveld gabbro, and by use of thermal expansion data and optimal phase boundary theory, estimated the temperature of formation of the four sets as approximately 1100, 850, 800, and 560°C. The presence of orthopyroxene lamellae that cut pigeonite lamellae of the first set and are deflected by them indicates that growth of the first set had ceased before the orthopyroxene lamellae formed and that different sets of lamellae do not simply reflect fluctuations in nucleation with continued growth, but discrete generations of lamellae. Why some lamellae should stop growing, resulting in the nucleation and growth of a new set of lamellae, remains a mystery; Robinson et al. (1977) suggested that partial dislocations resulting from strain-induced stacking faults (commonly found in pigeonite lamellae) accumulated on the augite–pigeonite interfaces, thus impeding step migration.

Some of the geometric differences between sets of lamellae in these complex crystals may be interpreted in relation to nucleation, but many questions remain unanswered. Thus the observation that lamellae of the third set (pigeonite "100") are thinner and more closely spaced than all other sets, both earlier and later, may be the result of a high nucleation rate, but the reason for the strong clustering of these lamellae is not obvious.

Following the creation of a nucleus of pigeonite, growth of the nucleus is achieved by the migration of atoms of Mg and Fe^{2+} to the crystallite and migration of Ca in the opposite direction. This migration, which may be regarded as an example of exchange diffusion, will now be examined briefly.

The crystal structures of augite and pigeonite are sufficiently similar that one may regard the exsolution reaction as a transformation of portions of the augite crystal to pigeonite. Thus, the coherent interfaces may be regarded as planes that migrate through the augite crystal, transforming augite octahedral sites to pigeonite octahedral sites, accompanied by changes in site composition. Evidence was presented above to suggest that nearly all of the required compositional changes may take place in the M2 sites, the $Mg:Fe^{2+}$ ratio of

M1 remaining virtually unchanged as augite M1 sites are transformed to pigeonite M1 sites. With regard to the Skaergaard augite examined by Nobugai *et al.* (1978), the position of point N2 relative to that of A2 in Fig. 1 provides a measure of the approximate net change of M2 site composition produced by the exsolution reaction. Thus, the Ca content of augite M2 increased from 69 to 87.5 percent, while the Mg content decreased from 9.5 to 3 percent, and Fe^{2+} from 21.5 to 9.5 percent. In other words, volumes of augite containing 100 M2 sites acquired a total of 18 Ca ions, and lost 6 Mg ions and 12 Fe ions.

A build-up of Ca and a depletion of Mg and Fe would be expected immediately adjacent to the lamellae during their growth, giving rise to concentration gradients within the augite crystal. Such gradients have been detected in an orthopyroxene by Lorimer and Champness (1973), but, in general, they have not been detected.

Of the various diffusion mechanisms that have been proposed for diffusion in ionic crystals, as reviewed by Shewmon (1963), that which postulates ionic jumps, aided by the presence of vacancies appears most likely. Accordingly, one may visualize a Mg ion in augite jumping from one M2 site to another in the direction of a nearby pigeonite lamella, in exchange for a Ca ion which simultaneously jumps in the opposite direction. This mechanism is facilitated if it involves three or four sites arranged in a ring, or if vacant sites are present. A small number of equilibrium vacancies will certainly be present in the pyroxene crystals. Vacant cation and anion sites may tend to occur adjacent to each other, and this may greatly facilitate the diffusion mechanism because octahedral cations that are surrounded by only five oxygen ions may escape more readily from that site. Although the composition of the M1 sites may not change appreciably as the reaction progresses, these sites may nevertheless be involved in that particular sites may accomodate various transient Mg and Fe ions, and perhaps occasionally a Ca, while the overall composition of these sites remains unchanged.

Regarding diffusion paths in augite, it is noteworthy that the M1 and M2 sites in this crystal lie within planes that are parallel to (100). Adjacent M2 sites within the planes are separated by 4.6 Å, and neighboring M2 sites in adjacent planes are separate from each other (through a layer of SiO_3 chains) by 5.0 Å. It appears likely therefore that the coefficients of Ca–Mg and Ca–Fe^{2+} exchange diffusion for a given temperature and composition will be greater in directions parallel to (100) than in directions normal to (100). This was anticipated by Poldervaart and Hess (1951) who suggested that the greater thickness of (001) lamellae, compared with (100) lamellae, is related to the ease of migration of ions parallel to the *c*-axis. In general, the size of crystals is governed by the ratio N/G, where N is nucleation rate and G is growth rate, with a small value for this ratio resulting in few crystals of large size and vice versa. Thus the common presence of few thick "001" lamellae and many thin "100" lamellae is consistent with a relatively large diffusion coefficient for directions parallel to (100), but the rate of nucleation and other factors are no doubt also important in determining the size of lamellae.

Kinetics of Exsolution

The question of why an exsolution reaction in augite ceases at some temperature in the cooling history of the crystal and the possible importance of the cooling rate in determining this temperature will now be qualitatively examined.

The rate of an exsolution reaction within a certain range of temperature may be expressed as the rate of production of exsolved pigeonite within an augite crystal of unit volume. Suppose such a reaction is in progress and only one set of lamellae is being produced. Suppose also that each lamella is growing at the same rate, and that no new lamellae are being created. Then the reaction rate may be expressed as a function of the rate of growth of one lamella.

The rate of growth may be limited by the mobility of ions or by their arrangement at the interface. Let us assume initially that growth is limited by mobility, i.e., by the diffusion coefficient for Ca–(Mg, Fe) exchange diffusion in the augite crystal surrounding the growing lamella. The following variables must be taken into consideration:

(a) the rate of cooling,
(b) the magnitude of the diffusion coefficient, and its dependence on temperature and Ca, Mg, Fe concentration, and
(c) concentration gradients within augite, and their change with time.

As a lamella grows, it extracts Mg^{2+} and Fe^{2+} from the surrounding augite in exchange for Ca^{2+}, and hence concentration gradients will be created, similar to those shown in Fig. 2. It is assumed that the concentration gradients (dc/dx, where c is the concentration and x is the distance normal to the interface) within augite immediately at the interface are small and unchanged with time. Then, by the diffusion equation

$$J = -\left(\frac{dc}{dx}\right)D, \tag{3}$$

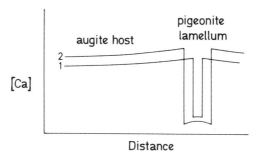

Fig. 2. Schematic gradients in the concentration of Ca ([Ca] = atomic Ca/(Ca + Mg + Fe)) across a growing lamella of pigeonite at time 1 and time 2. Mg and Fe gradients are represented by this figure in inverted form.

the flux of ions across the interface (J) is proportional to the diffusion coefficient (D). Hence a decline in the rate of growth and in the rate of the reaction is attributed to a decrease in the diffusion coefficient with decreasing temperature.

Diffusion coefficients for Ca–Mg and Ca–Fe exchange in augite are not yet available, but the self-diffusion of Ca in $CaSiO_3$, as reported by Lindner (1958), may be chosen as an illustrative example of the decline in ionic mobility with decreasing temperature. This choice also represents a simplification in that self-diffusion is independent of concentration.

The observed linear relationship between $\ln D$ and $1/T$ is shown in Fig. 3 as a plot of D against temperature in °C. The approximate change of D with time (t) may be visualized by supposing a uniform rate of cooling, whence the horizontal axis in Fig. 3 represents time. Now in order to illustrate the dramatic decrease in mobility of ions with falling temperature, the mean displacement of Ca (denoted $\bar{x}$) within $CaSiO_3$, as given by the equation

$$\bar{x} = \sqrt{2Dt} \tag{4}$$

was calculated for a time interval of 2000 years, and these distances in angstrom units are shown in Fig. 3.

A very rapid decrease in mobility of Ca, Mg and Fe with decreasing temperature, analogous to that shown in Fig. 3, is considered to be the cause for a rapid decrease in flux across the augite–pigeonite interface, resulting,

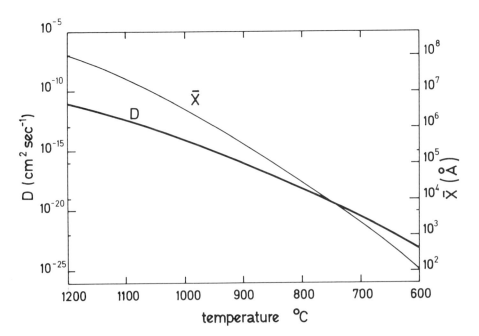

Fig. 3. Effect of temperature on diffusion coefficient for self-diffusion of Ca^{2+} in $CaSiO_3$ (D) (data from Lindner, 1958), and calculated mean displacement of Ca^{2+} ($\bar{x}$ in angstrom units) in 2000 years, as a function of temperature.

within a narrow range of temperature, in a rapid decline in the growth of lamellae and in the rate of the exsolution reaction. This narrow range of temperature, within which the rate of reaction decelerates to small or imperceptible values, is referred to as the freezing temperature.

The temperature estimate of 750°C for the Skaergaard augite, reported above, presumably represents an estimate of the freezing temperature. In view of the probable presence of small concentration gradients extending out from the lamellae–host interface, where equlibrium was presumably maintained during cooling, the temperature estimate may be slightly too high.

Although diffusion may be the rate-limiting process in the exsolution reaction, it is not necessarily rate-determining. This may be understood by imagining the rate of cooling, at a high temperature, to decrease to zero, whence the exsolution reaction would cease, regardless of the mobility of the ions. Only as the freezing temperature is approached would diffusion become rate-determining.

The above picture, of the reaction limited by diffusion and progressing steadily over an interval of temperature, would hold only while the lamellae are growing and have not yet reached the point when, for reasons not fully understood, growth becomes impeded, causing the nucleation of a new set of lamellae or the fatal termination of the reaction. Under certain conditions, therefore, the growth process may be the rate-limiting process.

Because the mobility of ions in crystals decreases very rapidly with decreasing temperature, a given rearrangement of ions can be achieved at progressively lower temperatures, provided correspondingly larger intervals of time are available. Thus volcanic and metamorphic rocks would be expected to show higher and lower freezing temperatures, respectively, when compared with shallow igneous bodies such as the Skaergaard. Estimates of freezing temperature may therefore be regarded as indicators of cooling rates. In agreement with this proposal is the presence of exsolution lamellae in metamorphic augite, as described by Jaffe *et al.* (1975), which must have formed below the temperature of metamorphism (about 700°C). In apparent disagreement is the calculation of Robinson *et al.* (1977), based on lattice fit, to indicate that exsolution in a Bushveld augite continued to about 560°C. However, these last-formed lamellae are very small and sparse and represent only a small fraction of the total product of the reaction. The Robinson *et al.* estimate for their most recent conspicuous set of lamellae is about 800°C, and it is likely that the Ca content of the augite would give a solvus temperature (Eq. (1)) close to 800°C. The Bushveld specimen examined by Boyd and Brown (1969) gives a solvus temperature (Eq. (1)) of 820°C.

Conclusion

The exsolution of pigeonite from crystals of augite in Skaergaard ferrogabbro 4430 continued on cooling to about 750°C, and Fe–Mg exchange between lamellae and host continued to very approximately 540°C. These temperature estimates were obtained by applying equations of equilibrium previously derived by Kretz (1982) to the host and lamellae analyses obtained by Nobugai et al. (1978).

The exsolution of pigeonite has brought about a change in the composition of the M2 sites of augite, amounting to an 18 (atomic) percent increase in Ca and a 6 and 12 percent decrease of Mg and Fe^{2+}, respectively. These estimates were calculated by comparing the host analyses, obtained by Nobugai et al. (1978) with grain analyses obtained by Brown (1957), and by assuming no change in M1 site composition, for which evidence was previously presented. These compositional changes were evidently brought about by Ca–(Mg, Fe) exchange diffusion in augite, which is considered to be the dominant rate-limiting process in the exsolution reaction.

The solvus–temperature estimate for augite (the freezing temperature) may be a function of the rate of cooling, and may, with "calibration" form a useful rate-of-cooling indicator.

References

Boyd, F. R., and Brown, G. M. (1969) Electron-probe study of pyroxene exsolution, *Mineral. Soc. Amer. Spec. Pap.* No. 2, 211–216.

Brown, G. M. (1957) Pyroxenes from the early and middle stages of fractionation of the Skaergaard Intrusion, East Greenland, *Mineral. Mag.* **31**, 511–543.

Buchanan, D. L. (1979) A combined transmission electron microscope and electron microprobe study of Bushveld pyroxenes from the Bethal area, *J. Petrol.* **20**, 327–354.

Burke, J. (1965) *The Kinetics of Phase Transformations in Metals.* Pergamon Press, Oxford.

Champness, P. E., and Lorimer, G. W. (1973) Precipitation (exsolution) in an orthopyroxene, *J. Mater. Sci.* **8**, 467–474.

Coleman, L. C. (1978) Solidus and subsolidus compositional relationships in some coexisting Skaergaard pyroxenes, *Contrib. Mineral. Petrol.* **66**, 221–227.

Copley, P. A., Champness, P. E., and Lorimer, G. W. (1974) Electron petrography of exsolution textures in an iron-rich clinopyroxene, *J. Petrol.* **15**, 41–57.

Davidson, L. R. (1968) Variation in ferrous iron-magnesium distribution coefficients of metamorphic pyroxenes from Quairading, Western Australia, *Contrib. Mineral. Petrol.* **19**, 239–259.

Hafner, S. S., Virgo, D., and Warburton, D. (1971) Cation distribution and cooling history of clinopyroxene from Oceanus Procellarum, *Proc. Second Lunar Sci. Conf., Geochim. Cosmochim. Acta. Suppl.* **2**, 91–108.

Jaffe, H. W., Robinson, P., and Tracy, R. J. (1975) Orientation of pigeonite exsolution lamellae in metamorphic augite: Correlation with composition and calculated optimal phase boundaries, *Amer. Mineral.* **60**, 9–28.

Kretz, R. (1982) Transfer and exchange equilibria in a portion of the pyroxene quadrilateral as deduced from natural and experimental data, *Geochim. Cosmochim. Acta* **46**, 411–422.

Lindner, R. (1958) Use of radioisotopes for the study of self-diffusion in oxide systems, *Proc. 2nd Int. Conf. Peaceful Uses Atomic Energy*, 116–119.

Lindsley, D. H., and Dixon, S. A. (1976) Diopside–enstatite equilibria at 850° to 1400°C, 5 to 35 kb, *Amer. J. Sci.* **276**, 1285–1301.

Lorimer, G. W., and Champness, P. E. (1973) Combined electron microscopy and analysis of an orthopyroxene, *Amer. Mineral.* **58**, 243–248.

McCallister, R. H., Finger, L. W., and Ohashi, Y. (1976) Intracrystalline Fe^{2+}–Mg equilibria in three natural Ca-rich clinopyroxenes, *Amer. Mineral.* **61**, 671–676.

McCallister, R. H., and Yund, R. A. (1977) Coherent exsolution in Fe-free pyroxenes, *Amer. Mineral.* **62**, 721–726.

Mueller, R. F. (1962) Energetics of certain silicate solid solutions, *Geochim. Cosmochim. Acta.* **26**, 265–275.

Nakajima, Y., and Hafner, S. S. (1980) Exsolution in augite from the Skaergaard Intrusion, *Contrib. Mineral. Petrol.* **72**, 101–110.

Nobugai, K., and Morimoto, N. (1979) Formation mechanism of pigeonite lamellae in Skaergaard augite, *Phys. Chem. Minerals* **4**, 361–371.

Nobugai, K., Tokonami, M., and Morimoto, N. (1978) A study of subsolidus relations of the Skaergaard pyroxenes by analytical electron microscopy, *Contrib. Mineral. Petrol.* **67**, 111–117.

Poldervaart, A., and Hess, H. H. (1951) Pyroxenes in the crystallization of basaltic magma, *J. Geol.* **59**, 472–489.

Raghaven, V., and Cohen, M. (1975) Solid-state phase transformation, in *Treatise on Solid State Chemistry*, Vol. 5, edited by N. B. Hannay. Plenum Press, New York.

Robinson, P., Ross, M., Nord, G. L., Jr., Smyth, J. R., and Jaffe, H. (1977) Exsolution lamellae in augite and pigeonite: fossil indicators of lattice parameters at high temperature and pressure, *Amer. Mineral.* **62**, 857–873.

Yund, M., and Huebner, J. S. (1979) Temperature-composition relationships between naturally occurring augite, pigeonite, and orthopyroxenes at one bar pressure. *Amer. Mineral.* **64**, 1133–1155.

Shewmon, P. G. (1963) *Diffusion in Solids*. McGraw-Hill, New York.

Virgo, D., and Hafner, S. S. (1969) Fe^{2+}, Mg order–disorder in heated orthopyroxenes, *Mineral Soc. Amer. Spec. Pap.* No. 2, 67–81.

Yund, R. A., and McCallister, R. H. (1970) Kinetics and mechanisms of exsolution, *Chem. Geol.* **6**, 5–30.

Chapter 3

Intracrystalline Cation Distribution in Natural Clinopyroxenes of Tholeiitic, Transitional, and Alkaline Basaltic Rocks

A. Dal Negro, S. Carbonin, G. M. Molin, A. Cundari, and E. M. Piccirillo

Introduction

Thermodynamic studies of Mueller (1962) and Thompson (1969) and the crystallographic work of Ghose (1965) gave new impetus to a rapidly developing branch of mineralogy. Orthopyroxenes were studied by Virgo and Hafner (1969) and Saxena and Ghose (1971) using Mössbauer technique. Site-occupancy data have been used in many mineralogical and petrological problems, e.g., to estimate the thermodynamic solution properties of pyroxenes and to understand the cooling history of rocks. A parallel development in the mineralogy of clinopyroxene did not take place, particularly because the Mössbauer technique proved unsatisfactory for calcic pyroxenes. Recognizing this, we undertook a detailed X-ray crystallographic study of the clinopyroxenes and succeeded to determine crystal-structural parameters for Fe–Mg site occupancies of the nonequivalent M1 and M2 sites. Until now there are no published site-occupancy data on a series of clinopyroxene of intermediate composition. This paper attempts to fill this important gap in the data on clinopyroxenes. Besides our concern for the crystal-chemical and thermodynamic aspects of such a study, we were also motivated by the possible use of pyroxene crystallographic parameters in the classification and characterization of magmatic rocks. With the advent of automation in X-ray crystallography, it is possible to study a number of crystals from one or more rock samples in a relatively short time and provide petrologically useful information.

   Important findings reported in this work are:

(a) It is possible to estimate Fe–Mg site occupancies on M1 and M2 sites in clinopyroxenes of intermediate compositions through crystal structure refinements as done by Clark *et al.* (1969), Takeda (1972), Takeda *et al.* (1974), Ghose *et al.* (1975), and others.
(b) There are distinct correlations between crystal-structural parameters (mainly generated as a function of Ca concentration in the M2 site), and concentrations of trivalent and tetravalent cations in the M1 site and Si in

the T site. As a result of these relationships, it is possible to completely determine the M1 and M2 site occupancy by Al^{3+}, Fe^{3+}, Ti^{4+}, Cr^{3+}, Fe^{2+}, Mg, Ca, and Na. The composition-structural parametric relationships make it possible to predict the bulk composition and site occupancies in crystals that have not been analyzed chemically.

(c) A new type of M2 site in the clinopyroxenes of intermediate compositions is reported. The presence of this new site, called M2', is a function of the pyroxene chemistry and it may have some bearing on the process of unmixing (exsolution) in calcic pyroxenes.

(d) Basaltic rocks may be distinguished on the basis of crystallographic parameters of the associated pyroxenes.

Petrologic Outlines of the Host Rocks

The investigated pyroxenes occur in volcanic rocks of the central Ethiopian Plateau ("Trap Series"). These volcanics are mainly represented by basaltic rocks related to both fissural (Aiba, Alaji, Ashangi *p.p.* formations) and

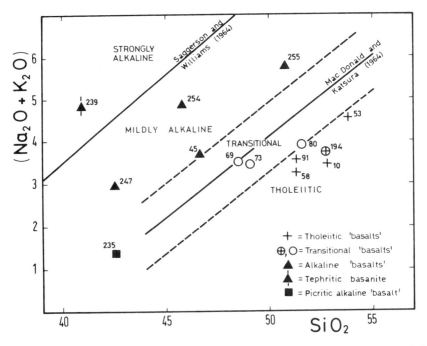

Fig. 1. Distribution of pyroxene-bearing lavas in terms of SiO_2 versus $\sum(Na_2O + K_2O)$ wt%. Field of transitional basalts between dashed lines according to Zanettin *et al.* (1974) and Piccirillo *et al.* (1979).

central (Termaber formation) type activities (Zanettin *et al.*, 1974, 1978, 1980, Appendix I).

The alkali versus silica diagram (Fig. 1) shows that the lavas display a wide spectrum of compositions, varying from strongly alkaline to tholeiitic (major element chemistry in Zanettin *et al.* (1976)).

Petrological data clearly indicate that these rocks are related to different primary magmas originated from variously depleted mantle sources and are affected by crystal fractionation processes during their ascent to surface (Piccirillo *et al.*, 1979; see also Brotzu *et al.*, 1981).

The *tholeiitic basalts* are distinctive for the presence of relatively low-Ca augite accompanied by orthopyroxene (91) and/or pigeonite (10, 53, 58), absence of olivine, and late crystallization of opaques. The *alkaline basalts* (45, 235, 247, 254, 255) are distinguished by the occurrence of Mg-rich olivine, often abundant in groundmass, Ca-rich augite, and early crystallization of opaques. The *transitional basalts* (69, 73, 80, 194), instead, are intermediate between the preceding ones: i.e., they show groundmass Mg-olivine, moderately Ca-rich augite, late crystallization of opaques, etc. Note that the transitional basalt 194 shows pigeonite in the groundmass, which is distinctive tholeiitic affinity.

Rock name, mineral assemblage, sample location, and other chemical parameters of the investigated samples are given in Appendix I.

Optical Features and Chemical Composition of the Pyroxenes

The pyroxene phenocrysts show a weak to moderate zoning which tends to become more pronounced in those of the alkaline rock-types. Generally, zoning is mainly confined to a thin rim whose thickness varies on average from 0.01 mm (tholeiitic and transitional basalts) to 0.03 mm (alkaline basalts). Zoning in the pyroxene microphenocrysts is often very weak or absent.

The colour of pyroxenes generally varies with the rock-type. It normally ranges from pale green to light green in the tholeiitic basalts, from pale brown to brownish (rim) in the transitional basalts, and from greenish brown or brownish violet or pale violet to intense violet or deep green (rim) in the alkaline basalts. Exsolution was carefully sought under the microscope, but not observed.

Notably, the size of pyroxene phenocrysts in the thoeliitic and transitional basalts is generally smaller (average: 0.6 mm; maximum: 3–4 mm in length) than those in the alkaline ones (average: 1.4 mm; maximum: 5–6 mm in length, excluding the pyroxenes in a picritic basalt 235 which are up to 1.8 cm in length).

Mineral Separation

Single crystals were usually separated by hand-picking and selected under a microscope from a rock section with a thickness of about 100 μ, by using various types of needles. This separation has been particularly useful to isolate those crystals with no appreciable zoning and homogeneous in color. Also, this simple technique is effective for separating pyroxenes belonging to different stages of crystallization.

In some cases the pyroxenes with weak zoning (mainly from tholeiitic and transitional basalts) were removed by using a Frantz isodynamic separator and heavy liquids, after crushing a rock chip.

Electron Microprobe Analyses and Chemical Variation

Chemical analyses of the same single crystal used for crystal structure refinements were obtained from polished thin sections by means of an automated JXA-5A electron microprobe operated at 15 kV (accelerating voltage) and 0.02–0.03 μA (specimen current). The results were corrected following the method of Mason et al. (1969) and are considered accurate to within 2–3% for major elements and better than 9% for minor elements.

Several analyses were carried out on core and rim of each single crystal in order to obtain average chemical compositions and to reduce the uncertainty in the estimation of the Fe^{3+} content by the charge-balance method (Finger, 1972; Papike et al., 1974).

The chemical variability among single analyses of each crystal is generally relatively low, averaging 0.020 (range 0.005–0.055) atoms/6 oxygens for Si, Fe^{2+}, Fe^{3+}, Mg and 0.010 (range 0.008–0.025) atoms/6 oxygens for Ca. This indicates that the selected single crystals may be considered homogeneous within the experimental error. Mean chemical analyses and structural formulae are given in Table 1.

In the conventional pyroxene system $Mg–Ca–Fe^{2+} + Fe^{3+} + Mn$ (Fig. 2) the clinopyroxene from the alkaline and strongly alkaline basalts forms a Ca-enrichment trend and those from transitional and tholeiitic basalts plot on constant or decreasing Ca-control lines, respectively.

Crystallographic Procedures

Intensity data were collected by using a computer-controlled Philips PW 1100 four-circle diffractometer with Mo $K\alpha$ radiation monochromatized by a flat graphite crystal. X-ray data collection has been carried out following the same procedure for all samples listed in Table 2. The intensities of the reflections with $\theta \leqslant 30°$ were collected by using the ω-scan method; the equivalent hkl–$h\bar{k}l$ pairs were scanned. The intensities were corrected for absorption

Table 1. Chemical composition of single crystals of clinopyroxene as determined by electron-probe analysis.

wt%	S.53	S.58	S.91	S.194	S.80	S.80A	S.80B	S.80C	S.69B	S.45A	S.45B	S.255	S.255A	S.255B	S.255C	S.235A	S.235B	S.235C	S.239	S.247	S.254A
SiO_2	51.02	50.46	51.43	53.15	52.50	53.73	52.22	53.28	51.25	49.98	51.11	48.93	49.39	50.75	51.68	49.71	47.99	42.57	47.44	46.73	47.36
TiO_2	1.15	1.24	0.87	0.70	0.68	0.68	0.63	0.63	1.53	1.47	1.60	1.37	1.15	0.84	0.58	1.13	2.03	4.38	2.12	2.38	1.89
Al_2O_3	1.92	2.76	2.58	2.14	1.79	1.37	2.11	1.39	2.45	2.98	2.88	4.04	4.18	3.57	2.69	4.00	5.33	8.58	5.90	6.83	5.88
Cr_2O_3	0.16	0.03	0.07	n.d.	0.48	0.92	0.73	0.72	n.d.	0.13	0.22	0.00	0.33	0.12	0.31	0.50	0.08	0.00	0.00	0.14	0.05
Fe_2O_3[a]	2.05	2.61	1.61	0.79	1.69	0.63	2.03	0.96	0.12	5.68	2.01	4.15	4.80	2.78	3.06	3.98	5.17	6.54	5.40	4.18	5.58
FeO[a]	9.56	9.57	5.96	6.73	3.09	4.15	4.20	4.01	9.05	1.88	5.70	4.53	2.30	4.50	3.13	1.97	2.13	2.48	3.27	3.13	2.09
MnO	0.28	0.30	0.19	0.18	0.17	0.15	0.16	0.13	0.18	0.15	0.14	0.18	0.13	0.16	0.14	0.09	0.11	0.20	0.19	0.12	0.10
MgO	16.02	15.67	17.47	17.06	18.41	18.62	17.56	18.44	14.78	15.15	15.48	14.37	15.29	15.24	16.54	15.78	14.15	11.69	12.51	13.21	13.54
CaO	17.28	17.41	18.86	19.44	20.20	20.30	20.20	20.26	19.79	21.13	21.39	21.32	22.04	21.43	21.55	21.66	22.48	22.82	22.52	22.48	22.54
Na_2O	0.29	0.28	0.18	0.25	0.31	0.31	0.32	0.31	0.31	0.35	0.37	0.43	0.46	0.56	0.41	0.53	0.63	0.45	0.86	0.50	0.65
Σ[b]	99.73	100.33	99.22	100.44	99.32	100.86	100.16	100.13	99.46	98.90	100.90	99.32	100.07	99.95	100.09	99.35	100.10	99.71	100.21	99.70	99.68
Si	1.906	1.878	1.903	1.942	1.923	1.941	1.910	1.940	1.916	1.878	1.876	1.830	1.823	1.875	1.894	1.837	1.777	1.608	1.766	1.741	1.763
Al^{IV}	0.085	0.121	0.097	0.058	0.077	0.059	0.090	0.059	0.084	0.122	0.124	0.170	0.177	0.125	0.106	0.163	0.223	0.382	0.234	0.259	0.237
σ	1.991	1.999	2.000	2.000	2.000	2.000	2.000	1.999	2.000	2.000	2.000	2.000	2.000	2.000	2.000	2.000	2.000	1.990	2.000	2.000	2.000
Al^{VI}	—	—	0.016	0.034	0.001	—	0.001	—	0.024	0.008	—	0.008	0.005	0.031	0.011	0.011	0.009	—	0.025	0.041	0.021
Ti	0.033	0.035	0.024	0.019	0.019	0.018	0.017	0.017	0.043	0.043	0.044	0.039	0.032	0.024	0.016	0.031	0.057	0.124	0.060	0.067	0.052
Fe^{3+}	0.058	0.073	0.045	0.022	0.047	0.016	0.056	0.026	0.003	0.053	0.055	0.117	0.133	0.077	0.085	0.111	0.145	0.186	0.151	0.117	0.157
Cr	0.005	0.001	0.002	—	0.014	0.026	0.021	0.021	—	0.004	0.006	—	0.010	0.004	0.008	0.016	0.002	0.001	—	0.004	0.001
Mg	0.893	0.869	0.963	0.935	1.005	1.002	0.957	1.000	0.826	0.839	0.847	0.801	0.841	0.840	0.903	0.870	0.781	0.658	0.694	0.734	0.752
Fe^{2+}	0.299	0.298	0.184	0.203	0.095	0.126	0.129	0.122	0.283	0.177	0.175	0.142	0.071	0.139	0.096	0.061	0.066	0.078	0.102	0.098	0.065
Mn	0.009	0.009	0.006	0.006	0.005	0.005	0.005	0.004	0.006	0.004	0.004	0.006	0.004	0.004	0.005	0.003	0.003	0.006	0.006	0.004	0.003
σ	1.297	1.285	1.240	1.219	1.186	1.193	1.186	1.190	1.185	1.128	1.131	1.113	1.096	1.119	1.124	1.102	1.065	1.052	1.038	1.065	1.051
Ca	0.691	0.694	0.747	0.764	0.793	0.786	0.791	0.791	0.793	0.846	0.841	0.854	0.871	0.848	0.846	0.858	0.892	0.924	0.899	0.897	0.900
Na	0.022	0.020	0.013	0.018	0.022	0.021	0.023	0.022	0.022	0.026	0.026	0.033	0.033	0.033	0.030	0.038	0.045	0.033	0.062	0.036	0.048
Σ	4.001	3.998	4.000	4.001	4.001	4.000	4.000	4.002	4.000	4.000	3.998	4.000	4.000	4.000	4.000	3.998	3.999	4.007	3.999	3.998	3.999

[a] Total Fe as FeO partitioned according to Papike et al. (1974).
[b] Includes K_2O = 0.00; n.d. = not determined.

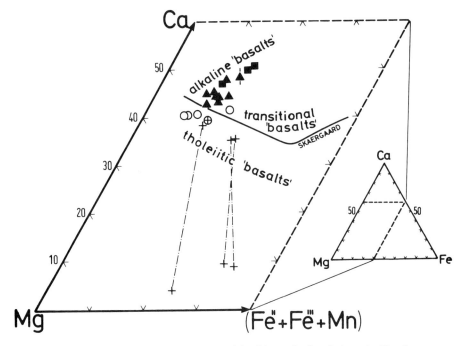

Fig. 2. Composition of clinopyroxenes used in this work. Symbols as in Fig. 1.

following the semiempirical method of North *et al.* (1968) and the values of equivalent pairs were averaged.

The X-ray data were processed by a program written for the PW 1100 diffractometers by Honstra and Stubbe (1972).[1]

All refinements were carried out in space group $C2/c$ starting with the atomic coordinates of diopside (Clark *et al.*, 1969). A locally rewritten version of the full-matrix least-squares program ORFLS (Busing *et al.*, 1962) was used. The refinements were carried out without chemical constraints. In particular, the ionized atomic scattering factors (international tables for X-ray crystallography (1974) and Tokonami (1965)) were used and all the structural sites were considered fully occupied. The choice of the scattering factors was: Ca^{2+} and Na^+ for M2; Mg^{2+} and Fe^{2+} for M1; $Si^{2.5+}$ for T. Partly ionized curves were adopted in view of the significant difference between F_0 and F_c observed for $\sin(\theta/\lambda) = 0-0.30$. The best fitting for the experimental data was obtained assuming 2.5 positive charges for Si; 1.5 negative charges for O, and complete ionization for other cations.

In the first stages of the refinements, isotropic temperature factors were used. The final cycles were performed allowing all the parameters (atomic coordinates, anisotropic temperature factors, M1, M2 site occupancies, scale

---

[1] Tables listing crystal-structure factors, coordinates, and anisotropic thermal factors are available from A. Dal Negro.

Table 2. Crystallographic data on clinopyroxenes.

	S.10	S.53	S.58	S.91	S.194	S.80	S.80A
a (Å)	9.750(1)	9.745(1)	9.739(1)	9.735(1)	9.732(1)	9.738(1)	9.735(1)
b (Å)	8.931(1)	8.927(1)	8.921(1)	8.910(1)	8.905(1)	8.916(1)	8.912(1)
c (Å)	5.258(1)	5.261(1)	5.261(1)	5.260(1)	5.258(1)	5.256(1)	5.254(1)
β	106.34(1)	106.58(1)	106.51(1)	106.47(1)	106.48(1)	106.37(1)	106.35(1)
V (Å³)	439.36	438.64	436.76	437.52	436.96	437.85	437.39
No. meas. refl.	650	652	649	650	648	647	647
No. obs. refl.	474	554	597	514	544	586	612
$I > n\sigma(I)$	$n = 2$	$n = 3$	$n = 5$	$n = 2$	$n = 1$	$n = 5$	$n = 5$
$R^a_{obs.}$	3.3	2.8	3.2	3.2	4.1	2.8	3.2
$R^b_{obs.}$	3.0	2.4	2.8	2.6	3.7	2.2	2.6

	S.80B	S.80C	S.69A	S.69B	S.73A	S.73B	S.45A
a (Å)	9.738(1)	9.741(1)	9.747(1)	9.746(1)	9.740(1)	9.740(1)	9.755(1)
b (Å)	8.915(1)	8.915(1)	8.918(1)	8.910(1)	8.916(1)	8.916(1)	8.916(1)
c (Å)	5.258(1)	5.257(1)	5.265(1)	5.268(1)	5.255(1)	5.258(1)	5.269(1)
β	106.36(1)	106.40(1)	106.38(1)	106.36(1)	106.31(1)	106.33(1)	106.23(1)
V (Å³)	437.99	437.95	439.08	438.93	437.99	438.19	440.01
No. meas. refl.	649	650	648	646	647	647	651
No. obs. refl.	555	534	547	560	569	545	541
$I > n\sigma(I)$	$n = 4$	$n = 5$	$n = 5$	$n = 4$	$n = 4$	$n = 3$	$n = 5$
$R^a_{obs.}$	2.7	2.6	2.5	2.6	2.7	2.7	2.1
$R^b_{obs.}$	2.0	2.1	2.0	2.2	2.2	2.2	1.7

Table 2 (continued).

	S.45B	S.255	S.255A	S.255B	S.255C	S.255D	S.235A
a (Å)	9.751(1)	9.748(1)	9.743(1)	9.744(1)	9.743(1)	9.739(1)	9.739(1)
b (Å)	8.914(1)	8.909(1)	8.901(1)	8.908(1)	8.912(1)	8.916(1)	8.901(1)
c (Å)	5.267(1)	5.269(1)	5.268(1)	5.264(1)	5.262(1)	5.253(1)	5.267(1)
β	106.21(1)	106.13(1)	106.16(1)	106.13(1)	106.21(1)	106.10(1)	106.18(1)
V (Å^3)	439.61	439.57	438.80	438.93	438.73	438.24	438.49
No. meas. refl.	651	650	647	647	647	648	647
No. obs. refl.	595	625	550	598	560.	588	598
$l > n\sigma(I)$	$n = 5$	$n = 5$	$n = 5$	$n = 5$	$n = 5$	$n = 5$	$n = 5$
$R^a_{obs.}$	2.6	2.6	2.0	2.2	2.3	2.2	2.1
$R^b_{obs.}$	2.3	2.3	1.6	1.9	1.8	1.8	1.8

	S.235B	S.235C	S.239	S.247	S.254A	S.254B
a (Å)	9.749(1)	9.769(1)	9.754(1)	9.744(1)	9.751(1)	9.747(1)
b (Å)	8.896(1)	8.888(1)	8.896(1)	8.883(1)	8.899(1)	8.904(1)
c (Å)	5.277(1)	5.301(1)	5.284(1)	5.282(1)	5.279(1)	5.272(1)
β	106.09(1)	106.02(1)	105.98(1)	106.12(1)	106.02(1)	106.04(1)
V (Å^3)	439.73	442.39	440.78	439.21	440.29	439.73
No. meas. refl.	650	650	650	647	650	649
No. obs. refl.	572	544	579	584	557	506
$l > n\sigma(I)$	$n = 4$	$n = 2$	$n = 5$	$n = 5$	$n = 5$	$n = 2$
$R^a_{obs.}$	1.7	2.1	1.6	1.8	2.2	2.6
$R^b_{obs.}$	1.5	1.9	1.5	1.6	2.1	2.5

[a] R value obtained before considering M2' contribution.
[b] R value obtained considering M2' contribution.

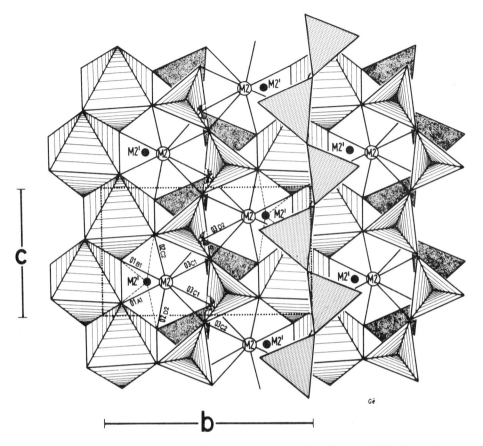

Fig. 3. The crystal structure of C2/c clinopyroxene projected on to (100) plane. Atom nomenclature after Burnham *et al.* (1967). The geometry of the new M2′ site as determined in this paper has been added to the diagram.

factors and secondary extinction coefficient (Zachariasen, 1963)) to vary until the shifts were less than the least-squares difference of the corresponding parameters.

When the anisotropic refinements reached convergence, the Fourier synthesis difference was computed. In all the samples the highest residual electron density peak occurred at about 0.6 Å from the M2 sites as shown by Rossi *et al.* (1978). This new site (M2′, see Fig. 3) may be linked with 01 and 02 at distances varying from 1.82 Å to 2.30 Å and was assumed to be occupied by Mg. Its occupancy was unconstrained. Further least-squares refinements by varying alternatively the M2′ occupancy and all the other parameters were carried out. The temperature factor of M2′ was fixed equal to the equivalent isotropic temperature factor of M2. A lower value for R was obtained at the end of the refinements (Table 2). It should be noted that the electronic contribution of M2′ is considered as indicative only, given that the resolution of the data is 0.7 Å. It should also be noted that this residue of electron density occurs in all natural clinopyroxenes investigated in this work. It has not been observed in synthetic diopside, but it occurs in the latter with increasing proportions of enstatite in solid solution (E. Bruno *et al.*, personal communication, 1981). The M2′ site occupancy tends to increase with decreasing Ca in the clinopyroxene. The electronic contribution of this new site is about equal to the Mg occupancy estimated for M2.

The position of the new M2′ site is distinct from that of M2 whose thermal motion is just about spherical and is not likely to extend in the direction of the M2′ site. Thus we believe that the M2′ site, recognized in all the natural pyroxenes studied here, is sufficiently distinct and is not due to a distortion of the regular M2 site.

Crystal Chemistry

Polyhedron M2

The crystal chemical characteristics of this polyhedron largely depend on the Ca occupancy and, distinctly, for $Ca \geqslant 0.80$ and $Ca < 0.80$ atoms (atomic proportions), respectively. In the first case, the substantial occupancy of M2 by Ca produces a linear increase of M2-O2 (2.305–2.352 Å) and M2-O1 (2.338–2.368 Å) (see Table 3). Concurrently, there is a contraction of $M2-O3_{C1}$ (2.592–2.557 Å) and $M2-O3_{C2}$ (2.739–2.703 Å). In the second case, entry of Mg and Fe^{2+} in M2, to the exclusion of Ca, determines a contraction of M2-O2 (2.290–2.251 Å) and M2-O1 (2.340–2.297 Å) and a corresponding extension of $M2-O3_{C1}$ (2.603–2.630 Å) and $M2-O3_{C2}$ (2.748–2.780 Å). In the latter case the polyhedron shows the strongest distortion and may be moving towards a change in symmetry. A measure of this distortion is $\Delta M2$, which is given in terms of M-O bond lengths by

$$\Delta M2 = M2\text{-}O3_{C2} - (\overline{M2\text{-}O3_{C1} + M2\text{-}O2 + M2\text{-}O1}). \tag{1}$$

The corresponding volume variation for the polyhedron is sympathetic but less sensitive.

The occupancy of M2 (± 0.020 atoms) may be deduced from the correlation between $\Delta M2$ and $(Ca + Na)$ ($r = -0.987$; Fig. 4) and between M2-O1 and Ca ($r = 0.986$; Fig. 5). Mg and Fe^{2+} may be obtained from the solution of the equations

$$10\ Mg + 24\ Fe^{2+} = \sum e_{M2} - (10\ Na + 18\ Ca),$$
$$Mg + Fe^{2+} = 1 - (Na + Ca), \tag{2}$$

where 10 and 24 are the electrons assigned to Mg and Fe^{2+}, respectively, $\sum e_{M2}$ is the sum of the electrons corresponding to M2 + M2′ cations (see

Table 3. Polyhedral geometry in clinopyroxenes.

	S.10	S.53	S.58	S.91	S.194	S.80	S.80A	S.80B	S.80C
M(1) site									
M(1)-O(2)$_{C1,D1}$	2.053(3)	2.051(2)	2.048(2)	2.042(2)	2.045(3)	2.048(2)	2.047(2)	2.046(2)	2.049(2)
M(1)-O(1)$_{A2,B2}$	2.064(2)	2.061(2)	2.059(2)	2.057(2)	2.048(3)	2.057(2)	2.055(2)	2.056(2)	2.054(2)
M(1)-O(1)$_{A1,B1}$	2.139(3)	2.144(2)	2.142(2)	2.140(2)	2.139(3)	2.135(2)	2.135(2)	2.134(2)	2.135(2)
Mean of 6	2.085	2.085	2.083	2.080	2.077	2.080	2.079	2.079	2.079
Volume (Å^3)	12.00	12.00	11.96	11.90	11.86	11.91	11.89	11.88	11.90
M(2) site									
M(2)-O(2)$_{C2,D2}$	2.276(3)	2.251(2)	2.256(2)	2.269(2)	2.278(3)	2.283(2)	2.284(2)	2.286(2)	2.282(2)
M(2)-O(1)$_{A1,B1}$	2.311(2)	2.297(2)	2.300(2)	2.310(2)	2.318(3)	2.327(2)	2.325(2)	2.326(2)	2.327(2)
M(2)-O(3)$_{C1,D1}$	2.614(3)	2.630(2)	2.626(2)	2.607(2)	2.600(3)	2.597(2)	2.597(2)	2.598(2)	2.597(2)
M(2)-O(3)$_{C2,D2}$	2.770(3)	2.780(2)	2.769(2)	2.759(2)	2.758(3)	2.748(2)	2.746(2)	2.748(2)	2.750(2)
Mean of 8	2.492	2.489	2.488	2.486	2.488	2.489	2.488	2.489	2.489
ΔM(2)	0.369	0.387	0.375	0.364	0.359	0.346	0.344	0.345	0.348
Volume (Å^3)	25.41	25.22	25.19	25.22	25.28	25.34	25.34	25.37	25.34
M(2') site									
M(2')-O(1)$_{A1,B1}$	1.83	1.82	1.83	1.82	1.85	1.85	1.84	1.83	1.84
M(2')-O(2)$_{C2,D2}$	2.26	2.24	2.24	2.25	2.25	2.26	2.26	2.26	2.26
T site									
T-O(2)	1.592(3)	1.595(2)	1.595(2)	1.594(2)	1.587(3)	1.591(2)	1.590(2)	1.591(2)	1.589(2)
T-O(1)	1.614(2)	1.612(2)	1.613(2)	1.613(2)	1.614(3)	1.608(2)	1.609(2)	1.611(2)	1.610(2)
T-O(3)$_{A1}$	1.658(3)	1.663(2)	1.663(2)	1.661(2)	1.659(3)	1.664(2)	1.660(2)	1.665(2)	1.665(2)
T-O(3)$_{A2}$	1.680(3)	1.673(2)	1.674(2)	1.679(2)	1.681(3)	1.679(2)	1.681(2)	1.680(2)	1.678(2)
Mean of nonbrg.	1.603	1.603	1.604	1.603	1.600	1.599	1.600	1.601	1.599
Mean of brg.	1.669	1.668	1.668	1.670	1.670	1.671	1.670	1.672	1.671
Mean of 4	1.636	1.636	1.636	1.636	1.635	1.635	1.635	1.637	1.635
T-T	3.104(1)	3.102(2)	3.100(1)	3.103(1)	3.104(2)	3.104(1)	3.103(1)	3.105(1)	3.105(1)
Volume (Å^3)	2.23	2.23	2.23	2.23	2.23	2.23	2.23	2.23	2.23
O(3)-T-O(3)	105.21(8)	105.45(6)	105.42(7)	105.23(8)	105.14(10)	104.94(6)	105.02(7)	104.93(6)	104.95(6)

Table 3 (continued).

	S.69A	S.69B	S.73A	S.73B	S.45A	S.45B	S.255	S.255A	S.255B
M(1) site									
M(1)-O(2)$_{C1,D1}$	2.043(2)	2.042(2)	2.049(2)	2.048(2)	2.043(2)	2.043(2)	2.038(2)	2.038(1)	2.039(1)
M(1)-O(1)$_{A2,B2}$	2.060(2)	2.058(2)	2.055(2)	2.057(2)	2.061(1)	2.058(1)	2.058(2)	2.055(1)	2.055(1)
M(1)-O(1)$_{A1,B1}$	2.139(2)	2.138(2)	2.135(2)	2.135(2)	2.137(1)	2.134(2)	2.132(2)	2.128(1)	2.130(1)
Mean of 6	2.081	2.079	2.080	2.080	2.080	2.078	2.076	2.074	2.075
Volume (Å^3)	11.92	11.89	11.90	11.91	11.91	11.88	11.83	11.79	11.81
M(2) site									
M(2)-O(2)$_{C2,D2}$	2.285(2)	2.290(2)	2.293(2)	2.287(2)	2.305(2)	2.308(2)	2.317(2)	2.320(1)	2.314(2)
M(2)-O(1)$_{A1,B1}$	2.323(2)	2.330(2)	2.335(2)	2.329(2)	2.338(1)	2.340(2)	2.346(2)	2.348(1)	2.344(1)
M(2)-O(3)$_{C1,D1}$	2.603(2)	2.596(2)	2.591(2)	2.597(2)	2.592(2)	2.588(2)	2.582(2)	2.576(1)	2.580(2)
M(2)-O(3)$_{C2,D2}$	2.753(2)	2.748(2)	2.744(2)	2.746(2)	2.739(2)	2.739(2)	2.729(2)	2.727(1)	2.731(1)
Mean of 8	2.491	2.491	2.491	2.490	2.493	2.494	2.493	2.493	2.492
ΔM(2)	0.349	0.342	0.338	0.342	0.327	0.327	0.314	0.312	0.318
Volume (Å^3)	25.40	25.43	25.45	25.38	25.55	25.56	25.57	25.57	25.54
M(2') site									
M(2')-O(1)$_{A1,B1}$	1.84	1.86	1.85	1.83	1.85	1.85	1.84	1.85	1.85
M(2')-O(2)$_{C2,D2}$	2.26	2.26	2.27	2.27	2.28	2.28	2.29	2.29	2.28
T site									
T-O(2)	1.595(2)	1.596(2)	1.589(2)	1.590(2)	1.597(1)	1.596(2)	1.597(2)	1.597(1)	1.597(1)
T-O(1)	1.613(2)	1.613(2)	1.607(2)	1.610(2)	1.613(1)	1.613(2)	1.615(2)	1.615(1)	1.615(1)
T-O(3)$_{A1}$	1.664(2)	1.663(2)	1.662(2)	1.664(2)	1.666(1)	1.662(2)	1.667(2)	1.666(1)	1.665(1)
T-O(3)$_{A2}$	1.681(2)	1.682(2)	1.680(2)	1.680(2)	1.685(1)	1.686(2)	1.686(2)	1.684(1)	1.684(1)
Mean of nonbrg.	1.604	1.604	1.598	1.600	1.605	1.604	1.606	1.606	1.606
Mean of brg.	1.672	1.672	1.671	1.672	1.675	1.674	1.676	1.675	1.674
Mean of 4	1.638	1.638	1.634	1.636	1.640	1.639	1.641	1.640	1.640
T-T	3.107(1)	3.108(1)	3.105(1)	3.105(1)	3.110(1)	3.109(1)	3.110(1)	3.109(1)	3.108(1)
Volume (Å^3)	2.24	2.24	2.22	2.23	2.25	2.24	2.25	2.25	2.25
O(3)-T-O(3)	105.11(6)	105.17(6)	104.91(6)	104.94(6)	104.97(5)	104.95(6)	104.82(7)	104.87(5)	104.80(5)

Table 3 (continued).

	S.255C	S.255D	S.235A	S.235B	S.235C	S.239	S.247	S.254A	S.254B
M(1) site									
M(1)-O(2)$_{C1,D1}$	2.044(1)	2.048(2)	2.036(2)	2.030(1)	2.021(2)	2.026(1)	2.023(1)	2.029(2)	2.037(2)
M(1)-O(1)$_{A2,B2}$	2.056(1)	2.058(2)	2.053(1)	2.055(1)	2.058(1)	2.056(1)	2.049(1)	2.054(2)	2.054(2)
M(1)-O(1)$_{A1,B1}$	2.131(1)	2.129(2)	2.125(2)	2.127(1)	2.128(1)	2.126(1)	2.123(1)	2.125(2)	2.127(2)
Mean of 6	2.077	2.078	2.071	2.071	2.069	2.069	2.065	2.069	2.073
Volume (Å^3)	11.85	11.88	11.75	11.73	11.70	11.71	11.64	11.71	11.78
M(2) site									
M(2)-O(2)$_{C2,D2}$	2.308(2)	2.314(2)	2.324(2)	2.336(1)	2.351(1)	2.352(1)	2.338(1)	2.346(2)	2.341(2)
M(2)-O(1)$_{A1,B1}$	2.341(1)	2.345(2)	2.351(2)	2.358(1)	2.368(1)	2.368(1)	2.361(1)	2.365(2)	2.362(2)
M(2)-O(3)$_{C2,D1}$	2.582(1)	2.580(2)	2.571(2)	2.567(1)	2.559(1)	2.557(1)	2.560(1)	2.569(2)	2.565(2)
M(2)-O(3)$_{C2,D2}$	2.738(1)	2.731(2)	2.724(2)	2.718(1)	2.703(1)	2.712(1)	2.711(1)	2.710(2)	2.721(2)
Mean of 8	2.492	2.492	2.492	2.495	2.495	2.497	2.492	2.497	2.497
ΔM(2)	0.328	0.318	0.309	0.298	0.277	0.286	0.291	0.283	0.298
Volume (Å^3)	25.52	25.55	25.56	25.65	25.73	25.76	25.59	25.68	25.74
M(2') site									
M(2')-O(1)$_{A1,B1}$	1.85	1.85	1.85	1.86	1.86	1.90	1.88	1.87	1.88
M(2')-O(2)$_{C2,D2}$	2.28	2.29	2.29	2.30	2.30	2.30	2.29	2.30	2.30
T site									
T-O(2)	1.592(2)	1.589(2)	1.597(2)	1.602(1)	1.614(2)	1.605(1)	1.606(1)	1.604(2)	1.598(2)
T-O(1)	1.612(1)	1.605(1)	1.614(2)	1.620(1)	1.630(1)	1.622(1)	1.623(1)	1.622(2)	1.614(2)
T-O(3)$_{A1}$	1.665(1)	1.666(2)	1.668(2)	1.669(1)	1.675(1)	1.671(1)	1.670(1)	1.672(2)	1.666(2)
T-O(3)$_{A2}$	1.682(1)	1.682(2)	1.684(2)	1.687(1)	1.696(1)	1.689(1)	1.689(1)	1.689(2)	1.688(2)
Mean of nonbrg.	1.602	1.597	1.605	1.611	1.622	1.613	1.614	1.613	1.606
Mean of brg.	1.673	1.674	1.676	1.678	1.685	1.680	1.679	1.680	1.677
Mean of 4	1.638	1.635	1.641	1.644	1.654	1.646	1.647	1.647	1.641
T-T	3.108(1)	3.106(1)	3.110(1)	3.118(1)	3.124(1)	3.117(1)	3.115(1)	3.115(1)	3.113(1)
Volume (Å^3)	2.24	2.23	2.25	2.26	2.30	2.27	2.28	2.27	2.25
O(3)-T-O(3)	104.85(5)	104.57(6)	104.72(6)	104.83(4)	104.86(5)	104.75(4)	104.82(4)	104.60(5)	104.76(6)

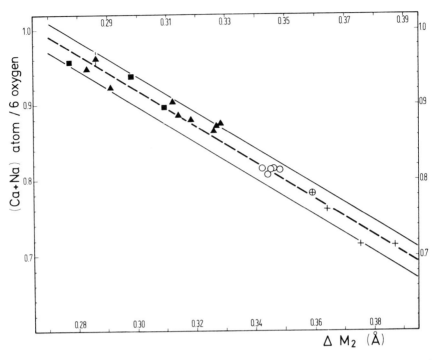

Fig. 4. Plot of ΔM2 against $\sum(Ca + Na)$ cations in clinopyroxenes. Symbols as in Fig. 1.

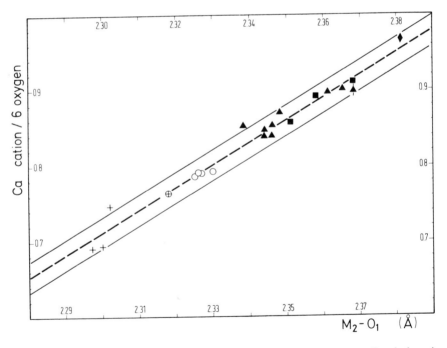

Fig. 5. Plot of M2-O1 bond distance against Ca cation in clinopyroxene. Symbols as in Fig. 1. Solid diamond: cpx from a clinopyroxenite ejecta from Vesuvius.

Table 4. Electrons of M1, M2 and M2' as calculated by site occupancy.

S.No.	1 e_{M1}	2 e_{M2}[a]	3 e_{M2}[b]	4 $e_{M2'}$	5 e_{M2}[c]	6 e_{M1+M2}[d]	7 e_{M1+M2}[e]
10	12.541	18.594	18.330	0.576	18.750	31.291	—
53	12.373	18.816	18.540	0.576	18.966	31.339	30.958
58	12.303	18.648	18.408	0.600	18.828	31.131	31.071
91	11.792	17.896	17.368	0.672	17.968	29.760	29.429
194	11.554	17.984	17.528	0.576	18.044	29.598	29.480
80	10.973	17.832	17.368	0.672	17.936	28.909	28.666
80A	10.973	17.904	17.384	0.768	18.028	29.001	28.755
80B	11.302	18.040	17.568	0.648	18.128	29.430	29.350
80C	11.099	17.960	17.520	0.624	18.052	29.151	28.849
69A	12.079	18.240	17.784	0.624	18.324	30.403	—
69B	12.359	18.224	17.776	0.600	18.300	30.659	30.767
73A	11.078	17.688	17.280	0.624	17.796	28.874	—
73B	11.302	17.992	17.584	0.576	18.076	29.378	—
45A	12.310	18.144	17.760	0.552	18.228	30.538	30.375
45B	12.114	18.032	17.696	0.456	18.092	30.206	30.451
255	12.646	18.160	17.808	0.504	18.236	30.882	30.731
255A	12.226	18.088	17.776	0.504	18.184	30.410	30.109
255B	12.168	18.110	17.800	0.456	18.183	30.351	30.019
255C	11.645	17.880	17.520	0.552	17.976	29.621	29.498
255D	11.064	17.704	17.384	0.504	17.796	28.860	—
235A	11.805	17.800	17.500	0.456	17.878	29.683	29.614
235B	12.548	17.872	17.656	0.360	17.994	30.492	30.462
235C	13.479	17.912	17.712	0.312	17.968	31.447	31.858
239	13.528	17.720	17.608	0.168	17.748	31.276	31.152
247	12.639	17.816	17.616	0.312	17.872	30.511	30.701
254A	12.660	17.912	17.736	0.264	17.956	30.616	30.607
254B	12.149	17.768	17.584	0.264	17.808	29.957	—

[a] Electrons computed by M2 site refinement without considering M2'.
[b] Electrons computed by M2 and M2' site refinement.
[c] Average electrons: (col.2 + col.3 + col.4)/2. The electrons for M2 + M2' result from the average of the electrons of M2 obtained by refining M2 occupancy without considering M2', and those calculated as the sum of the electrons of M2 and M2' after the last cycles of refinement. This average reduces the error in the electron computation, owing to the partial overlap existing between the electron distribution of M2 and M2'.
[d] Electrons calculated from structural data: col.1 + col.5.
[e] Electrons calculated from microprobe analysis (Table 1).

Table 4), Na, Ca, Mg and Fe^{2+} are atomic proportions, respectively; Mn is calculated as Fe^{2+}.

Polyhedron M1

A notable variation for this polyhedron is given by the M1-O2 bond distance (2.021–2.051 Å), which reflects the occupancy by ions smaller than Mg and

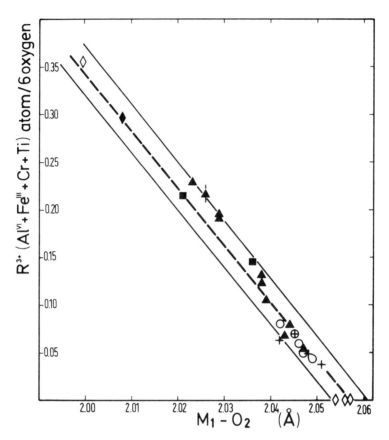

Fig. 6. Plot of M1-O2 bond distance against the sum of R^{3+} cations. Symbols as in Fig. 1; open diamonds = synthetic pyroxenes (E. Bruno *et al.*, personal communication, 1981), solid diamond = cpx from a clinopyroxenite ejecta from Vesuvius.

Fe^{2+}, i.e., Ti^{4+}, Fe^{3+}, Al^{3+}, and Cr^{3+}. The occupancy by the latter ions produces a strong angular distortion of the polyhedron.

The strong negative correlation between M1-O2 and $R^{3+} = Al^{3+} + Cr^{3+} + Fe^{3+} + Ti^{4+}$ ($r = -0.989$) (see Fig. 6) may be used to derive a value for R^{3+} (± 0.020). The occupancy of Mg and Fe^{2+} may then be obtained from the solution of the equations:

$$10\,Mg + 24\,Fe^{2+} + \sum e_{R^{3+}} = \sum e_{M1},$$

$$2.081\,Mg + 2.126\,Fe^{2+} + (\overline{M1\text{-}O})_{R^{3+}} = \overline{M1\text{-}O}, \tag{3}$$

$$Mg + Fe^{2+} + R^{3+} = 1.000,$$

where 10 and 24 are the electrons assigned to Mg and Fe^{2+}, respectively,

$\sum e_{R^{3+}}$ is the sum of electrons corresponding to R^{3+} cations, and $\sum e_{M1}$ is the sum of electrons corresponding to M1 cations; 2.081 Å is the $\overline{M1-O}$ in an octahedron coordinated by Mg (E. Bruno et al., personal communication, 1981) and 2.126 the $\overline{M1-O}$ in an octahedron coordinated by Fe^{2+} (Ungaretti et al., 1978); $(\overline{M1=O})_{R^{3+}} = 1.930 \, Al^{VI} + 2.030 \, Fe^{3+} + 1.99 \, Ti^{4+}$ (Ungaretti et al., 1978).

For the Ethiopian clinopyroxenes the following relationships are obtained from the chemical analyses

$$Al^{VI} : Fe^{3+} : Ti^{4+} = 1 : 2 : 7 \quad (Ca < 0.80 \text{ atoms}), \tag{4}$$

$$Al^{VI} : Fe^{3+} : Ti^{4+} = 1 : 5.5 : 3.5 \quad (Ca \geqslant 0.80 \text{ atoms}). \tag{5}$$

The proportions of Al^{VI}, Fe^{3+} and Ti^{4+} should satisfy, within experimental error, the following equations:

$$R^{3+} = -5.975(M1\text{-}O2) + 12.293, \tag{6}$$

$$AR^{3+} = 3 \, Al^{IV} + 1 \, Na, \tag{7}$$

where A is the sum of the R^{3+} charges.

Polyhedron T

A main feature of this polyhedron is the negative correlation between $\overline{T-O}$ nonbridging and Si $(r = -0.993)$, illustrated in Fig. 7. In the above correlation the data on the fassaites from Allende (Dowty and Clark, 1973), Quebec (Peacor, 1967), and Ador (Hazen and Finger, 1977), the Takasima augite (Takeda, 1972), and α and γ synthetic diopsides (E. Bruno et al., personal communcation, 1981) are included. Notably, the regression curve is close to that obtained by Hazen and Finger (1977).

Discussion

The existence of a new site M2′ is of special interest in the interpretation of pyroxene topology and appears to be supported by the crystal-structure data on the synthetic, $C2/c$, $ZnSiO_3$ pyroxene provided by Morimoto et al. (1975). In the latter pyroxene Zn in the M2 site is situated far from O3 and close to O1 and O2 (Zn2-O1 and Zn2-O2 are 1.933 and 2.031 Å, respectively), forming an irregular tetrahedral coordination of O1 and O2 around Zn2. The M2′ of the natural pyroxenes investigated show identical atomic coordinates to those

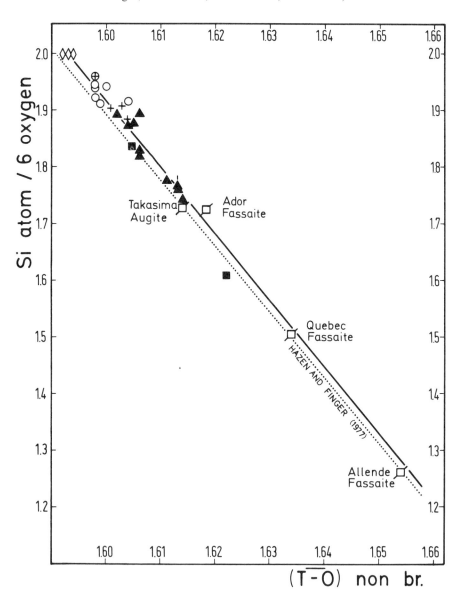

Fig. 7. Plot of nonbridging T-O bond distance against Si^{4+}. Symbols as in Fig. 1; open diamonds = synthetic pyroxenes (E. Bruno *et al.*, personal communication, 1981); barred squares = fassaites and augite after Dowty and Clark (1973), Hazen and Finger (1977), Peacor (1967), and Takeda (1972).

reported for Zn ($x = 0.0$; $y = 0.2361$; $z = 0.25$; Morimoto *et al.* (1975), Table 3). From the Fourier synthesis, the M2'-O1 and M2'-O2 vary within 1.8–1.9 Å and 2.2–2.3 Å, respectively (Table 3). The evidence for the existence of the M2' site decreases with increasing Ca in the investigated clinopyroxenes and disappears above Ca > 0.95 atoms.

The O1 and O2 bond lengths of M2 are particularly unfavourable to Mg occupancy, relative to Fe^{2+}. It is expected that M2' is preferentially occupied by Mg, while Fe^{2+} should occupy M2. It seems, therefore, possible that the "splitting" of the M2 site to M2' may represent a precursor to the unmixing of the (Ca, Mg, Fe^{2+}) pyroxenes where Fe^{2+} or Mg substitute for Ca in M2.

The presence of M2' in synthetic pyroxenes in the diopside–enstatite join at 1 bar, obviously results from a high-temperature crystallization (over 1100°C), indicating that this structural feature forms at a high-temperature stage in the pyroxene crystallization history.

An important result of this work is the calculation of the site occupancies for all the cations in the M1, M2, and T sites by using sets of Eqs. (2) and (3) in combination with the chemical analysis of the pyroxenes. Results of the site occupancies calculated in this manner (Eqs. (2), (3), (6), and (7)) are presented in Table 5. The Fe^{3+} can be simply estimated using R^{3+} (Eq. (6)) and Al^{VI}, Ti, and Cr as determined by probe analysis (see Table 1). The site-occupancy data of the Ethiopian pyroxenes are treated in the next section.

From the various correlations established in this work (Figs. 4–7) it has become possible to estimate the site occupancies in different sites independently of the chemical analysis. In other words, for volcanic pyroxenes with compositions which fall within the range of the compositions used in this work, it is possible to completely determine the site occupancies and, therefore, the complete chemical composition of a pyroxene. This could be done by using the following equations:

$$Ca = 2.975(M2\text{-}O1) - 6.130, \tag{8}$$

$$(Ca + Na) = -2.3959\Delta M2 + 1.6394, \tag{9}$$

$$R^{3+} = -5.975(M1\text{-}O2) + 12.293, \tag{6}$$

$$Si = -11.495(T\text{-}O)_{nonbr} + 20.306. \tag{10}$$

Site occupancies as determined by using Eqs. (6) and (8)–(10) are presented in Table 6. As expected, the reproducibility of the data from linear regression relationships (Eqs. (6), and (8)–(10)) is quite satisfactory. One may, therefore, confidently use these relationships in predicting site occupancies from crystallographic data even if the pyroxene cannot be chemically analyzed. Table 6 includes some pyroxenes whose compositions have not been determined.

The charge balance of the investigated pyroxenes generally approximates 12.00, ranging from 11.92 to 12.02. It should be noted that the best approximation to the ideal value of 12.00 is given by pyroxenes with Ca > 0.80 atoms, whereas the low-Ca pyroxenes show a charge deficiency up to 0.08.

The compositions of the pyroxenes studied in the present work lie on two compositional trends. The subcalcic trend which begins around Ca ≈ 0.80 atoms and continues parallel to other basaltic trends (e.g., the Skaergaard trend in Fig. 2). The second trend is the calcic trend (Ca > 0.80 atoms).

Table 5. Cation site-occupancy data using chemical data for R^{3+} cations, Ca, Na, and Mn.

	S.53	S.58	S.91	S.194	S.80	S.80A	S.80B	S.80C	S.69B	S.45A	S.45B	S.255	S.255A	S.255B	S.255C	S.235A	S.235B	S.235C	S.239	S.247	S.254A
M(1) site																					
R^{3+}	0.038	0.049	0.062	0.070	0.034	0.050	0.059	0.044	0.080	0.068	0.054	0.124	0.131	0.106	0.080	0.146	0.068	0.215	0.217	0.229	0.196
Mg	0.820	0.824	0.845	0.853	0.919	0.921	0.900	0.918	0.786	0.811	0.810	0.792	0.826	0.809	0.872	0.850	0.779	0.700	0.694	0.723	0.758
Fe^{2+}	0.142	0.127	0.093	0.077	0.047	0.029	0.041	0.038	0.134	0.121	0.136	0.084	0.043	0.085	0.048	0.004	0.148	0.085	0.089	0.048	0.046
M(2) site																					
Ca	0.691	0.694	0.747	0.764	0.793	0.786	0.791	0.791	0.793	0.846	0.840	0.854	0.871	0.848	0.846	0.858	0.892	0.924	0.899	0.897	0.900
Na	0.022	0.020	0.013	0.018	0.022	0.021	0.023	0.022	0.022	0.026	0.026	0.033	0.033	0.033	0.030	0.038	0.042	0.031	0.062	0.036	0.043
Mg	0.030	0.039	0.106	0.076	0.064	0.081	0.046	0.079	0.040	0.012	0.032	0.009	0.002	0.014	0.031	0.064	0.000	0.000	0.000	0.013	0.000
Fe^{2+}	0.250	0.238	0.128	0.136	0.116	0.107	0.135	0.104	0.139	0.106	0.098	0.098	0.090	0.101	0.088	0.037	0.063	0.039	0.033	0.050	0.054
Mn	0.007	0.009	0.006	0.006	0.005	0.005	0.005	0.004	0.006	0.004	0.004	0.006	0.004	0.004	0.005	0.003	0.003	0.006	0.006	0.004	0.003
$K_D{}^a$	0.021	0.025	0.091	0.050	0.028	0.018	0.016	0.031	0.049	0.025	0.055	0.010	0.012	0.015	0.019	0.009	0.0	0.0	0.0	0.017	0.0

$^a K_D = \{(Fe^{2+}/Mg)_{M1}/(Fe^{2+}/Mg)_{M2}\}$.

Table 6. Cation site-occupancy data using crystal-structural relationships established in this paper.

	S.10[a]	S.53	S.58	S.91	S.194	S.80	S.80A	S.80B	S.80C
T site									
Si	1.880	1.880	1.869	1.880	1.915	1.926	1.915	1.903	1.926
Al^{IV}	0.120	0.120	0.131	0.120	0.085	0.074	0.085	0.097	0.074
M(1) site									
R^{3+}	0.036	0.030	0.047	0.063	0.084	0.035	0.043	0.053	0.047
Mg	0.804	0.821	0.816	0.855	0.866	0.919	0.917	0.886	0.903
Fe^{2+}	0.160	0.149	0.137	0.082	0.051	0.047	0.040	0.061	0.049
M(2) site									
Ca	0.745	0.704	0.713	0.742	0.766	0.793	0.787	0.790	0.793
Na	0.010	0.009	0.028	0.025	0.013	0.018	0.028	0.023	0.013
Mg	0.035	0.039	0.023	0.083	0.080	0.069	0.053	0.052	0.066
Fe^{2+} [b]	0.210	0.249	0.236	0.150	0.141	0.121	0.132	0.135	0.128
K_D	0.033	0.028	0.016	0.053	0.033	0.029	0.018	0.026	0.028

	S.69A[a]	S.69B	S.73A[a]	S.73B[a]	S.45A	S.45B	S.255	S.255A	S.255B
T site									
Si	1.869	1.869	1.938	1.915	1.857	1.846	1.846	1.846	1.846
Al^{IV}	0.131	0.131	0.062	0.085	0.143	0.154	0.154	0.154	0.154
M(1) site									
R^{3+}	0.057	0.083	0.038	0.045	0.078	0.072	0.126	0.132	0.121
Mg	0.829	0.798	0.908	0.890	0.813	0.829	0.776	0.804	0.811
Fe^{2+}	0.113	0.119	0.054	0.066	0.108	0.099	0.099	0.064	0.068
M(2) site									
Ca	0.781	0.802	0.817	0.799	0.826	0.846	0.849	0.855	0.843
Na	0.022	0.018	0.013	0.021	0.030	0.019	0.038	0.037	0.034
Mg	0.043	0.040	0.073	0.054	0.023	0.038	0.005	0.006	0.015
Fe^{2+} [b]	0.154	0.119	0.097	0.126	0.121	0.096	0.108	0.102	0.107
K_D	0.038	0.043	0.045	0.032	0.025	0.047	0.006	0.005	0.040

	S.255C	S.255D[a]	S.235A	S.235B	S.235C	S.239	S.247	S.254A	S.254B[a]
T site									
Si	1.892	1.949	1.857	1.788	1.662	1.765	1.754	1.765	1.846
Al^{IV}	0.108	0.051	0.143	0.212	0.338	0.235	0.246	0.235	0.154
M(1) site									
R^{3+}	0.086	0.060	0.147	0.170	0.216	0.218	0.229	0.192	0.139
Mg	0.859	0.908	0.830	0.770	0.689	0.685	0.746	0.756	0.807
Fe^{2+}	0.055	0.033	0.023	0.060	0.094	0.097	0.026	0.053	0.054
M(2) site									
Ca	0.834	0.846	0.864	0.885	0.915	0.915	0.894	0.906	0.897
Na	0.019	0.031	0.035	0.040	0.035	0.039	0.048	0.040	0.028
Mg	0.047	0.043	0.027	0.008	0.000	0.014	0.003	0.000	0.027
Fe^{2+} [b]	0.099	0.079	0.074	0.066	0.050	0.032	0.055	0.053	0.048
K_D	0.030	0.012	0.010	0.009	0.0	0.062	0.002	0.0	0.038

[a] Samples not analyzed by microprobe.

[b] Includes Mn.

The Subcalcic Pyroxene Trend

This general trend forms around the ternary immiscibility field. Thermody-namically this may be explained as a consequence of the binary (enstatite–diopside) miscibility gap extending into the ternary compositional field. It is for the first time that we have crystallographic data on a series of pyroxenes of intermediate compositions. Therefore, a crystal structural description of such trend is now possible. Pyroxenes along this trend (e.g., 58 and 80) may be compared by referring to Table 3.

The increase of Fe^{2+} produces a drastic reduction of the M2-O1 and M2-O2, and an increase (with entering of Al^{IV}) of the T-O1 and T-O2 bond distances. Although the increase of the T-O bonds should be expected with the replacement of Ca by Fe^{2+} in M2, as a consequence of the reduction of M2-O bonds, in the present case we find a distinct correlation of Al^{IV} with Ca (Fig. 8) and Fe^{2+}_{M2}. The reduction of M2-O is also accompanied by shorter M1-O1

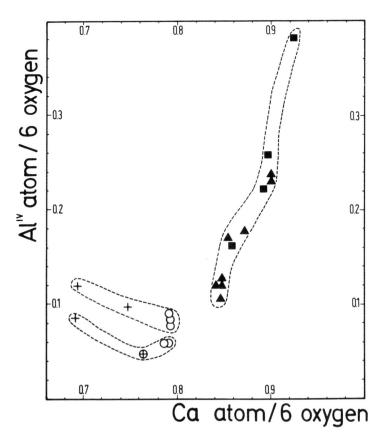

Fig. 8. Plot of tetrahedral Al ions against Ca to show the compositional variation in calcic and subcalcic pyroxenes. Symbols as in Fig. 1.

and M1-O2 due to trivalent cations. Notably T-O3 bridging bonds are reduced concurrently with the increase of M2-O3. Consequently for subcalcic pyroxenes only two of the T-O bond lengths increase, increasing the angle O3-T-O3 and producing a larger c_0 correlated to the Al content in the T site. Therefore, it can be concluded that the replacement of Ca by Fe^{2+} in M2 produces structural changes influencing the bond distances of M1 and T sites. An immediate consequence is the entering in these sites of trivalent cations and Al^{IV}, respectively. These structural variations have a very important bearing on the stability of Fe-rich pyroxenes, which may change significantly depending on Al^{3+} concentration in the environment.

The Calcic–Pyroxene Trend

The Ethiopian calcic–pyroxenes (Ca > 0.80 atoms) show an opposite effect of increasing Ca^{2+} on the tetrahedral site occupancy. As shown in Fig. 8, there is a positive correlation between Ca and the Al occupancy of the tetrahedral sites. While the entry of Al in the tetrahedral sites of subcalcic pyroxenes was determined by increased tetrahedral bond lengths, the increased preference of Al over Si to enter the tetrahedral sites of calcic pyroxenes is the result of a different change brought about by increasing Ca in M2. The M2-O3 bonds are shortened, thereby decreasing the effective negative charge of O3. Si may then be substituted by a trivalent cation of larger radius due to the increased T-O3 bond lengths. This is illustrated in Fig. 9 by two examples, 53 and 239, which differ in their Ca occupancy of M2, relative to Fe^{2+}.

Figure 10 shows yet another effect of increasing Ca occupancy of the M2 site. With the occupancy of the T site by Al, R^{3+} cations may substitute for Mg at the neighboring M1 site. The negative correlation between Mg in M1 and Al^{IV} (Fig. 10) supports this crystal-chemical effect.

Finally, Fig. 11 shows the correlation between the Fe occupancy of M2 site and the total concentration of (Ca + Na). As expected, there is a negative correlation. However, the slopes for the variation in the calcic and subcalcic pyroxenes are different. In the calcic pyroxenes, the larger slope indicates a greater influence of (Ca + Na) on Fe occupancy than in the subcalcic pyroxenes.

Applications

Thermodynamics and Cooling History

The thermodynamic interpretation of the data presented in this paper is rendered difficult because of the paucity of experimental data on the distribution of cations between the M1 and M2 sites. The intracrystalline Fe–Mg

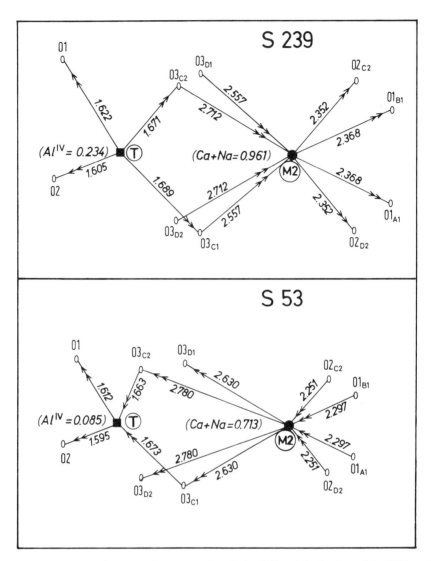

Fig. 9. Comparison between the geometries of the M2 polyhedra in calcic (239) and subcalcic (53) pyroxenes. Double arrows toward (away from) T and M2, respectively, indicate shortening (lengthening) of the bond distances.

exchange can be represented by the equation

$$Mg(M1) + Fe(M2) \rightleftharpoons Fe(M1) + Mg(M2).$$

(a)

At equilibrium, we have

$$K_a = \frac{\left(X_{Fe}^{M1} X_{Mg}^{M2}\right)}{\left(X_{Mg}^{M1} X_{Fe}^{M2}\right)} (K_\gamma),$$

(11)

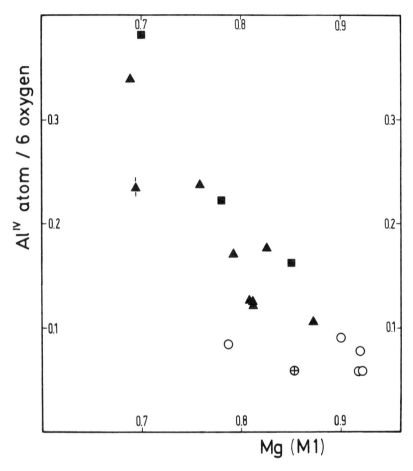

Fig. 10. Plot of the Mg site occupancy in M1 against tetrahedral Al. Symbols as in Fig. 1.

where K_a is the equilibrium constant, X_i the atomic fraction (Fe/(Fe + Mg)), and K_γ the nonideality term which includes the activity coefficients of Fe and Mg on the two sites and also the energy change of the ordered and disordered components (for example, in orthopyroxene the energy change ΔG between $Mg_{M1}Fe_{M2}Si_2O_6$ and $Mg_{M2}Fe_{M1}Si_2O_6$, see Thompson (1970)). For clinopyroxenes the expression for K_γ has not been developed yet. Since the experimental data on site occupancies in clinopyroxene is inadequate, we work with atomic ratios only, employing K_D which is the compositional term in Eq. (11).

K_D values for pyroxenes are listed in Table 5. The variation in these values may be due to temperature and concentrations of the R^{3+} cations and Ca. Figures 12 and 13 show that K_D decreases with increasing concentrations of R^{3+} cations and Ca, respectively. McCallister *et al.* (1976) determined site occupancies on two clinopyroxenes which were equilibrated at several temperatures. Their pyroxenes are compositionally more Al rich than ours. Their

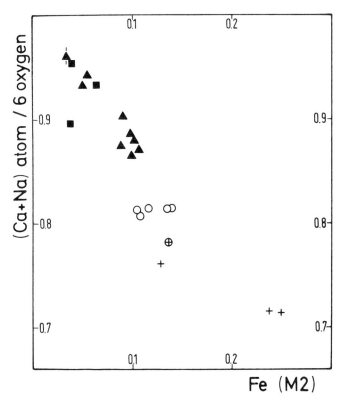

Fig. 11. Plot of the Fe occupancy of M2 against the sum of Ca and Na cations. Symbols as in Fig. 1.

data on site occupancies at 675 and 927°C were processed to yield the following two equations[2] which express K_D as a function of the concentrations of the R^{3+} and Ca:

$$\ln K_D(675°C) = -0.5609 - 1.7328(R^{3+}) - 2.3434(Ca), \qquad (12)$$

$$\ln K_D(927°C) = -1.2353 - 0.5201(R^{3+}) - 0.7181(Ca), \qquad (13)$$

where (R^{3+}) is the sum of octahedrally coordinated Al^{3+}, Ti^{4+}, Cr^{3+}, and Fe^{3+}, and (Ca) the sum of Ca, Mn, and Na ions.

Assuming that the coefficient in Eqs. (12) and (13) are linear with temperature, we obtain the following equation for determining the temperature of intracrystalline equilibrium:

$$T(°K) = \frac{5.465(R^{3+}) + 7.324(Ca) - 3.039}{-\ln K_D + 4.032(R^{3+}) + 5.383(Ca) - 3.767} \times 1000. \qquad (14)$$

[2] These and Eq. (14) are provided by Saxena (personal communication, 1981).

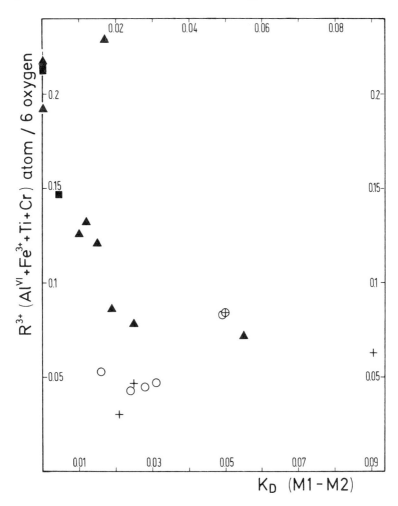

Fig. 12. Plot of the site distribution coefficient K_D(M1-M2) against the sum of R^{3+} cations. Symbols as in Fig. 1.

Equation (14) is to be used with caution. It is based on limited data on pyroxenes which contain a large concentration of octahedral Al. We could still perhaps be able to use Eq. (14) in obtaining some information on the cooling history of the various rocks under study. Note that McCallister *et al.*'s data at 802°C was left out in constructing this geothermometer. The K_D for the Kakanui pyroxene at this temperature calculated from Eq. (14) is 0.105 which is closer to their value of 0.096 than the value of 0.123 (McCallister *et al.*, 1976).

Temperatures of intracrystalline cation equilibration are listed in Table 7. Although these temperatures are subject to a large uncertainty, we notice a general correlation between the estimated temperature and texture of the

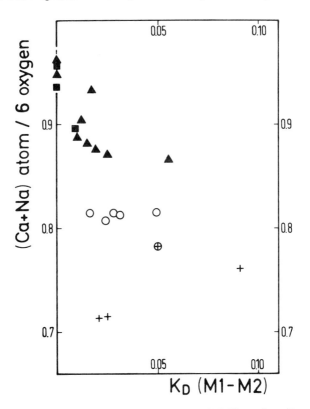

Fig. 13. Plot of the site distribution coefficient K_D(M1-M2) against Ca + Na cations. Symbols as in Fig. 1.

basalts. A high equilibration temperature (1148°C, sample 196[3]) corresponds to a pyroxene occurring in a fine-grained groundmass with abundant fresh glass, presumably formed by rapid cooling. On the other hand, relatively low equilibration temperatures (447°C, sample 80; 404°C, sample 255) are ob-

Table 7. Equilibrium temperatures in Ethiopian clinopyroxenes.[a]

S. No.	10[b]	53	58	91	194	80	80A	80B	80C	69A[b]	69B	73A[b]
T (°C)	424	307	346	692	560	447	420	374	469	505	579	548

73B[b]	45A	45B	255	255A	255B	255C	255D[b]	235A	247	254B[b]	196A
479	492	632	404	438	446	458	388	407	534	618	1148

[a] The temperatures are calculated using Eq. (14) based on experimental work of McCallister *et al.* (1976).
[b] The temperatures are calculated using K_D values obtained from crystallographic data.

[3] The microprobe and crystallographic data will be published elsewhere.

tained from pyroxenes in well-crystallized groundmasses, suggesting that they have formed under slow cooling conditions.

Temperatures below 400°C were measured for the two tholeiitic samples 53 and 58. These samples differ from others in the present study and from those used by McCallister *et al.* (1976) in that they contain low Ca and octahedral Al, respectively. If our compositional extrapolation is valid, these estimated temperatures show that ordering may continue in clinoyroxenes to much lower temperatures than in orthopyroxenes for which a cutoff temperature of about 450°C was suggested by Virgo and Hafner (1969), Mueller (1969), and Saxena (1973).

Other Petrologic Applications

Another important application of the crystallographic data concerns the classification of basaltic rocks. Generally, there is a sympathetic compositional variation between the pyroxene and the host rock. Such relationship may not apply when the pyroxene, for example, is xenocrystal (included in a magma of different origin or composition) or when the pyroxene forms at high pressure.

We have demonstrated that the chemistry of Ethiopian pyroxenes can be estimated from crystallographic data (Table 6). When a sufficient number of compositional varieties have been studied, we believe that a general relationship between crystal parameters (e.g., $\Delta M2$, bond lengths, the angle β) and compositions (multicomponent) can be established. It will then be possible to characterize basaltic types through the crystallographic study of pyroxenes, without necessarily determining the chemical composition of the host rock, which is sometimes deeply altered. This idea is supported by the data plotted in Figs. 12 and 13.

In Fig. 14 the pyroxene structural response to various magmatic environments of crystallization is illustrated in terms of the variation $\Delta M2$ versus β. This figure shows, for comparison, additional data not discussed in this paper. The Ethiopian pyroxenes form two distinct series, one related to basic–ultrabasic lavas and the other to alkaline–peralkaline rhyolites. The clinopyroxenes from the Euganean trachytes fall in an intermediate position between the two Ethiopian series. The clinopyroxenes from ultramafic nodules are also distinct in their relatively high β values. In general, the latter tends to decrease with decreasing Mg content in the bulk rocks, reflecting largely the substitution of Mg by Fe^{2+} in M1 sites. The positive correlation between β and $\Delta M2$ for the individual pyroxene series is due to the Mg and/or Fe^{2+} substitution for Ca in M2.

The M2-O1 versus M1-O2 variation (Fig. 15) illustrates the relationship between the site occupancy of M1 and M2 where Ca and R^{3+} group are negatively correlated. Consequently, the transition from tholeiitic through transitional to alkaline basaltic rocks is closely and distinctly imprinted in the crystal structure of the associated clinopyroxenes, providing a high-resolution discrimination for the various magmatic types to which they are related.

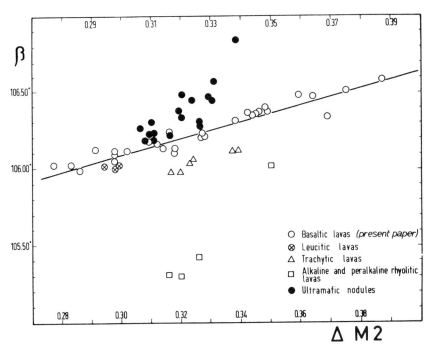

Fig. 14. Plot of the angle β against the distortion index ΔM2 to show the variation of pyroxenes from various rock-types.

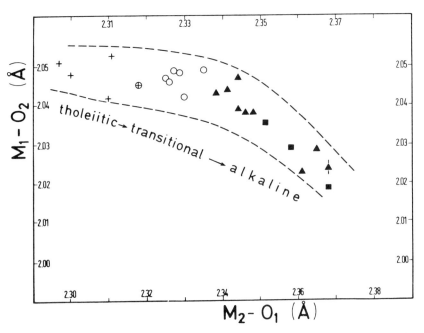

Fig. 15. Plot of M1-O2 against M2-O1 bond distances in clinopyroxenes from tholeiitic, transitional, and alkaline basalts Symbols as in Fig. 1.

Appendix I

S.No.	Rock Name[a]	S.I.[b]	Q[c]	Ne[c]	Mineral Assemblage[d]	Sample Location[e]	Volcanic Formation[f]
10	Tholeiitic basalt	30.4	4.7	—	pl + cpx + opx + pig + op	Socota area	Infrasedimentary (JR)
53	Andesi-basalt	17.8	8.0	—	pl + cpx + pig + op	Dessie area	Ashangi (Pre-OLG)
58	Andesi-basalt	24.5	3.3	—	pl + cpx + pig + op	Amba Costantino area	Ashangi (Pre-OLG)
91	Tholeiitic basalt	31.2	1.1	—	pl + cpx + opx + op	Borona area	Aiba (OLG)
194	Tholeiitic–transitional basalt	37.1	1.2	—	ol + cpx + pl + op + (pig)	Molale area	Alaji-Molale (MIO)
69	Transitional basalt	33.7	—	—	ol + cpx + pl + op	Amba Alaji area	Aiba (OLG)
73	Transitional basalt	33.8	—	—	ol + cpx + pl + op	Amba Aiba area	Aiba (OLG)
80	Transitional basalt	28.3	2.6	—	ol + cpx + pl + op	Amba Aiba area	Aiba (OLG)
45	Alkaline basalt	28.1	—	—	ol + cpx + op + pl	Ashangi Lake area	Ashangi (Pre-OLG)
255	Hawaiite	23.1	—	—	ol + cpx + pl + op + bi	Worra Ilu area	Termaber Guassa (O-M)
235	Picritic alkaline basalt	52.3	—	1.2	ol + cpx + op + (pl)	Abuna Josef area	Termaber Guassa (O-M)
247	Basanitic alkaline basalt	43.2	—	5.4	ol + cpx + op + pl	Guassa area	Termaber Guassa (O-M)
254	Basanitic alkaline basalt	32.8	—	7.1	ol + cpx + op + pl	Guassa area	Termaber Guassa (O-M)
239	Tephritic basanite	26.0	—	13.5	ol + op + pl + (ol)	Abuna Joseph area	Termaber Guassa (O-M)

[a] Nomenclature according to mineralogy (see the text) and chemical classifications of De La Roche *et al.* (1980) and G. Bellieni, E. M. Piccirillo, and B. Zanettin (1981, unpublished data).

[b] S.I. (solidification index) = $100 \, MgO/(MgO + FeO_{tot} + Na_2O + K_2O)$.

[c] CIPW normative quartz (Q) and nepheline (Ne) calculated assuming $Fe_2O_3 = 1.5$ wt%.

[d] pl = plagioclase; cpx = Ca-rich clinopyroxene; opx = orthopyroxene; pig = pigeonite; op = opaques; ol = olivine; bi = biotite.

[e] The precise sample location can be found in Zanettin *et al.* (1976).

[f] JR = Jurassic (Abbate *et al.*, 1969); Pre-OLG = Pre-oligocene; OLG = oligocene (32–28 m.y.); MIO = miocene (26–16 m.y.); O-M = oligocene–miocene (28–24 m.y.).

Acknowledgments

The authors are deeply indebted to Professor S. K. Saxena for his stimulating discussion, helpful suggestions, and improvement of the manuscript. Thanks are due to Professors E. Cannillo, G. Rossi, and L. Ungaretti, and Dr. M. C. Domeneghetti (Pavia University) for their helpful suggestions and to the "Centro per la Cristallografia Strutturale" (Pavia) of C.N.R. for the use of diffractometer facilities.

The authors also thank Professors A. Gregnanin, E. Justin-Visentin, and B. Zanettin for rock samples, and Mr. Mezzacasa for drafting the figures.

This research was financed by Italian C.N.R.: CT No. 79.00041.05 and CT No. 80.02585.05 (Padova), "Gruppo di Ricerche Geologiche e Petrografiche in Africa Orientale," and "Centro di Studio per i Problemi dell'Orogeno delle Alpi Orientali" (Padova).

References

Abbate, E., Facibeni, P., Gregnanin, A., Merla, G., and Sagri, M. (1969) Basalt flows and related sandstones in Socota area (Northern Ethiopia), *Boll. Soc. Geol. It.* **88**, 499–516.

Brotzu, P., Ganzerli-Valentini, M. T., Morbidelli, L., Piccirillo, E. M., and Traversa, G. (1981) Basaltic volcanism in the northern sector of the Main Ethiopian Rift, *J. Volc. Geoth. Res.*, **10**, 365–382.

Burnham, C. W., Clark, J. R., Papike, J. J., and Prewitt, C. T. (1967) A proposed crystallographic nomenclature for clinopyroxene structures, *Z. Krist.* **125**, 1–6.

Clark, J. R., Appleman, D. E., and Papike, J. J. (1969) Crystal-chemical characterization of clinopyroxenes based on eight new structure refinements, *Miner. Soc. Amer. Spec. Pap.* **2**, 31–50.

Dowty, E., and Clark, J. R. (1973) Crystal structure refinement and optical properties of a Ti^{3+} fassaite from the Allende meteorite, *Amer. Miner.* **58**, 230–242.

Finger, L. W. (1972) The uncertainty in the calculated ferric iron content of a microprobe analysis, *Carn. Inst. Wash.*, *Y. B.* **71**, 600–603.

Ghose, S. (1965) Mg^{2+}, Fe^{2+} order in an orthopyroxene, $Mg_{0.93}Fe_{1.07}S_{12}O_6$, *Z. Krist.* **122**, 81–99.

Ghose, S., Wan, C., and Okamura, F. P. (1975) Site preference and crystal chemistry of transition metal ions in pyroxenes and olivine (Abstract), *Acta Cryst. A* **31**, S76.

Hafner, S. S., and Virgo, D. (1970) Temperature-dependent cation distributions in lunar and terrestrial pyroxenes, *Proc. Apollo 11 Lun. Sci. Conf.* **3**, 2183–2198.

Hafner, S. S., Virgo, D., and Warburton, D. (1971) Cation distributions and cooling history of clinopyroxenes from Oceanus Procellarum, *Proc. 2nd Lun. Conf.* **1**, 91–108.

Hazen, R. M., and Finger, L. W. (1977) Crystal structure and compositional variation of Angra Dos Reis fassaite, *Earth Planetary Sci. Lett.* **35**, 357–362.

Honstra, J., and Stubbe, B. (1972) *PW 1100 Data Processing Program*, Research Laboratories, Eindhoven, Holland.

MacDonald, G., and Katsura, T. (1964) Chemical compositions of Hawaiian lavas, *J. Petrol.* **5**, 82–133.

Mason, P. K., Frost, M. T., and Reed, J. S. B. (1969) Computer programs for calculating correlations in quantitative X-ray microanalysis, *Nat. Phys. Lab.* (U.K.) *IMS Rep.* **2**.

McCallister, R. H., Finger, L. W., and Ohashi, Y. (1976) Intracrystalline Fe^{2+}-Mg equilibria in three natural Ca-rich clinopyroxenes, *Amer. Miner.* **61**, 671–676.

Morimoto, N., Nakajuma, Y., Syono, Y., Akimoto, S., and Matsui, Y. (1975) Crystal structures of pyroxene-type $ZnSiO_3$ and $ZnMgSi_2O_6$, *Acta Cryst.* B **31**, 1041–1049.

Mueller, R. F. (1962) Energetics of certain silicate solutions, *Geochim. Cosm. Acta* **26**, 581–598.

North, A. C. T., Phillips, D. C., and Mathews, F. S. (1968) A semi-empirical method of absorption correction, *Acta Cryst.* A **24**, 351–359.

Papike, J. J., Cameron, K., and Baldwin, K. (1974) Amphiboles and pyroxenes: characterization of other than quadrilateral components and estimates of ferric iron from microprobe data, *Geol. Soc. Amer.* **6**, 1053–1054.

Peacor, D. R. (1967) Refinement of the crystal structure of a pyroxene formula $M_IM_{II}(Si_{1.5}Al_{0.3})O_6$, *Amer. Miner.* **52**, 31–41.

Piccirillo, E. M., Justin-Visentin, E., Zanettin, B., Joron, J. L., and Treuil, M. (1979) Geodynamic evolution from plateau to rift: major and trace elment geochemistry of the central eastern Ethiopian plateau volcanics, *N. Jb. Geol. Paläont. Abh.* **158**, 139–179.

Rossi, G., Tazzoli, V., and Ungaretti, L. (1978) Crystal-chemical studies on sodic clinopyroxenes, *Proc. 11th Int. Mineralog. Ass. 11th Meeting*, Novosibirsk.

Saggerson, E. P., and Williams, L. A. J. (1964) Ngurumanite from southern Kenya and its bearing on the origin of rocks in the northern Tanganyka alkaline district, *J. Petrol.* **5**, 40–81.

Saxena, S. K. (1973) *Thermodynamics of Rock-Forming Crystalline Solutions*. Springer-Verlag, Berlin.

Saxena, S. K., and Ghose, S. (1971) Mg^{2+}–Fe^{2+} order–disorder and the thermodynamics of the orthopyroxene–crystalline solution, *Amer. Miner.* **56**, 532–559.

Takeda, H. (1972) Crystallographic studies of coexisting aluminan orthopyroxene and augite of high-pressure origin, *J. Geophys. Res.* **77**, 5798–5811.

Takeda, H., Miyamoto, M., and Reid, A. M. (1974) Crystal chemical control of element partitioning for coexisting chromite–ulvöspinel and pigeonite–augite in lunar rocks, *Proc. 5th Lun. Sci. Conf.* **1**, 727–741.

Thompson, J. B., Jr. (1969) Chemical reaction in crystals, *Amer. Miner.* **54**, 341–375.

Tokonami, M. (1965) Atomic scattering factor for O^{2-}, *Acta Cryst.* **19**, 486.

Ungaretti, L., Mazzi, F., Rossi, G., and Dal Negro, A. (1978) Crystal-chemical characterization of blue amphiboles, *Proc. 11th Int. Mineralog. Ass. 11th Meeting*, Novosibirsk.

Virgo, D., and Hafner, S. S. (1969) Fe^{2+}, Mg order–disorder in heated orthopyroxenes, *Miner. Soc. Amer. Spec. Pap.* **2**, 67–81.

Zachariasen, W. H. (1963) The secondary extinction correction, *Acta Cryst.* **16,** 1139–1144.

Zanettin, B., Gregnanin, A., Justin-Visentin, E., Mezzacasa, G., and Piccirillo, E. M. (1974) Petrochemistry of the volcanic series of the Central Eastern Ethiopian plateau and relationships between tectonics and magmatology, *Mem. Ist. Geol. Miner. Univ. Padova* **31,** 1–34.

Zanettin, B., Gregnanin, A., Justin-Visentin, E., Mezzacasa, G., and Piccirillo, E. M. (1976) New chemical analyses of the Tertiary volcanics from the Central Eastern Ethiopian plateau, *C.N.R., Istit. Miner. Petrol. Univ. Padova* **1,** 1–43.

Zanettin, B., Justin-Visentin, E., and Piccirillo, E. M. (1978) Volcanic succession, tectonics and magmatology in Central Ethiopia, *Memorie Accademia Patavina Scienze Lettere Arti,* **90,** 1–19.

Zanettin, B., Justin-Visentin, E., Nicoletti, M. and Piccirillo, E. M. (1980) Correlations among Ethiopian volcanic formations with special references to the chronological and stratigraphical problems of the "Trap series," in *Geodynamic Evolution of the Afro-Arabian Rift System,* Acc. Naz. Lincei, Roma, **47,** 231–252.

II. Melts, Fluids, and Solid-Fluid Equilibria

Chapter 4

The Densities and Structures of Silicate Melts

D. R. Gaskell

Introduction

As the density of a silicate melt is determined by the masses of the individual particles and by the manner in which the particles are arranged with respect to one another, the results of studies of the variations of melt density with composition and temperature are amenable to structural interpretation. Furthermore, as melt density, via its relationship with molar volume, is also a thermodynamic property, the interpretation of melt density in terms of the geometrical arrangements and configurations of ionic entities can be augmented by the interpretation of thermodynamic activities in terms of the extents and magnitudes of the interactions occurring among the various structural entities. This is particularly so in the consideration of ternary silicate melts. Recent X-ray diffraction studies of liquid alkali silicates (Waseda and Suito, 1977; Waseda and Toguri, 1977a) and alkaline earth silicates (Waseda and Toguri, 1977b) have confirmed that the tetrahedral coordination of silicon by oxygen, long detected in silicate glasses, persists in the melt, and hence the structures of melts can be approached by considering the ways in which silicate tetrahedra can be arranged with respect to one another.

Binary Silicate Melts

Early measurements of the densities and thermal expansivities of binary alkali (Bockris, Tomlinson, and White, 1956) and alkaline earth (Tomlinson, Heynes, and Bockris, 1958) silicates showed that melts within the composition range $0.88 < X_{SiO_2} < 1$ have zero expansivity. This behavior indicates that the expandable M^+-O^- and $M^{2+}-O^-$ interactions, formed by the breaking of double oxygen bonds in the silica, are completely contained within "cages" in a randomly damaged three-dimensional silica network. The sudden appearance, at ~ 12 mol% basic oxide, of a thermal expansivity indicates that a structure of discrete silicate anions becomes more stable than a three-dimensional network structure. Bockris $et\ al.$ (1956) postulated that, in the range $0.88 > X_{SiO_2} > 0.5$, discrete "pencil" polyanions are formed by face-to-face polymerization of $Si_3O_9^{6-}$, $Si_4O_{12}^{8-}$, or $Si_5O_{15}^{10-}$ metasilicate ring ions and, in the range $0.5 > X_{SiO_2} > 0.33$, chain ions of the general formula $Si_xO_{3x+1}^{(2x+2)-}$ are formed by linear polymerization of the SiO_4^{4-} tetrahedra. Volume calculations (Tomlinson, Heynes, and Bockris, 1958) suggested that the equilibrium structure for a given composition within the range $0.88 > X_{SiO_2} \geqslant 0.5$ would show a transition from three- to four- or five-membered rings as the silica content increases, i.e., that, although the $Si_3O_9^{6-}$ ion is stable in the vicinity of the metasilicate composition, in more siliceous melts polyions formed from polymerization of three-membered rings are likely to be less stable than polyions formed from four- and five-membered rings. This conclusion is also supported by consideration of the Si–O–Si bond angles required in polyions formed from the three types of rings, and the stability of the $Si_4O_{12}^{8-}$ ring has

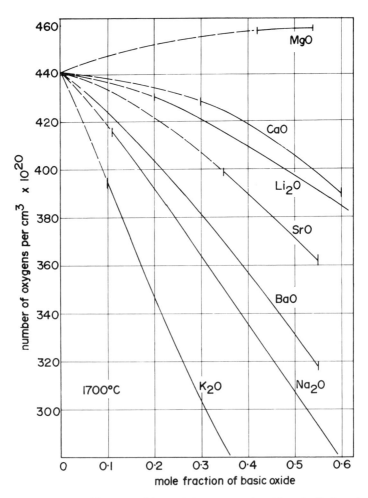

Fig. 1. The variation, with composition, of the oxygen densities of alkali and alkaline earth silicate melts at 1700°C.

been confirmed by examination of the trimethylsilyl derivatives of quenched silicates (Gotz and Masson, 1970, 1971; Masson, Jamieson, and Mason, 1974). The essentially coulombic nature of the interactions between alkali and alkaline earth cations and the silicate anions was demonstrated by the linear variation of the measured thermal expansivity of melts of a given silica content with the inverse of the oxygen attraction parameter.

The variations of the oxygen densities (the number of oxygen atoms per cubic centimeter of melt) with composition at 1700°C are shown in Fig. 1. Consideration of oxygen density, which eliminates the influence of the masses of the particles, shows that the addition of alkaline and alkaline earth oxides to silica causes a smooth dilation in the silicate structure, the extent of which can be correlated with cation size. In magnesium silicates the cation size is small enough that a contraction occurs in the silicate structure. Figure 2 shows the oxygen densities of metasilicate melts as a function of the cube of the radius of the cation. The linear variation in the case of $MgO \cdot SiO_2$, $CaO \cdot SiO_2$, and $SrO \cdot SiO_2$ suggests that these melts have similar structures, with the differences in molar volume being determined only by the differences among

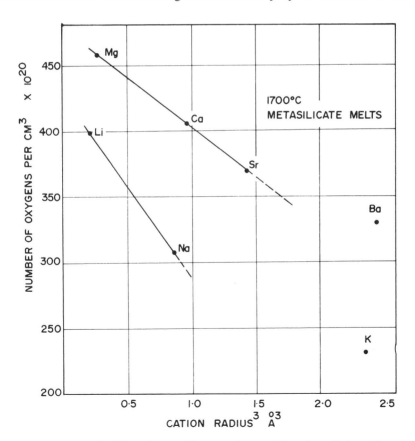

Fig. 2. The oxygen densities of metasilicate melts as a function of the cube of the radius of the cation.

the sizes of the cations. The larger-than-expected oxygen density of $BaO \cdot SiO_2$ suggests that the larger size of the Ba^{2+} ion causes a fundamental change in the structure. Similarly, by analogy with the alkaline earth silicates, it would appear that the large size of the K^+ ion causes the structure of $K_2O \cdot SiO_2$ to be fundamentally different from those of $Li_2O \cdot SiO_2$ and $Na_2O \cdot SiO_2$. As alkali silicates contain twice as many cations as do alkaline earth silicates, the oxygen densities of alkali silicates are lower than those of the corresponding alkaline earth silicates containing cations of similar size.

The densities of melts in three highly basic binary silicate systems have been measured. Figure 3 shows the results of the two most recent studies of melts in the system FeO–SiO_2 at 1410°C (Gaskell and Ward, 1967; Shiraishi, Ikeda, Tamura, and Saito, 1978), Fig. 4 shows the densities of melts in the system PbO–SiO_2 at 1100°C (Hino, Ejima, and Kameda, 1967), and Fig. 5 shows the densities of melts in a limited range of composition in the system MnO–SiO_2 at 1500°C (Segers, Fontana, and Winand, 1978). In each of these systems the melt density is greater than that calculated assuming ideal mixing of MO and SiO_2, i.e., assuming a linear variation of molar volume from MO to SiO_2. The densities of mechanical mixtures are drawn as broken lines in Figs. 3–5. This

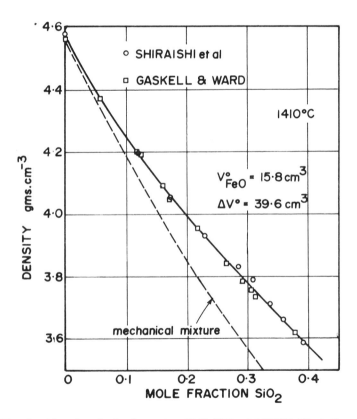

Fig. 3. The densities of melts in the system FeO–SiO_2 at 1410°C (Gaskell and Ward, 1967; Shiraishi, Ikeda, Tamura, and Saito, 1978).

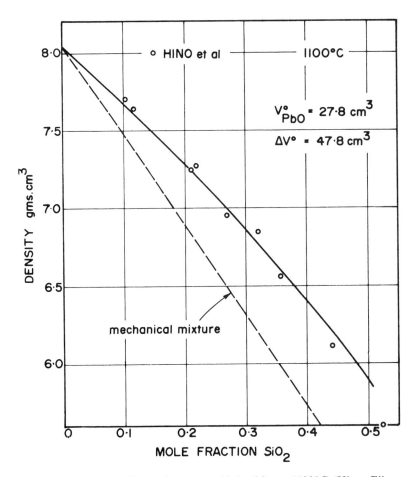

Fig. 4. The densities of melts in the system PbO–SiO$_2$ at 1100°C (Hino, Ejima, and Kameda, 1967).

behavior has been accounted for by Grau and Masson (1976) in terms of Masson's polymerization model of liquid silicates (Masson, 1965). The latter model postulates that basic binary silicate melts contain cations, free O^{2-} ions and an array of linear chain ions of the formula Si$_x$O$_{3x+1}^{(2x+2)-}$. Random mixing of the anions is assumed, which is equivalent to the assumption that the melt is a Raoultian mixture of the components MO and M_{x+1}Si$_x$O$_{3x+1}$, in which case the molar volume of the melt is given as

$$\tilde{V} = N_{MO}V^\circ_{MO} + \sum N_{M_{x+1}Si_xO_{3x+1}} V^\circ_{M_{x+1}Si_xO_{3x+1}}, \qquad (1)$$

where V°_{MO} and $V^\circ_{M_{x+1}Si_xO_{3x+1}}$ are the molar volumes of MO and pure monodisperse M_{x+1}Si$_x$O$_{3x+1}$. From the thermodynamic assumptions in the polymerization model it follows that

$$V^\circ_{M_{x+2}Si_{x+1}O_{3x+4}} - V^\circ_{M_{x+1}Si_xO_{3x+1}} = V^\circ_{M_2SiO_4} - V^\circ_{MO} = \Delta V^\circ \qquad (2)$$

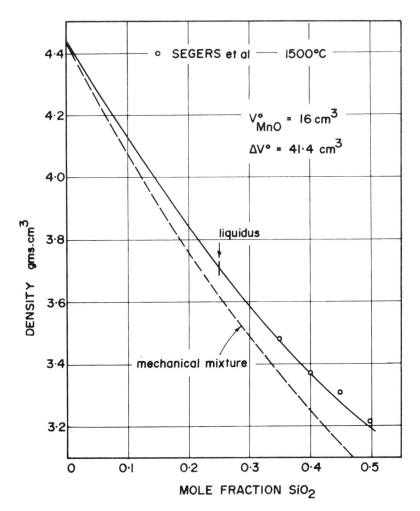

Fig. 5. The densities of melts in the system MnO–SiO_2 at 1500°C (Segers, Fontana, and Winand, 1978).

in which case

$$V^\circ_{M_{x+1}Si_xO_{3x+1}} = V^\circ_{MO} + x\Delta V^\circ. \tag{3}$$

Thus, Eq. (1) can be written as

$$\tilde{V} = V^\circ_{MO} + \Delta V^\circ \sum_1^\infty x N_{M_{x+1}Si_xO_{3x+1}}. \tag{4}$$

From the polymerization model (Masson, 1968) the summation in Eq. (4) is equal to $(1/X_{SiO_2} - 2)$ and hence Grau and Masson derived

$$\tilde{V} = V^\circ_{MO} + \frac{\Delta V^\circ}{(1/X_{SiO_2} - 2)}. \tag{5}$$

As a result of the assumption made in Eq. (2), i.e., that the volume of an SiO_3 group is constant, the molar volume given by Eq. (5) is independent of the degree of polymerization of the melt and, also, as only linear chain anion formation is considered, the molar volume becomes infinite at $X_{SiO_2} = 0.5$ where, in theory, the system contains a single infinitely long-chain metasilicate ion.

Grau and Masson derived the molecular weight (MW) of the melt as

$$MW = MW_{MO} + \frac{MW_{MSiO_3}}{(1/X_{SiO_2} - 2)} \tag{6}$$

and, hence, the melt density as

$$\rho = \frac{X_{MO}MW_{MO} + X_{SiO_2}MW_{SiO_2}}{X_{MO}V_{MO}^\circ + X_{SiO_2}(\Delta V^\circ - V_{MO}^\circ)} . \tag{7}$$

The full lines drawn in Figs. 3–5 are Eq. (7) with $V_{FeO}^\circ = 15.8$ cm^3 and $\Delta V_{FeSiO_3}^\circ = 39.6$ cm^3, $V_{PbO}^\circ = 27.8$ cm^3 and $\Delta V_{PbSiO_3}^\circ = 47.8$ cm^3, and $V_{MnO}^\circ = 16$ cm^3 and $\Delta V_{MnSiO_3}^\circ = 41.4$ cm^3. As is seen, the observed density behavior is well accounted for by the theory.

Ternary Silicate Melts

The thermodynamic properties of ternary silicate melts are usually considered in terms of the extent to which the behavior conforms with ideal silicate mixing (Richardson, 1956). In the ideal silicate mixing model it is postulated that random mixing of the cations occurs when two binary silicates containing the same mole fraction of SiO_2 are mixed. This single assumption, which allows the activities of the three oxide components of a ternary silicate melt to be calculated from knowledge of the mixing properties of the constituent binary silicates, predicts a linear variation of molar volume with composition in the systems $x(AO, BO) \cdot SiO_2$. Strictly, ideal silicate mixing is to be expected only when the free-energy change, ΔG°, for the exchange reaction

$$xAO \cdot SiO_2 + xBO = xBO \cdot SiO_2 + xAO \tag{8}$$

is zero. As this free-energy change is given as

$$\Delta G^\circ = (x + 1)(\Delta G_A^M - \Delta G_B^M),$$

where ΔG_A^M is the free energy of formation of $xAO \cdot SiO_2$ from AO and SiO_2 per mole of oxide and ΔG_B^M is the corresponding value for $xBO \cdot SiO_2$, it is seen that ideal silicate mixing requires that $\Delta G_A^M = \Delta G_B^M$, and, hence, that the same degree of polymerization occur in both $xAO \cdot SiO_2$ and $xBO \cdot SiO_2$. Thus, the more chemically similar the oxides AO and BO, the more nearly ideal will be the mixing in the ternary silicate melt.

Figure 6 shows the variations, with composition, of the molar volumes of melts in four pseudobinary sections of the system $CaO–FeO–SiO_2$ (Gaskell,

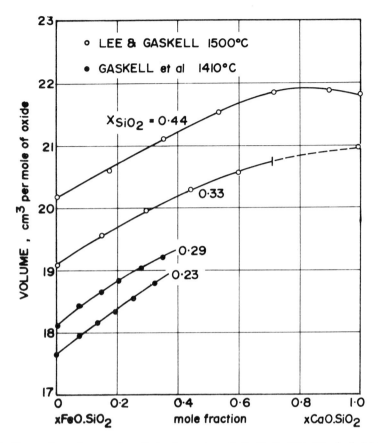

Fig. 6. The molar volumes of melts in the system FeO–CaO–SiO$_2$ (Gaskell, McLean, and Ward, 1969; Lee and Gaskell, 1974).

McLean, and Ward, 1969; Lee and Gaskell, 1974). In these systems the free-energy changes for the exchange reactions given by Eq. (8) and the deviations from ideal silicate mixing are large. This, together with the positive deviations from linearity of the molar volumes, can be interpreted by considering the types of oxygen surrounding the cation in the oxygen coordination sphere of the cation. In the terminal binary silicates the ratio of free oxygen ions to oxygens singly bonded to silicon, O^{2-}/O^-, in the oxygen coordination sphere of the cation is solely determined by the degree of polymerization of the silicate anions, and hence is greater in an iron silicate than in the corresponding calcium silicate (Masson, 1965). In a mixture of the two binary silicates any deviation of the ratio O^{2-}/O^- in the coordination sphere of a cation from the value for the pure binary silicate will cause a deviation in the value of the partial molar volume from that of the molar volume of the pure binary silicate. In Fig. 7 the excess molar volume in the system 1.273(FeO, CaO)SiO$_2$ increases linearly with composition up to about 40 percent replacement of Fe^{2+} by Ca^{2+}, and consequently, within this range of composition,

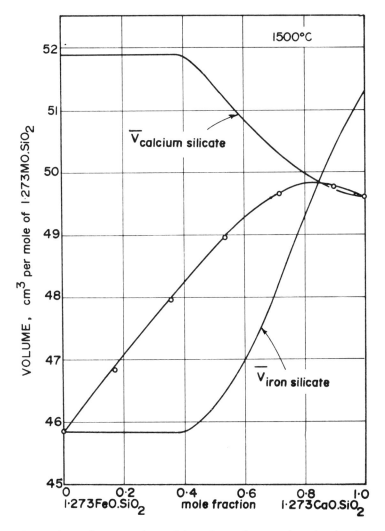

Fig. 7. The molar volumes and partial molar volumes of melts in the system 1.273(FeO, CaO) · SiO$_2$ at 1500°C (Lee and Gaskell, 1974).

the partial molar volumes of the pseudobinary components 1.273CaO · SiO$_2$ and 1.273FeO · SiO$_2$ are constant; the partial molar volume of the iron silicate equals the molar volume of the pure iron silicate and the partial molar volume of the calcium silicate is greater than the molar volume of the pure calcium silicate. Thus, within this range of composition, the ratio O^{2-}/O$^-$ in the coordination sphere of Fe^{2+} is the same as that in the pure iron silicate and the ratio O^{2-}/O$^-$ in the coordination sphere of Ca^{2+} is greater than that in the pure calcium silicate. The linear variation of the molar volume of the mixture with composition indicates that, within this range of composition, ideal mixing of these two types of groups occurs. When the extent of replacement of Fe^{2+} by Ca^{2+} exceeds some critical value, marked changes occur in the oxygen coordination spheres of both types of cation. After about 40 percent replacement the ratio O^{2-}/O^- in the coordination sphere of Ca^{2+} rapidly decreases toward the value for pure calcium silicate and the ratio in the coordination sphere of Fe^{2+} increases rapidly. Variation of the O^{2-}/O^- ratio with composition is caused by a variation in the overall degree of polymerization of the silicate anion population and/or by the occurrence of preferred ionic association, wherein one type of cation is preferentially coordinated with O^{2-} and the other type is preferentially coordinated with the O^- on the silicate ions. The observed behavior of the partial molar volumes of the pseudobinary components suggests that the O^{2-}/O^- ratio for the Fe^{2+} is increasing with increasing Ca^{2+} content due to preferred association of Fe^{2+} with O^{2-}, and the ratio O^{2-}/O^- for the Ca^{2+} is decreasing due to preferred association of Ca^{2+} with O^-. This interpretation predicts that the partial molar volume of FeO should decrease and the partial molar volume of CaO should increase with increasing extent of replacement of Fe^{2+} by Ca^{2+} beyond 40 percent. This ionic microsegregation is thermodynamic in origin, being determined by the free-energy change for the reaction expressed in Eq. (8), and the physical misfit between the two types of ionic groupings, produced by the microsegregation, causes a positive excess volume change on mixing in the pseudobinary system. In Fig. 7 this misfit is maximized at 59 percent replacement of Fe^{2+} by Ca^{2+}, the composition at which the excess molar volume has a maximum value.

The greater stabilities of phosphates than of the corresponding silicates causes a greater extent of ionic microsegregation in ternary phosphate melts, and, in the systems $Na_2O-FeO-P_2O_5$ and $CaO-FeO-P_2O_5$ microsegregation leads to macrosegregation in that two-liquid regions occur in which the systems separate into coexisting iron oxide-rich and calcium phosphate-rich or sodium phosphate-rich liquids (Muan and Osborn, 1965). In agreement with the greater stabilities of sodium phosphates than of calcium phosphates, the extent of the miscibility gap in the system $Na_2O-FeO-P_2O_5$ is greater than in the system $CaO-FeO-P_2O_5$, and also, in agreement with the lower stabilities of silicates than of phosphates, the addition of SiO_2 to the ternary phosphates decreases the extents of the miscibility gaps (Muan and Osborn, 1965). The influence of microsegregation on the molar volumes of four pseudobinary phosphate melts is illustrated in Fig. 8 which shows the data of Boyer et al. (1967). As is seen, significant positive excess volumes occur, which, in agreement with the relative stabilities of the component binary phosphates, are greater in the system $ZnO-Na_2O-P_2O_5$ than in the system $ZnO-CaO-P_2O_5$.

It is to be expected that this type of ionic segregation will occur in all basic ternary silicates containing an alkali or alkaline earth oxide and a transition metal oxide. Figure 9 shows the variation, with composition, of the molar volumes of melts in four pseudobinary sections of the system $CaO-MnO-SiO_2$ at 1500°C (Segers, Fontana, and Winand, 1978). The differences between the stabilities of the pseudobinary components in these systems are considerably less than in the corresponding $CaO-FeO-SiO_2$ systems and the deviations

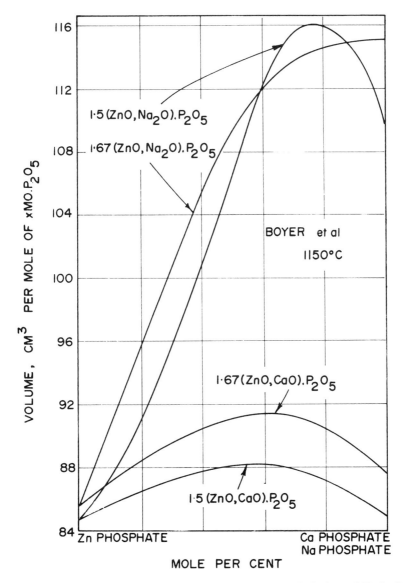

Fig. 8. The molar volumes of melts in the systems $ZnO-Na_2O-P_2O_5$ and $ZnO-CaO-P_2O_5$ at $1150°C$ (Boyer, Fray, and Meadowcraft, 1967).

from ideal silicate mixing are correspondingly less. Figure 9 shows that Mn-rich melts exhibit ideal behavior with very small positive deviations occurring in Ca-rich melts. The expansivities of melts in the systems $1.273(FeO, CaO) \cdot SiO_2$ and $1.222(MnO, CaO) \cdot SiO_2$ are shown in Fig. 10. The minimum in the thermal expansivity indicates that the inherent expansivities of the ionic interactions are counteracted by thermal relaxation of the ionic microsegregation.

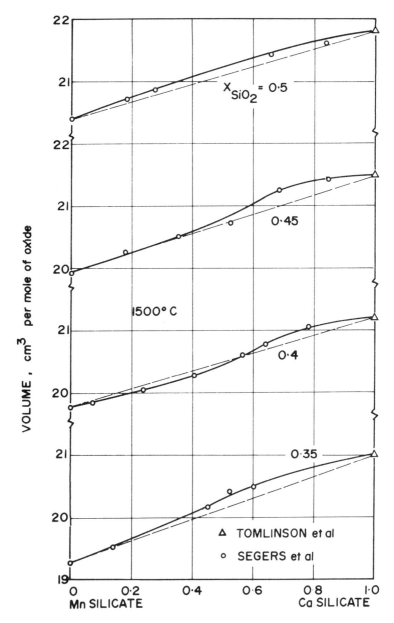

Fig. 9. The molar volumes of melts in the system $CaO–MnO–SiO_2$ at 1500°C (Segers, Fontana, and Winand, 1978).

Figure 11 shows the molar volumes of melts in the systems $(PbO, Li_2O) \cdot SiO_2$ (Ejima, Hino, and Kameda, 1970), $(PbO, Na_2O) \cdot SiO_2$ (Hino, Ejima, and Kameda, 1968), and $(PbO, K_2O) \cdot SiO_2$ at 1200°C (Hino, Ejima, and Kameda, 1969). Also included in this figure are the molar volumes of the three alkali metasilicates measured by Bockris *et al.* (1956). Good agreement exists be-

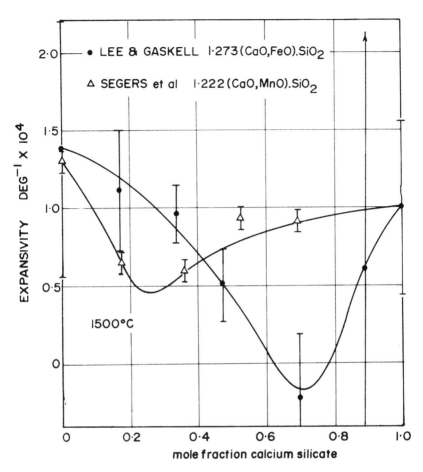

Fig. 10. The thermal expansivities of melts in the systems 1.273(FeO, CaO) · SiO₂ and 1.222(MnO, CaO) · SiO₂ at 1500°C (Segers, Fontana, and Winand, 1978; Lee and Gaskell, 1974).

tween the molar volume of $K_2O \cdot SiO_2$ obtained by extrapolating the data of Hino *et al.* (1969) and that of Bockris *et al.* (1956), but significant differences occur between the corresponding values for $Na_2O \cdot SiO_2$ and $Li_2O \cdot SiO_2$. Thus, although the molar volumes in the system $(PbO, K_2O) \cdot SiO_2$ clearly show very slight positive deviations from linearity, it could be argued that the molar volumes in the system $(PbO, Li_2O)SiO_2$ show either a linear variation or a distinct minimum at about 80 percent $Li_2O \cdot SiO_2$. The system $(PbO, Na_2O) \cdot SiO_2$ shows the expected maximum.

The correspondence found between the solution thermodynamics and molar volume behavior in the systems $MnO–CaO–SiO_2$, $FeO–CaO–SiO_2$, and $PbO–Alk_2O–SiO_2$ is not found in mixtures of $PbO \cdot SiO_2$ with alkaline earth metasilicates. Although these systems show the expected deviations of the thermodynamic properties from ideal silicate mixing (Ouchi and Kato, 1977),

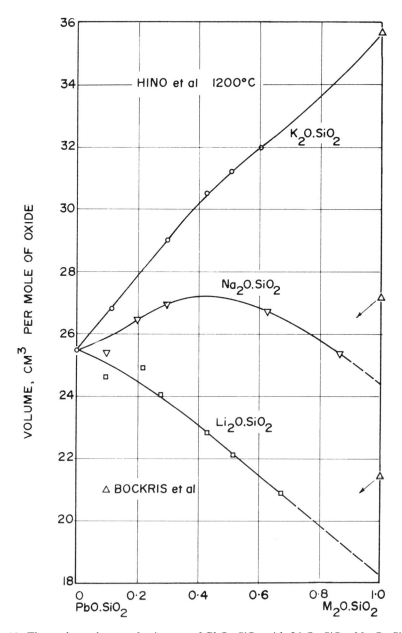

Fig. 11. The molar volumes of mixtures of PbO · SiO$_2$ with Li$_2$O · SiO$_2$, Na$_2$O · SiO$_2$, and K$_2$O · SiO$_2$ at 1200°C (Ejima, Hino, and Kameda, 1970; Hino, Ejima, and Kameda, 1968; Hino, Ejima, and Kameda, 1969).

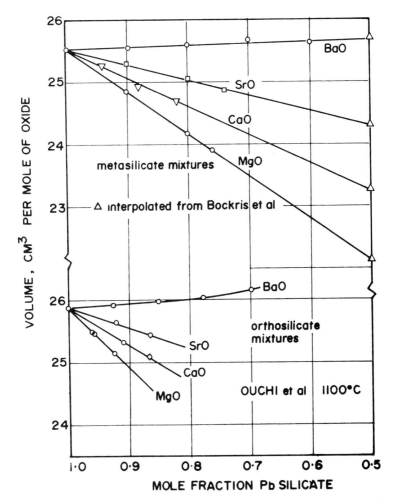

Fig. 12. The molar volumes of orthosilicate and metasilicate mixtures containing PbO and alkaline earth oxides at 1100°C (Ouchi, Yoshida, and Kato, 1977).

linear variations of the molar volume with composition are found (Ouchi, Yoshida, and Kato, 1977). Figure 12 shows the molar volumes in the systems $(PbO, MgO) \cdot SiO_2$, $(PbO, CaO) \cdot SiO_2$, $(PbO, SrO) \cdot SiO_2$, and $(PbO, BaO) \cdot SiO_2$ and the corresponding orthosilicate mixtures at 1100°C (Ouchi, Yoshida, and Kato, 1977). In the case of the metasilicate mixtures, the lines are drawn between the molar volume of $PbO \cdot SiO_2$ and the molar volumes of the alkali earth metasilicates measured by Tomlinson *et al.* (1958). The experimental data points for the ternary silicates lie exactly on these lines and the molar volumes of the orthosilicate mixtures are also linear within the restricted ranges of composition studied.

As has been stated, the extent of deviation from ideal silicate mixing is determined, in part, by the chemical dissimilarity between the two basic oxide

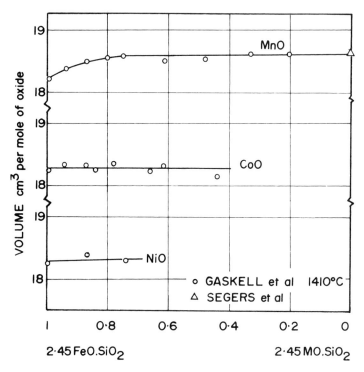

Fig. 13. The molar volumes of melts in the systems $2.45(FeO, MnO) \cdot SiO_2$, $2.45(FeO, CoO) \cdot SiO_2$ and $2.45(FeO, NiO) \cdot SiO_2$ at $1410°C$ (Gaskell, McLean, and Ward, 1969; Segers, Fontana, and Winand, 1978).

components. Ideal silicate mixing has been found in melts in the system $2(FeO, MnO) \cdot SiO_2$ (Song and Gaskell, 1979) and small deviations occur in melts in the system $2(FeO, CoO) \cdot SiO_2$ (Belton, Suito, and Gaskell, 1973). The molar volume of melts in the systems $2.45(FeO, MnO) \cdot SiO_2$, $2.45(FeO, CoO) \cdot SiO_2$, and $2.45(FeO, NiO) \cdot SiO_2$ are shown in Fig. 13 (Gaskell, McLean, and Ward, 1969). Grau and Masson applied Eq. (5) to ideal mixing in pseudobinary sections of ternary silicate melts, and, from the data in Fig. 13, they estimated 16.0 cm^3 and 14.5 cm^3 as the values of $V°_{MnO}$ and $V°_{CoO}$, respectively, and 41.3 cm^3 and 41.0 cm^3 as the values of $\Delta V°_{MnSiO_3}$ and $\Delta V°_{CoSiO_3}$, respectively. It is of interest to note the virtually exact agreement with the values of $V°_{MnO}$ and $\Delta V°_{MnSiO_3}$ obtained independently from the later data for binary manganese silicates presented in Fig. 5.

In highly acidic ternary silicates the concentrations of free oxygen ions are sufficiently low that ionic microsegregation is negligible and it is considered that the small deviations from ideal silicate mixing which have been found in $K_2O-Na_2O-SiO_2$ melts (Belton, Gaskell, and Choudary, 1974) are due to the difference in the oxygen coordination numbers of the cations caused by the difference in cation size. This behavior is in agreement with the observed linear variations of molar volume with composition in the systems $0.667(Li_2O,$

$Na_2O) \cdot SiO_2$ and $0.667(Li_2O, K_2O) \cdot SiO_2$ at 1400°C (Bockris, Tomlinson, and White, 1956).

Conclusions

Consideration of the densities of binary silicate melts, together with the behavior of other structure-sensitive physical properties and thermodynamic properties, yields a self-consistent, although, as yet, incomplete, picture of the ionic constitution of these melts. It is firmly established that the structures are based on the polymerization of SiO_4 units to form a size range of discrete polysilicate anions. The metasilicate composition, $MO \cdot SiO_2$, is critical as, in melts more siliceous than the metasilicate, the occurrence of pencil ions, formed by face-to-face polymerization of the metasilicate ring ions $Si_3O_9^{6-}$ and/or $Si_4O_{12}^{8-}$ is postulated, and in melts more basic than the metasilicate, the occurrence of chain ions formed by polymerization of SiO_4 units is postulated. The current problem is to determine the point at which, on the addition of SiO_2 to a basic metal oxide, ring ions first begin to make a significant appearance in the melts and to determine the nature of the gradual changeover from linear to ring ion polymerization as the metasilicate composition is traversed. Trimethylsilylation studies of glassy silicates have shown that quite complex ring ions can form. For example, pentagonal ring ions containing up to four internal –O–Si–O– bridges have been detected in iron blast furnace slag (Masson, Jamieson, and Mason, 1974).

The constitution of basic ternary silicate melts is influenced by ionic microsegregation to an extent which is determined by the difference between the stabilities of the constituent binary silicate systems and by the silica content of the melt. The structural misfit caused by the ionic microsegregation gives rise to positive excess molar volumes in the ternary melts. In any given ternary system the effect of microsegregation will be maximized at that composition which contains equivalent concentrations of the two types of cation and equivalent numbers of O^{2-} and O^{-}. With increasing silica content the concentration of O^{2-} decreases and with decreasing silica content the concentration of O^{-} decreases such that, in both cases, the extent and influence of microsegregation decrease. In sufficiently siliceous melts the concentration of O^{2-} becomes small enough that microsegregation is eliminated, in which case the volume change on mixing and the thermodynamic behavior are determined mainly by the difference between the sizes of the cations.

Acknowledgment

This paper is a contribution from the Laboratory for Research on the Structure of Matter, University of Pennsylvania, which is supported by the National Science Foundation under Grant DMR76-80994A02.

References

Belton, G. R., Suito, H., and Gaskell, D. R. (1973) Free energies of mixing in liquid iron–cobalt orthosilicates at 1450°C, *Metall. Trans.* **4**, 2541.

Belton, G. R., Gaskell, D. R., and Choudary, U. V. (1974) Thermodynamics of mixing in molten sodium–potassium silicates, in *Physical Chemistry of Process Metallurgy; The Richardson Conference*, p. 247. Institute of Mining and Metallurgy, London.

Bockris, J. O'M., Tomlinson, J. W., and White, J. L. (1956) The structure of liquid silicates; partial molar volumes and expansivities, *Trans. Faraday Soc.* **52**, 299.

Boyer, A. J. G., Fray, D. J., and Meadowcroft, T. R. (1967) The surface tensions and molar volumes of the binary phosphates of sodium, lithium, calcium and zinc, *J. Phys. Chem.* **71**, 1442.

Ejima, T., Hino, M., and Kameda, M. (1970) Surface tension, density and viscosity of $PbO-SiO_2-Li_2O$ ternary melts, *Nippon Kinzoku Gakkaishi* **34**, 546.

Gaskell, D. R., McLean, A., and Ward, R. G. (1969) Densities and structures of ternary silicate melts, *Trans. Faraday Soc.* **65**, 1498.

Gaskell, D. R., and Ward, R. G. (1967) Density of iron oxide–silica melts, *Trans. Metall. Soc. AIME* **239**, 249.

Gotz, J., and Masson, C. R. (1970) Trimethylsilyl derivatives for the study of silicate structures, Part I, A direct method of trimethyl silylation, *J. Chem. Soc. A*, 2683.

Gotz, J., and Masson, C. R. (1971) Trimethylsilyl derivatives for the study of silicate structures, Part II, Orthosilicate, pyrosilicate and ring structures, *J. Chem. Soc. A*, 686.

Grau, A. E., and Masson, C. R. (1976) Densities and molar volumes of silicate melts, *Can. Metall. Quart.* **15**, 367.

Hino, M., Ejima, T., and Kameda, M. (1967) Surface tension and density of liquid lead silicate, *Nippon Kinzoku Gakkaishi* **31**, 113.

Hino, M., Ejima, T., and Kameda, M. (1968) Surface tension, density and viscosity of $PbO-Na_2O-SiO_2$ ternary melts, *Nippon Kinzoku Gakkaishi* **32**, 809.

Hino, M., Ejima, T., and Kameda, M. (1969) Surface tension, density and viscosity of $PbO-K_2O- SiO_2$ ternary melts, *Nippon Kinzoku Gakkaishi* **33**, 617.

Lee, Y. E., and Gaskell, D. R. (1974) The densities and structures of melts in the system $CaO-FeO-SiO_2$, *Metall. Trans.* **5**, 853.

Masson, C. R. (1965) An approach to the problem of ionic distributions in liquid silicates, *Proc. Roy. Soc. London* **287A**, 201.

Masson, C. R. (1968) Ionic equilibria in liquid silicates, *J. Amer. Ceram. Soc.* **51**, 134.

Masson, C. R., Jamieson, W. D., and Mason, F. G. (1974) Ionic constitution of metallurgical slags, in *Physical Chemistry of Process Metallurgy; The Richardson Conference*, p. 223. Institute of Mining and Metallurgy, London.

Muan, A., and Osborn, E. F. (1965) *Phase Equilibria among Oxides in Steelmaking*, p. 158. Addison-Wesley Publishing Co., Reading, Mass.

Ouchi, Y., and Kato, E. (1977) Activities of PbO in binary and ternary silicate melts, *Nippon Kinzoku Gakkaishi* **41**, 855.

Ouchi, Y., Yoshida, T., and Kato, E. (1977) Densities of ternary lead silicate melts, *Nippon Kinzoku Gakkaishi* **41**, 865.

Richardson, F. D. (1956) Activities in ternary melts, *Trans. Faraday Soc.* **52**, 1312.

Segers, L., Fontana, A., and Winand, R. (1978) Poids specifiques et volumes molaires de melanges d'oxydes fondus du systeme $CaO-SiO_2-MnO$, *Electrochem. Acta* **23**, 1275.

Shiraishi, Y., Ikeda, K., Tamura, A., and Saito, T. (1978) On the viscosity and density of molten $FeO-SiO_2$ system, *Trans. Japan Inst. Metals* **19**, 264.

Song, K. S., and Gaskell, D. R. (1979) The free energies of mixing in melts in the systems $2FeO \cdot SiO_2-2MnO \cdot SiO_2$ and $2.33FeO \cdot TiO_2-2.33MnO \cdot TiO_2$, *Metall. Trans.* **10B**, 15.

Tomlinson, J. W., Heynes, M. S. R., and Bockris, J. O'M. (1958) The structure of liquid silicates, Part 2, Molar volumes and expansivities, *Trans. Faraday Soc.* **54**, 1822.

Waseda, Y., and Suito, H. (1977) The structure of molten alkali silicates, *Trans. Iron and Steel Inst. Japan* **17**, 82.

Waseda, Y., and Toguri, J. M. (1977a) Temperature dependence of the structure of molten silicates $M_2O \cdot 2SiO_2$ and $M_2O \cdot SiO_2$ (M = Li, Na and K), *Trans. Iron and Steel Inst. Japan* **17**, 601.

Waseda, Y., and Toguri, J. M. (1977b) The structure of molten binary silicate systems $CaO-SiO_2$ and $MgO-SiO_2$, *Metall. Trans.* **8B**, 563.

Chapter 5

The Thermodynamics of Supercritical Fluid Systems

K. I. Shmulovich, V. M. Shmonov, and V. A. Zharikov

Introduction

The aim of the present paper is to discuss briefly the results of investigations into the thermodynamics of a model of natural fluids. These investigations have been conducted over the past decade in the laboratory of hydrothermal systems at the Institute of Experimental Mineralogy, USSR Academy of Sciences. Scientists have long been interested in hydrothermal solutions due to their importance in the formation of many ore bodies and of virtually all metamorphic, metasomatic, and magmatic rocks in the earth's crust. One should be aware, however, that natural processes involving a fluid phase are very complex, involving too many problems to be covered adequately in one paper.

Experiments have been conducted on a model fluid to determine the equilibrium thermodynamic properties of the components over the rather narrow temperature range of 400–700°C. These are the temperatures at which the most active processes of postmagmatic contact metamorphism, regional metamorphism (below the sanidine and granulite facies), metasomatism, and pegmatite formation occur. However, these properties are not well known as geochemical data has, in the past, come from experimental work undertaken in the chemical industry in which interest is normally restricted to a range of pressure and temperature up to only a few hundred atmospheres and a few hundred degrees Celsius. By the time we began our investigations into the thermodynamics of fluid systems, the properties of H_2O had already been studied over a fairly large range of P, T (Burnham et al., 1969). Therefore, we concentrated on CO_2—the second most important component of a natural fluid.

Pure Species

Models based on only a single component (H_2O, CO_2, or H_2) agree rather poorly with analyses of inclusions believed to represent the mineral-forming fluids and other estimates of equilibrium fluid compositions. In spite of this, these simple models are still applied to the petrological processes as, for instance, in estimating metamorphic reactions at $P_{total} \approx P_{H_2O}$.

By applying the principle of corresponding states, Breedveld and Prausnitz (1973) derived generalized diagrams for isothermal and isobaric densities, fugacity coefficients, enthalpy, and internal energy at values of P^* (P/P^{crit}) up to 2000 and of T^* (T/T^{crit}) up to 50. However, it was necessary to test experimentally both the validity of this principle and the reliability of their results. $P-V-T$ relations were determined for pure carbon dioxide within the range 200–800 MPa, 400–800°C (Shmonov and Shmulovich, 1974). The measurements were obtained by using the displacement method. Two runs were made—one with, and one without, a metal cylinder in the "gradient-free" zone—in order to determine the difference in the amount of gas; the volume of the piezometer was not measured. A similar apparatus was used earlier to measure the $P-V-T$ relations for CO_2 in the same pressure range, from 50 to 400°C (Tzyklis et al., 1969). A corrected function method was applied to the data from Vukalovich and Altunin (1965) (obtained at P to 60 MPa), Jüza et al. (1965) (at P to 400 MPa, $T = 50$–475°C), Michels et al. (1935) (at P to 300 MPa, $T = 150$°C), Tzyklis et al. (1969), and our own work. Thus, we were able to compile very detailed tables of the thermodynamic functions of CO_2 in steps of 10°C, 10 MPa up to 1000°C, 1000 MPa (Shmulovich and Shmonov, 1978). The average errors in the tabulated values were 1.1% for the fugacity coefficients, 0.4–0.5% for the entropy, and 0.35% for both the molar volume and the Gibbs free energy increment (this increased to 0.5–0.7% at low temperatures). A comparison of these calculated values of V_{CO_2} with those given in the $\rho^*(P^*, T^*)$ plot of Breedveld and Prausnitz (1973) revealed that the use of the principle of corresponding states introduced an error of 10% at $T^* \approx 3$, $P^* \approx 30$; that error decreased to 4% at $P^* = 140$.

The above figures give an idea of the accuracy obtainable with the principle of corresponding states. Analogous results were obtained by use of the empirical two-constant equations of state (the van der Waals, Redlich-Kwong, and other equations) since substances that obey these equations of state are thermodynamically similar and, thus, are amenable to use of the principle of corresponding states. However, relative errors inherent in the gas volume (density) estimates are not totally representative. The picture will be more complete if these errors are put in terms of absolute increments in the Gibbs free energy with pressure: a value of $5\%(G_n^{T,P} = 1 - G_n^{T,P-1})$ for H_2O and CO_2 will be, respectively, 0.47 and 0.55 kcal/mol at 500°C, 100 MPa; 0.7 and 1 kcal/mol at 500°C, 1000 MPa; 0.8 and 0.9 kcal/mol at 1000°C, 100 MPa; and 1.2 and 1.5 kcal/mol at 1000°C, 1000 MPa.

The use of more sophisticated and complicated forms of the theory of corresponding states, or of the two-constant equations, will lead to greater accuracy. Melnik (1978) applied an analytical graphic correction to generalized experimental diagrams, and so calculated thermodynamic properties of Ar, H_2, N_2, O_2, CH_4, CO, CO_2, HCl, H_2O, H_2S, NH_3, and SO_2 in the range

400–1500°K and 50–1500 MPa. Touret and Bottinga (1979) used the Redlich–Kwong equation, with the two constants generalized to functions of temperature and pressure, to extrapolate properties of CO_2 to 1200°C, 2000 MPa. The estimates of Melnik and of Touret and Bottinga agree well with values obtained by interpolation between the static $P–V–T$ data of Shmulovich and Shmonov (1978) and the shock-wave compression data of Zubarev and Telegin (1962).

Over the past few years, studies of the $P–V–T$ properties of pure components at high temperatures and pressures have been concentrated in the ultrahigh region attained in dynamic experiments, while static measurements were made mainly on compositionally complex mixtures.

The major qualitative result of the studies on the thermodynamics of pure components in a natural fluid is clearly demonstrated when the fugacities of CO_2 and H_2O are compared. Their behavior is totally different: over the geologically important temperature range, the fugacity of CO_2 increases exponentially with pressure, while H_2O becomes a linear function of pressure, and the coefficient of proportionality differs only slightly from unity. These relationships are demonstrated in Fig. 1. As a result, with increasing depth of metamorphism, carbonation reactions increase in intensity due to the drastic rise in the chemical potential of CO_2 relative to that of H_2O.

To sum up, the basic concept of Korzhinskii (1940), according to which mineral parageneses undergo successive carbonation with increasing depth of metamorphism, and from which the "mineral depth facies" scheme has been developed, has a sound physical basis: the cause of this phenomenon lies in the changing inherent properties of fluid components.

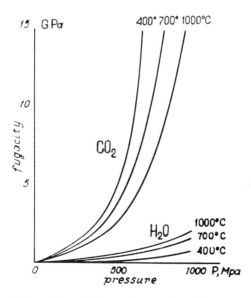

Fig. 1. Variations of fugacities of CO_2 and of H_2O with pressure. f_{CO_2} values from Shmulovich and Shmonov (1978). f_{H_2O} values from Burnham *et al.* (1969).

Binary Systems

The increasing complexity of natural fluid models eventually resulted in the development of binary systems of the H_2O–nonpolar gas type. Although such systems are complicated by the reactivity of the components, we will confine ourselves to their thermodynamic properties and will not discuss problems that would normally arise in chemically interacting systems.

H_2O–CO_2

This is the most important system petrologically and the best studied in the high-temperature region. Franck and Tödhaide (1965) published tables of compressibility for binary solutions up to 750°C and 200 MPa, from which Ryzhenko and Malinin (1971) calculated activity coefficients and fugacities of components in the mixtures. Greenwood (1969) measured P–V–T relations in the H_2O–CO_2 system up to 800°C and 50 MPa. By approximating the P–V–T–X surface with a polynomial, he compiled tables of compressibility (Z) and activity–concentration diagrams for the P, T range in question. Recently, Gehrig (1980) reported P–V–T measurements in the system H_2O–CO_2 at 400, 450, and 500°C, up to 60 MPa, and calculated the activity coefficients of components. In the present study, measurements were obtained by using the apparatus of Zakirov (1977) at 400°C up to 100 MPa (Shmulovich et al., 1979), and at 400 and 500°C up to 500 MPa. To improve the accuracy of the measurements at low pressures, the piezometer was lined with gold; a piston separated the hot and cold zones, and a special pressure cell, sensitive to 0.01 MPa, was used. The pressure measurements were accurate to 0.2%.

In Fig. 2, the available data on volumes of mixing, $V^E = V^{mix} - V_{CO_2}(X_{CO_2} - V_{H_2O}) \cdot X_{H_2O}$, are plotted against P up to 100 MPa at $X_{CO_2} = 0.4$. It can be seen that data from Greenwood, Gehrig, and our own studies are fairly consistent, but that the results of Franck and Tödhaide give systematically lower V^E values. $(\partial V^E/\partial X_n)_{P,T}$ is also important in determining activity coefficients of components in homogeneous mixtures. The coefficients for H_2O and CO_2 were calculated from our own and published experimental data, or, where there were no data or it was not sufficiently accurate, they were estimated from theoretical equations of state. The results are given by Shmulovich et al. (1980b). The standard state of a component was taken as the pure fluid ($a_n = X_n = 1$) at temperature and pressure of interest throughout the calculations. The activity coefficient of component n, γ_n, was calculated from

$$RT\ln\gamma_n = \int_0^P \left(V^E - (1 - X_n)\left(\frac{\partial V^E}{\partial X_n}\right)_{P,T} \right) \cdot dP. \qquad (1)$$

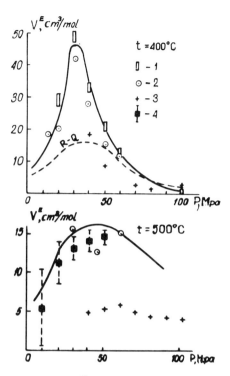

Fig. 2. Excess volumes of mixing, V^E, plotted against P in the system H_2O–CO_2, with $X_{CO_2} = 0.4$. Symbols: 1—authors' data; 2—Gehrig (1980); 3—Franck and Tödhaide (1959); 4—Greenwood (1969). Solid line is calculated from Eqs. (2)–(5), with $\xi = 0.85$. Dashed line is calculated from the Redlich–Kwong equation of de Santis *et al.* (1974).

No correction has been made for any nonideality of the pure component, i.e., its fugacity coefficient is assumed to be unity.

At pressures up to 10 MPa, an equation of state with a second virial coefficient was employed to estimate V^E for Eq. (1). The cross-coefficient B_{H_2O,CO_2} was taken from the Lennard–Jones potential tables (Hirchfelder *et al.*, 1954). At 100°C the experimental and calculated values show a very good agreement (Coan and King, 1971). In the region of maximum absolute values of V^E—that is, from 10 to 100 MPa—the V^E–X_{CO_2} relations were interpolated and refined mainly by using the values calculated from the equation of state:

$$Z^{mix} = Z^{hs} + \sum_{i=2}^{5} \left(B_i^{LD}(T) - B^{hs} \right) \cdot \rho^{i-1}, \qquad (2)$$

where Z^{mix} represents the compressibility of the mixture, Z^{hs} represents the compressibility of the hard-sphere model, approximated by the Carnahan–Stirling equation, B_i^{LD} are the virial coefficients for the (6–12) Lennard–Jones paired "effective" potential, and ρ represents the density. The parameters of this equation are the diameter of the hard spheres (σ) and the depth of the potential well (ϵ), which were calculated from $P-V-T$ data on the pure components by minimizing the nonlinear functional:

$$\min \sum_{n=1} \left(Z^{\exp}(V_n, T) - Z^{\mathrm{mod}}(\sigma_n, \epsilon_n) \right). \tag{3}$$

By using the obtained values of σ_n and ϵ_n, and rearranging in the usual way, σ_{mix} and ϵ_{mix} were calculated from

$$\sigma_{\mathrm{mix}} = \sigma_{H_2O} \cdot X_{H_2O}^2 + X_{H_2O} \cdot X_{CO_2}(\sigma_{H_2O} + \sigma_{CO_2}) + \sigma_{CO_2} \cdot X_{CO_2}^2, \tag{4}$$

$$\epsilon_{\mathrm{mix}} = \epsilon_{H_2O} \cdot X_{H_2O}^2 + 2 \cdot X_{H_2O} \cdot X_{CO_2} \xi \sqrt{\epsilon_{H_2O} \cdot \epsilon_{CO_2}} + \epsilon_{CO_2} \cdot X_{CO_2}^2, \tag{5}$$

where ξ is the correction factor calculated from our measurements (Shmulovich et al., 1979). The best results were obtained with $\xi = 0.85$.

In earlier works, the values obtained for the activity of components at pressures above 100 MPa by using Eq. (2) and the Redlich–Kwong equation, were not substantiated by experiments. We have undertaken a special $P-V-T$ study in the system H_2O-CO_2 over the range 100–500 MPa at 400 and 500°C. Experiments were run on an external heating apparatus, using the displacement method. A sealed, thin-walled, gold capsule containing the H_2O-CO_2 mixture in weighed proportions was placed in the piezometer; the CO_2 pressure and the specified temperature were set. By discharging portions of the CO_2 into the receiving gas system and measuring both the quantity of the discharged gas and the drop in pressure, the values of $V^{\mathrm{mix}} - V_{CO_2}$ in the capsule were determined. This procedure eliminated some of the errors inherent in the calculation of V^E because V_{CO_2} was determined on the same device. The detailed techniques and results were published in Shmonov (1977), Shmonov and Shmulovich (1978), and Shmulovich et al. (1980b).

In Fig. 3, V^E, measured on the 400 and 500°C isotherms, is plotted against pressure measured in steps of 50 MPa. Although very small, the values of V^E are consistently positive within the wide range of pressures and compositions considered. However, values of V^E calculated from Eq. (2), with $P > 200$ MPa, are negative. The more accurate measurements of the properties of dense gaseous mixtures obtained by using the above method clearly require that theoretical models allow for certain subtle effects due to nonspherical symmetry of molecules, orientational contributions to the potential function, and possibly, the dependence of σ on ρ ("soft" spheres).

At pressures above 200 MPa, activity coefficients were calculated by fitting the Margules two-constant equation to the experimental values of V^E (Shmulovich et al., 1980b). The results are listed in Table 1.

The values of V^E calculated by using the modified Redlich–Kwong equation, although considerably different from experimental values at low pressures, agreed rather well with them at high pressures. Also, activities of components calculated by Flowers (1979), using de Santis' modification of the Redlich–Kwong equation (de Santis et al., 1974), agreed rather well with the activities ($a_n = \gamma_n \cdot X_n$) given in Table 1. The consistency of results obtained by using the Redlich–Kwong equation, in which the constant of attraction is

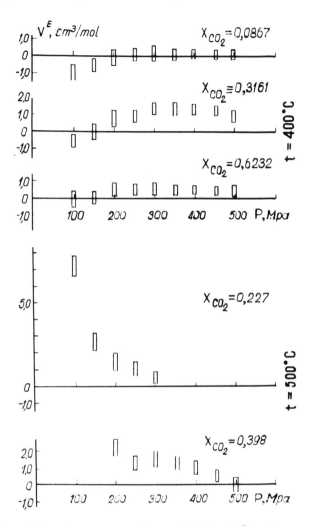

Fig. 3. Experimental results on excess volumes of mixing, V^E, in the system H_2O–CO_2 at high pressures. Vertical bars indicate range of uncertainty of measurements.

temperature dependent, may be due to the fact that the errors involved in calculating the fugacity coefficients for the pure components and for components in the binary system compensate each other.

The nonideality of solutions in the H_2O–CO_2 system can also be evaluated by a method based on the experimentally established differences between equilibrium conditions in carbonation or hydration reactions involving a binary gas system and those involving the pure one-component fluid. Although this method is far less accurate than P–V–T measurements, its advantages are simplicity and coverage of an extremely wide range of P, T. Also, it bears a relatively high resemblance to natural processes, in that the

Table 1. The activity coefficients of components in the system H_2O–CO_2.

X_{CO_2}	0		0.1		0.2		0.4		0.6		0.8		0.9		1.0	
P, MPa	H_2O	CO_2	H_2O	CO_2	H_2O	CO_2	H_2O	CO_2	H_2O	CO_2	H_2O	CO_2	H_2O	CO_2	H_2O	CO_2
400°C																
30	1	1.8	1.01	1.53	1.03	1.35	1.11	1.14	1.21	1.04	1.30	1.00	1.33	1.00	1.35	1
50	1	3.2	1.02	2.35	1.06	1.83	1.22	1.30	1.46	1.08	1.71	1.01	1.81	1.00	1.86	1
100	1	3.9	1.02	2.86	1.06	2.21	1.25	1.50	1.60	1.18	2.14	1.04	2.50	1.01	2.93	1
200	1	3.3	1.01	2.72	1.04	2.27	1.20	1.64	1.56	1.26	2.33	1.06	3.03	1.01	4.12	1
300	1	3.8	1.01	2.98	1.05	2.40	1.23	1.66	1.62	1.26	2.43	1.06	3.13	1.01	4.18	1
500	1	4.9	1.02	3.55	1.07	2.67	1.29	1.70	1.76	1.25	2.62	1.06	3.32	1.01	4.29	1
500°C																
30	1	1.2	1.00	1.14	1.01	1.10	1.03	1.05	1.06	1.02	1.09	1.00	1.11	1.00	1.12	1
50	1	1.5	1.01	1.35	1.02	1.23	1.08	1.09	1.14	1.02	1.19	1.00	1.21	1.00	1.20	1
100	1	2.5	1.01	1.91	1.05	1.54	1.18	1.17	1.33	1.03	1.42	1.00	1.41	1.00	1.35	1
200	1	3.4	1.02	2.40	1.07	1.82	1.25	1.26	1.48	1.06	1.66	1.00	1.68	1.00	1.64	1
300	1	3.7	1.02	2.60	1.07	1.96	1.26	1.33	1.53	1.09	1.80	1.01	1.90	1.00	1.94	1
500	1	4.0	1.02	2.84	1.07	2.15	1.27	1.44	1.59	1.14	2.02	1.02	2.25	1.00	2.48	1
600°C																
30	1	1.06	1.00	1.04	1.00	1.02	1.01	1.00	1.02	1.00	1.02	1.00	1.01	1.00	1.00	1
50	1	1.14	1.00	1.04	1.01	1.05	1.02	1.01	1.04	1.00	1.04	1.00	1.03	1.00	1.01	1
100	1	1.41	1.00	1.26	1.02	1.16	1.06	1.05	1.11	1.00	1.12	1.00	1.11	1.00	1.08	1
200	1	1.78	1.01	1.52	1.03	1.34	1.11	1.13	1.20	1.03	1.29	1.00	1.31	1.00	1.32	1
300	1	1.91	1.01	1.63	1.03	1.43	1.12	1.18	1.24	1.06	1.39	1.00	1.46	1.00	1.53	1
500	1	1.99	1.01	1.70	1.03	1.49	1.12	1.22	1.27	1.08	1.46	1.00	1.58	1.00	1.70	1

*Institute of Experimental Mineralogy, Academy of Sciences of the USSR, 142432 Chernogolovka, USSR.

method allows for the effects of the solubility of the mineral components on the activities of the volatiles in the fluid. Using this method, Walter (1963) calculated a_{CO_2} from phase equilibria in the system $MgO-CO_2-H_2O$ at 650–750°C, 100 MPa. We have used the grossular carbonation reaction to find a_{CO_2} for 900–1100°C, 100–600 MPa.

$$CaCO_{3(Cc)} + CaSiO_{3(Woll)} + CaAl_2Si_2O_{8(An)} = Ca_3Al_2Si_3O_{12(Gros)} + CO_2.$$

(6)

This reaction is characterized by an anomalously low change in entropy and, hence, extremely accurate estimates of the equilibrium concentration of CO_2 may be made (Zharikov et al., 1977). a_{H_2O} was calculated at 450–600°C, 600 MPa from the analcime dehydration reaction in the system H_2O-CO_2 (Likhoïjdov et al., 1977). Although the accuracies of the a_n values obtained from the $P-V-T$ measurements and from mineral equilibria are not comparable, nearly all the estimates of a_n obtained by the latter method were consistently 20–30% higher than the corresponding $P-V-T$ values.

Other Systems of the H_2O–Nonpolar Gas Type

Marakushev and Perchuk (1975) showed that with depth metamorphic rocks become increasingly rich in "reduced" species (H_2, CO, CH_4), owing to a marked drop in P_{CO_2}. Therefore, it is important to know the activity–concentration relationships in such systems in order to calculate accurately the composition of metamorphic fluids.

$P-V^E-T$ relations were calculated for systems of the H_2O–nonpolar gas type by using the virial equation of state with five coefficients for the Lennard–Jones "effective" potential (Eqs. (2)–(5)). The results were correlated with the experimental findings in the systems H_2O-CH_4, H_2O-N_2 (Basaev et al., 1974), H_2O-Ar (Lentz and Franck, 1969), and H_2O-Xe (Franck et al., 1974). As Eqs. (2), (4), and (5) satisfactorily describe the experimental results for the systems studied, no correction factor was needed, i.e., in Eq. (4), $\xi = 1$. The activities of nonpolar gases, as calculated from the equation of state and as graphically derived from the experimental data, were accurate to within experimental uncertainty for the same values of P, T, and X_n (Shmulovich et al., 1980a). Consequently, the generalized plot of $a_n - X_n$ (Fig. 4) can be used to estimate, to a first approximation, activities for the systems H_2O–nonpolar gas to 100–200 MPa. The only exceptions are H_2O-CO_2, where V^E and a_{CO_2} are lower than in other systems, and H_2O-H_2. Values of V^E as calculated in the system H_2O-H_2 were approximately 1.5 times those for $H_2O \approx (CH_4, N_2,$ Ar, Xe) at 400–500°C and to 200 MPa. The calculated activities of components in the system H_2O-H_2 are considerably higher than in similar systems (see Fig. 5). This must be due to the strongly differing parameters of the potential of intermolecular interaction. According to Fig. 5, the calculated and experimental a_{H_2} (Shaw, 1963) agree fairly well.

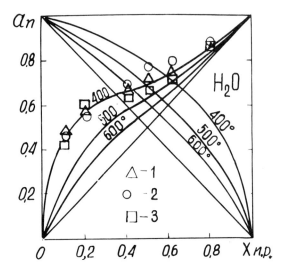

Fig. 4. Generalized activity–concentration diagram for systems of the H_2O–nonpolar gas type. Experimental results: (Symbols) 1—H_2O–N_2; 2—H_2O–CH_4; 3—H_2O–Ar. Curves have been calculated using Eqs. (2)–(5).

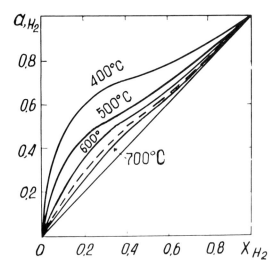

Fig. 5. Activity–concentration diagram for H_2O–H_2 at 100 MPa. Solid lines are a_{H_2} calculated from V^E (Eqs. (2)–(5)). Dashed line represents data of Shaw (1963) for 700°C, 80 MPa.

H_2O–Electrolyte Systems

The thermodynamic and transport characteristics of H_2O–electrolyte systems have been studied in considerable detail. Here we will consider only one aspect: the effect of the electrolyte on the activity of H_2O in supercritical solutions involved in high-pressure mineral reactions. Barnes and Ernst (1963) studied the hydration reaction

$$MgO + H_2O = Mg(OH)_2 \qquad (7)$$

and recorded the displacement of the equilibrium temperature caused by the addition of the electrolyte NaOH. They found that values of a_{H_2O} along the P–T curve of this reaction were slightly lower than X_{H_2O}, up to 650°C, 200 MPa. Urusova (1971) obtained values of a_{H_2O} to 400°C, 30 MPa from the steam pressure and showed that for solutions involving salts of the NaCl and KCl type $a_{H_2O} > X_{H_2O}$ at $T > T_{H_2O}^{crit}$. Hence, in solutions of low to medium density, the systems H_2O–electrolyte are similar to H_2O–nonpolar gas in that they deviate from Raoult's law with the same sign. This has been supported by the P–V–T measurements of Bach *et al.* (1977) on the system H_2O–HCl. Recent studies (Frantz and Marshall, 1979) have shown that above 500°C and up to 200 MPa virtually all of the most common halide salts—NaCl, KCl, $CaCl_2$, $MgCl_2$, and $FeCl_2$—occur in the fluid phase mostly as neutral molecules (ion pairs). Therefore, it would not seem likely that the activity of H_2O would drop drastically as a result of ion solvation. The rise in density with increasing pressure along the P–T curve of reaction (7) will result in greater dissociation of salts, and thus in a suppression of the activity of H_2O; that is, the electrolyte will have greater effects on a_{H_2O} at higher pressures.

Attempts have been made to extend the results of Barnes and Ernst (1963) to higher pressures (up to 400 MPa) and to more concentrated solutions of electrolytes (NaCl up to 20M). The experimental technique was essentially unchanged—all runs were performed in sealed Pt capsules. The results of runs at 400 MPa are shown in Fig. 6. Circles denote that either periclase or brucite

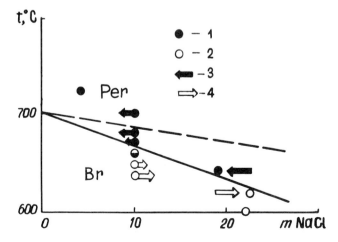

Fig. 6. Dependence of the equilibrium temperatures of brucite = periclase + H_2O on NaCl concentration of fluid, at 400 MPa. Symbols: 1—Per unchanged; 2—Br unchanged; 3—Per → Br; 4—Br → Per.

was unaltered. However, where the brucite–periclase transition was complete, the results were recalculated to the new concentrations of NaCl, allowing for the amount of water released or absorbed in the reaction. The initial and final concentrations are indicated by the ends of symbols 3 and 4. It was assumed that the brucite contained no Cl, as brucite lattice parameters were identical at 600°C, 600 MPa in runs with 10 and $20M$ NaCl solutions and in the runs with pure water. In Fig. 6, the equilibrium temperatures of the highly concentrated solutions ($\approx 20M$) have been displaced by as much as 40°C from the position calculated assuming an ideal solution.

The results of Barnes and Ernst (1963) and of our own work are summarized in Fig. 7 as a plot of activity of H_2O against electrolyte concentration ($a_{H_2O}-C_m$). The hatched region represents the change in the activity of H_2O with electrolyte concentration along the $P-T$ curve of the periclase–brucite equilibrium over the range 550–700°C, 100–400 MPa. The excellent agreement between the NaOH and NaCl curves implies that individual differences between the $I-I$ electrolytes may be insignificant. Also, an increase in pressure either does not markedly affect the $a_{H_2O}-C_m$ relationship along the $P-T$ curve of Eq. (5), or its effect is nullified by the opposing effect of increasing temperature. The curves for the activity of H_2O in solutions of certain strong electrolytes at 25°C are also presented in this diagram, as they represent the limiting case of total dissociation and maximum hydration. To sum up this brief discussion of H_2O–electrolyte systems, we should note that no matter how small the negative deviations from Raoult's law might be at high temperatures, they need to be considered, as they may lower the equilibrium temperature in the actual concentration region (obtained from analyses of

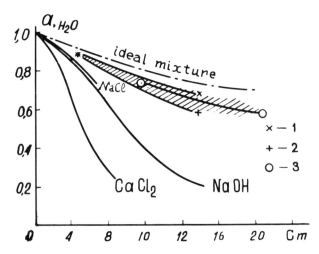

Fig. 7. Activity–concentration diagram for H_2O in electrolyte solutions in equilibrium with brucite and periclase. Symbols: 1 and 2—NaOH solution at 100 MPa and 200 MPa, respectively (after Barnes and Ernst (1963)); 3—authors' results for NaCl solution at 400 MPa.

fluid inclusions in minerals) by 40–60°C from values predicted by assuming ideal solutions.

Ternary Systems

The experimental study of ternary systems of the H_2O–nonpolar gas–electrolyte type only began in recent years. It was not until 1980 that the P–V–T relations and values of V_{CO_2} were determined in an H_2O–CO_2–NaCl system at 400, 450, and 500°C, at pressures up to 60 MPa (Gehrig, 1980).

Takenouchi and Kennedy (1965) conducted studies on the solution of CO_2 in NaCl–H_2O mixtures up to 450°C, 140 MPa. It might be inferred from their results that a solubility limit for CO_2 in these solutions will persist to higher temperatures. Hence, the differences in a_{CO_2} in a salt mixture compared to a_{H_2O} in pure water will also persist.

Special measurements of the solubilities of gases in water and in salt solutions (Malinin, 1979; Naimot et al., 1979) have given values of the coefficient, K_{sh}, obtained from the Sechenov equation

$$\log(S^0/S^{sl}) = K_{sh} \cdot m, \qquad (8)$$

where S^0 and S^{sl} are the solubilities of gas in water and in salt solutions of molality m, respectively. These measurements indicate a minimum in the K_{sh}–T curve. At temperatures in the range 100–150°C to 300°C, K_{sh} increased considerably at all electrolyte concentrations studied (up to $5M$). However, an isobaric increase of temperature and the related decrease of density of the solution will invariably result in association of the ions of the electrolyte and, thus, in a reduction of their "salting-out effect" on the nonpolar gas. Consequently, a maximum value of K_{sh} will be reached with increasing temperature. To evaluate the salting-out effect of the electrolyte in systems where the fluids are supercritical, we studied the CO_2 equilibrium concentrations in reactions (6) and (9) at 500–700°C, up to 100 MPa.

$$2CaCO_{3(Cc)} + SiO_{2(Q)} + CaAl_2Si_2O_{8(An)} = Ca_3Al_2Si_3O_{12(Gros)} + 2CO_2. \quad (9)$$

Experiments were run for 10 days in sealed Pt capsules with P_{O_2} held at about that of the NNO buffer; the CO_2 source was $Ag_2(COO)_2$. The equilibrium position was established from the direction of the reaction as determined by X-ray quantitative analysis. The starting materials were synthetic minerals. Grossular was obtained from a grossular gel treated at 700°C with $P_{H_2O} = 200$ MPa. The calcite–wollastonite–anorthite mixture was obtained from the grossular gel treated at 700°C with $P_{H_2O} = P_{CO_2}$ and $P_{Total} = 200$ MPa. The principal results were obtained using $4.6M$ solutions of the electrolytes KCl and $CaCl_2$ (Shmulovich and Kotova, 1980).

At temperatures and pressures for which a binary salt-free solution is supercritical, the conventional definition of the salting-out effect is no longer valid, as $S^0 = \infty$ in Eq. (8). Therefore, the supercritical salting-out effect may be defined as the decrease of the concentration of the nonpolar component as the electrolyte is added to the solution; a corresponding increase of the activity coefficient of the nonpolar component occurs so that the chemical potential remains constant. The supercritical salting-out coefficient was calculated from

$$\gamma^*_{CO_2} = \left(\gamma^0_{CO_2} \cdot X^0_{CO_2}\right) / \left(\gamma^{0(sl)}_{CO_2} \cdot X^{sl}_{CO_2}\right) \tag{10}$$

where the superscript 0 refers to the salt-free system; sl to the system involving the electrolyte; and $\gamma^{0(sl)}_{CO_2}$ is the activity coefficient of CO_2 in the binary salt-free system, with $X_{CO_2} = X^{sl}_{CO_2}$. The $\gamma^0_{CO_2}/\gamma^{0(sl)}_{CO_2}$ ratio corrects for the dependence of the activity coefficient of CO_2 on concentration in the salt-free system. Figure 8 presents some of the calculated values of $\gamma^*_{CO_2}$ and the data of Malinin (1979) on K_{sh}. According to this diagram, the maximum value of $\gamma^*_{CO_2}$ lies in the region of $T^c_{H_2O}$, and its value decreases above 700°C, 100 MPa, where $\rho \approx 0.36$ g/cm^3. The rise in density appears to increase $\gamma^*_{CO_2}$: for a 4.6M CaCl$_2$ solution at 925°C, 500 MPa, where $\rho_{mix} \approx 0.7$ g/cm^3, log$\gamma^*_{CO_2} \approx 0.3$. The supercritical salting-out effect may even persist up to silicate melt temperatures, contributing to the thermodynamic properties of those components that separate in the fluid phase during the crystallization of abyssal intrusions.

The changes in a_{H_2O} in a supercritical ternary gas solution appear, at a first approximation, to be the sum of the changes in the H_2O–nonpolar gas and the H_2O–electrolyte systems. It should be noted that the changes in a_{H_2O} in systems involving an electrolyte or a nonpolar gas occur in the opposite directions: a_{H_2O} decreases when an electrolyte is added and increases upon

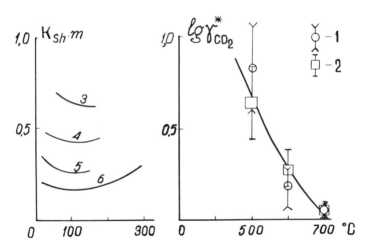

Fig. 8. Variations of the activity of CO_2 in electrolyte solutions. On right: variation of coefficient of supercritical salting-out for CO_2 with temperature for (symbols) 1—4.6M KCl; 2—4.6M CaCl$_2$. Vertical bars indicate experimental uncertainty. On left: variation of Sechenov's constant, K_{sh}, with temperature (Malinin, 1979) for curve 3—5M CaCl$_2$; curve 4—3M CaCl$_2$; curve 5—4M NaCl; curve 6—1M CaCl$_2$.

addition of a nonpolar gas. No particular studies of these problems have as yet been conducted.

Conclusions

The $P-T$ curves of reactions between minerals and fluids in the H_2O-CO_2-NaCl system should be calculated with due regard for certain factors that influence the activity coefficients of the components: (a) nonideality of the H_2O-CO_2 system, especially in the temperature range 400–600°C; (b) decreases of a_{H_2O} in systems involving electrolytes; and (c) the salting-out effect of electrolytes in CO_2. Therefore, the fugacities of CO_2 and H_2O in a three-component fluid model should be calculated from

$$f_{CO_2} = P \cdot \psi_{CO_2} \cdot \gamma_{CO_2(H_2O)}\gamma^*_{CO2(m)} \cdot X_{CO_2}, \tag{11}$$

$$f_{H_2O} = P \cdot \psi_{H_2O} \cdot \gamma_{H_2O(CO_2)} \cdot \gamma^*_{H_2O(m)} \cdot X_{H_2O}, \tag{12}$$

where ψ_n represents the fugacity coefficients of the pure components (Burnham et al., 1969, Shmulovich and Shmonov, 1978); $\gamma_{n_1(n_2)}$ represents the activity coefficients of components in a binary mixture (see Table 1); $\gamma^*_{CO_2(m)}$ represents the coefficient of the supercritical salting-out effect (see Fig. 8); and $\gamma^*_{H_2O(m)} = a_{H_2O}/X_{H_2O}$ (see Fig. 7). The dependence of $\gamma^*_{CO_2(m)}$ on the concentration of electrolytes is currently under investigation.

The calculation of fugacities also requires data on the concentration of the principal components in the fluid. In experimental mineral equilibrium studies, X_n is normally known or can easily be calculated. However, difficulties arise when the conditions of formation of natural paragenetic assemblages must be determined. This is essentially an inverse problem: X_n is to be determined from the known fugacities at a given P and T. For most parageneses containing a carbonate mineral, X_{CO_2} cannot be determined with reasonable accuracy because the value $(\partial T/\partial X_{CO_2})_P$ of the boundary reactions is extremely low—approximately $10°C/0.1X_{CO_2}$—over a wide range of fluid compositions.

Cationic isomorphism in solid phases of variable composition yields values for the formation temperatures of paragenetic assemblages, which may be used to calculate the X_{CO_2} of coexisting fluids. However, because the temperatures cannot be determined with sufficient accuracy, X_{CO_2} can be estimated only to within ± 0.4, which is not a useful result. As an alternative, a limited set of "informative" paragenetic assemblages, which are stable over a narrow range of X_{CO_2}, can be used to determine fluid compositions. For parageneses such as talc + quartz + grossular at temperatures of 700–750°C, and for some parageneses from the greenschist facies, it is possible to evaluate X_{CO_2} if T and P_{Total} are known, and X_{H_2O} by assuming $X_{H_2O} \approx 1 - X_{CO_2}$. Data on anionic isomorphism would make it possible to evaluate the equilibrium concentration of electrolytes, but sufficient information is not available at the present time.

References

Bach, R. W., Friedrichs, and Rau, H. (1977) $P-V-T$ relations for HCl–H_2O mixtures up to 500°C and 1500 bars, *H. Temp.–H. Pressure B* **9**, 305–312.

Barnes, H. H., and Ernst, W. G. (1963) Ideality and ionization in hydrothermal fluids: the system MgO–H_2O–NaOH, *Amer. J. Sci.* **261**, 129–150.

Basaev, A. R., Skripka, V. G., and Namiot, A. U. (1974) Volume properties of mixtures water vapor with methane and nitrogene at high temperatures and pressures, *J. Phys. Chem.* **12**, 1631–1674 (in Russian).

Breedveld, G. J. F., and Pransnitz, J. M. (1973) Thermodynamic properties of supercritical fluids and their mixtures at very high pressures, *AIChE J.* **19**, 783–796.

Burnham, C. W., Holloway, J. R., and Davis, N. F. (1969) Thermodynamic properties of water to 1000°C and 10000 bars, *Geol. Soc. Amer. Spec. Pap.* No. 132.

Coan, C. R., and King, A. D. J. (1971) Solubility of water in compressed carbon dioxide, nitrons oxide and ethare; Evidence of hydration of carbon dioxide and nitrons oxide in He gas phase. *J. Am. Chem. Soc.* **93**, 1857–1862.

de Santis, R., Breedveld, G., and Pranshitz, J. M. (1974) Thermodynamic properties of aqueous gas mixtures at advanced pressures, *Ind. Eng. Chem. Pros. Des. Dev.* **13**, 374–377.

Flowers, G. C. (1979) Correction of Holloway's (1977), "Adaptation of the modified Redlich–Kwong equation of state for calculation of the fugacities of molecular species in supercritical fluids of geological interest," *Contrib. Miner. Petrol.* **69**, 315–318.

Franck, E. U., and Tödhaide, K. (1959) Thermische Eigenschaften überkritischer Mischungen von Kohlendioxid und Wasser bis zu 750°C und 2000 Atm., *Z. Phys. Chem., Neue Folge* **22**, 232–245.

Frantz, J. D., and Marshall, W. L. (1979) Electrical conductance studies of $MgCl_2$–H_2O and $CaCl_2$–H_2O solution, *Geophys. Lab. Carnegie Inst. Washington Yearbook* **1978–1979**, 591–599, 603–606, and 586–591.

Gehrig, M. (1980) Phasengleichgewichte und $P-V-T$-Daten ternarer Mischungen aus Wasser, Kohlendioxid und Natriumchlorid bis 3 kbar und 550°C, Thes. Diss. Hochschule Verlag, Freiburg.

Greenwood, H. J. (1969) The compressibility of gaseous mixtures of carbon dioxide and water between 0 and 500 bars pressures and 450° and 800°C, *Amer. J. Sci.* **267-A**, 191–208.

Hirchfelder, J. O., Curtiss, C. F., and Bird, B. B. (1954) *Molecular Theory of Gases and Liquids*. Wiley, New York.

Jüza, J., Kmonicek, V., and Sifner, O. (1965) Measurements of the specific volume of carbon dioxide in the range of 700 to 4000 bar and 50 to 475°C. *Physica*, **31**, 1735–1744.

Korzhinskii, D. S. (1940) The factors of mineral equilibria and mineralogical depth facies, *Akad. Nauk USSR, M., Inst. Geol. Nauk* **12**, 100 (in Russian).

Lentz, H., and Franck, E. U. (1969) Das System Wasser-Argon bei hohen Drucken und Temperaturen, *Bericht. de. Buseny.* **73**, 28–35.

Likhoïjdov, G. G., Ivanov, I. P., and Shmulovich, K. J. (1977) The stability of analcime and activity of H_2O in the system $NaAlSi_2O_2-H_2O-CO_2$, *Int. Geol. Rev.* **19**.

Malinin, S. D. (1979) *Physical Chemistry of Hydrothermal Systems with Carbon Dioxide*, Nauka, Moscow.

Marakushev, A. A., and Perchuk, L. L. (1975) Thermodynamical calculations of gaseous and gaseous-mineral equilibria in application to the problem of fluid formation, *Geodynamic Studies*, No. 3, pp. 46–66, 1625–1639. Nauka, Moscow.

Melnik, U. P. (1978) *Thermodynamic Properties of Gases under the Conditions of Deep-Seated Petrohenesis*, Naukova Dumka, Kiev (in Russian).

Michels, A., Michels, C., and Wonters, H. (1935) Isotherms of carbon dioxide between 70 and 3000 atmospheres (Amagat densities between 200 and 600), *Proc. Roy. Soc. Ser. A* **153**, 214–224.

Namiot, A. U., Skripka, V. G., and Ashman, K. D. (1979) The influence of aqueous salts on solubility of methane at temperatures from 50° to 350°C, *Geokhymia* No. 1, 147–148 (in Russian).

Ryzenko, B. N., and Malinin, S. D. (1971) The fugacity rule in the system CO_2-H_2O, CO_2-CH_4, CO_2-N_2, CO_2-H_2, *Geokhymia* No. 8, 899–913 (in Russian).

Shaw, H. R. (1963) Hydrogen–water vapor mixtures: Control of hydrothermal atmospheres by hydrogen osmosis, *Science* **139**, 1220–1222.

Shmonov, V. M. (1977) The apparatus for measuring PVT properties of gases up to 1000°K and 8000 bars, *Ocherki Phys.-Chem. Petrol.* **1**, 236–245 (in Russian).

Shmonov, V. M., and Shmulovich, K. I. (1974) Molar volumes and equations of state for CO_2 between 100°C–1000°C and 2000–10000 bars, *Dokl. Akad. Sci. USSR* **217**, 935–938.

Shmonov, V. M., and Shmulovich, K. I. (1978) Measurement of PVT properties for the system H_2O-CO_2 at 500°C and pressures up to 5 kbars, in *Experiment and Technique of High Gaseous and Solid-Phase Pressures*, edited by I. P. Ivanov and I. A. Litvin, pp. 133–137. Nauka, Moscow (in Russian).

Shmulovich, K. I., and Shmonov, V. M. (1978) *Tables of Thermodynamic Properties of Gases and Liquids*, Vol. 3, *Carbon Dioxide*. Isdatelstvo Standartov, Moscow (in Russian).

Shmulovich, K. I., Shmonov, V. M., and Zakirov, I. W. (1979) PVT-measurements in a hydrothermal system at high pressures and temperatures, in *The Methods of Experimental Studies of Hydrothermal Equilibria*, edited by A. A. Godovikov, pp. 81–89. Nauka, Novosibirsk (in Russian).

Shmulovich, K. I., Masur, V. A., Kalinichev, A. G., and Khodorevskaja, I. (1980a) Relations PVT and activity–concentration in the systems H_2O–nonpolar gas type, *Geokhimia* No. 11, 1625–1639 (in Russian).

Shmulovich, K. I., Shmonov, V. M., Masur, V. A., and Kalinichev, A. G. (1980b) Relations PVT, activity–concentration in the system H_2O-CO_2, (homogeneous solution), *Geokhimiya*, No. 12, 1807–1824 (in Russian).

Shmulovich, K. I., and Kotova, N. P. (1980) The influence of electrolytes on the activity of CO_2 in supercritical aqueous solutions, *Dokl. Acad. Nauk USSR* **253**, 952–956 (in Russian).

Takenouchi, S., and Kennedy, G. C. (1965) The solubility of carbon dioxide in NaCl solutions at high temperatures and pressures, *Amer. J. Sci.* **263**, 445.

Touret, J., and Bottinga, Y. (1979) Equation d'etat pour le CO_2; application aux idusions carboni quus, *Bull. Mineral.* **102**, 577–583.

Tzyklis, D. S., Linshits, L. R., and Zimmerman, S. S. (1969) Measurement and computation of molar volumes for carbon dioxide, *J. Phys. Chem.* **43**, 1919–1926 (in Russian).

Urusova, M. A. (1971) Activity water in solutions of alkali-halide salts at advanced temperatures, *Isv. Acad. Nauk USSR, Ser. Chem.* **6**, 1145–1149 (in Russian).

Vukalovich, M. P., and Altunin, V. V. (1965) *Thermophysical Properties of Carbon Dioxide*. Atomisdat, Moscow (in Russian).

Walter, L. S. (1963) Data on the fugacity of CO_2 in mixtures of CO_2 and H_2O, *Amer. J. Sci.* **261**, 151–156.

Zakirov, I. V. (1977) Experimental apparatus and method for measuring CO_2 compressibility to 2500 bars and 1100°K, *Ocherki Physico-Chemical Petrology*, Vol. 7, pp. 28–33 (in Russian).

Zharikov, V. A., Shmulovich, K. I., and Bulatov, V. K. (1977) Experimental studies in the system $CaO-MgO-Al_2O_5-SiO_2-CO_2-H_2O$ and conditions of high-temperature metamorphism, *Tectonophysics* **43**, 145–162.

Zubarev, V. N., and Telegin, G. C. (1962) Shock compression of liquid nitrogen and solid carbon dioxide, *Dokl. Acad. Nauk USSR* **142**, 309–312.

Chapter 6

Thermodynamics of Crystal–Fluid Equilibria, with Applications to the System $NaAlSi_3O_8$–$CaAl_2Si_2O_8$–SiO_2–$NaCl$–$CaCl_2$–H_2O

J. G. Blencoe, G. A. Merkel, and M. K. Seil

Introduction

From fundamental thermodynamic principles it can be stated that, with a specific set of components to express the compositions of phases, there is only one "true" $\Delta G_f = f(P, T, X)$ function for an isostructural binary or multi-component system, and for each bulk composition in such a system at constant P and T there is a unique set of chemical potentials for the components (Denbigh, 1968, pp. 270–278 and 303–305).[1] However, it is also true that the choice of standard state(s) and the manner in which the compositions of phases are expressed profoundly affect the values of standard-state chemical potentials and activity coefficients for the components.

It is a major purpose of this paper to assess various standard states and thermochemical parameters for representing the chemical potentials of *undissociated* solute components in supercritical, *ideally dilute* aqueous fluids (vapors). As we offer our evaluations (1) special attention will be focused on the advantages and limitations of employing pseudobinary mole fractions to represent the concentrations of undissociated solute components, and (2) presentation of thermodynamic theory will be directed toward a specific and practical goal: to develop a thermodynamically rigorous procedure for calculating equilibria between components that are "shared" by coexisting nonideal binary crystalline solutions and ideally dilute aqueous vapors. Thermodynamic theory described in this chapter is also applicable to crystal–melt–vapor and melt–vapor equilibria, but our commentary and thermodynamic treatments will only deal with crystal–vapor equilibria; specifically, equilibria involving $NaAlSi_3O_8$–$CaAl_2Si_2O_8$ plagioclases and coexisting H_2O-rich fluids.

Due to the paltry quantities of experimental data on the mixing properties of solute components in aqueous fluids, it is very difficult to discern the P–T–X limits of ideally dilute (Henrian) behavior of these components in any particular system, but it may be stated generally that this behavior is favored by low pressure, high temperature, and increased dilution. Therefore, applications of theory presented herein will be restricted to plagioclase–fluid

[1] Notation employed in this paper is defined in Table 1.

equilibria under a set of $P-T$ conditions (2 kbars, 700°C) and concentrations of solute components $[\sum X_i(\text{solutes}) < 0.05]$ for which the assumption of Henrian behavior is reasonably justified.

In their analyses of plagioclase–fluid ion-exchange equilibria, previous investigators have used molalities or pseudobinary mole fractions to represent the concentrations of salt components in $NaCl-CaCl_2-H_2O$ fluids. This is legitimate, *provided* that there is a proper accounting for the thermodynamic effects of variable X_w in the fluids. However, we suggest that it is more correct and straightforward to employ ternary mole fractions to specify the proportions of aqueous salt components. This approach is harmonious with a Raoultian "molten solute" standard state for aqueous salt components in plagioclase–fluid ion-exchange equilibria, and we believe that this solute standard state has general applicability in thermodynamic treatments of crystal–vapor and melt–vapor equilibria.

Notation and Standard States

Most of the abbreviations and symbols employed in this paper are listed in Table 1. The definitions, categorical positioning, and utility of entries in this listing are self-explanatory, or will become intelligible in the context of subsequent discussions. However, at this point we notify the reader that only sparing use is made of the superscript v. To avoid a proliferation of this particular denotation, we declare now that all thermochemical quantities pertaining to $NaCl$, $CaCl_2$, and H_2O are understood to refer to these components as they "exist" either (1) in the molten state ($NaCl$ melt, $CaCl_2$ melt), or (2) dissolved in supercritical aqueous fluids (vapors). On the other hand, affixation of superscript v to quantities related to aqueous feldspar components is necessary to avoid confusion with the crystalline feldspar components of $NaAlSi_3O_8-CaAl_2Si_2O_8$ plagioclases.

For all components listed in Table 1—which include crystalline, solvent, and solute components—we have at constant P and T

$$\mu_i = \mu_i^\circ + RT \ln a_i, \tag{1}$$

where $a_i = \gamma_i X_i$. Standard states adopted herein for these components, each referring to system P and T, are as follows:

(1) *Crystalline components* (ab = $NaAlSi_3O_8$, an = $CaAl_2Si_2O_8$, qz = SiO_2): The standard state of a crystalline component is 1 mole of the pure crystalline component with the same structure as the crystalline phase under consideration (Blencoe, 1979).

(2) *Solvent component* ($w = H_2O$): The standard state of a solvent component is 1 mole of the pure component in the supercritical fluid (vapor) state.

(3) *Solute components* ($NaCl$, $CaCl_2$, ab = $NaAlSi_3O_8$, an = $CaAl_2Si_2O_8$, qz = SiO_2): The standard state of a solute component is 1 mole of the pure, metastable liquid (molten) component.

Therefore, in Eq. (1), the standard-state free energy of a component, μ_i°, is taken as the *free energy of formation* (from the elements) of 1 mole of the pure component at system P and T, and $a_i = 1.0$ for the component under these conditions.

The standard states listed above can be described as Raoultian because, for each component, $\gamma_i \to 1$ as $X_i \to 1$. However, there are two other types of standard states in common use: a 1 molal standard state (applied solely to solute components) in which $\gamma_i \to 1$ as $m_i \to 0$, and a Henrian standard state in which $\gamma_i \to 1$ as $X_i \to 0$. In subsequent discussions there will be repeated references to these standard states or "conventions" (Denbigh, 1968), and for the sake of brevity they will be referred to as the R (Raoultian), M (1 molal), and H (Henrian) standard states, respectively. The relationships between the respective activity coefficients—hereinafter designated $\gamma_{i,R}$ (R standard state), $\gamma_{i,M}$ (M standard state), and $\gamma_{i,H}$ (H standard state)—are derived in Appendix A. Additionally, graphical representations of the standard-state chemical potentials for each "convention" are illustrated in Fig. 1. The reasoning behind our uncustomary selection of an R, "molten solute" standard state for aqueous solute components will become evident in subsequent discussions of plagioclase–fluid equilibria in the system $NaAlSi_3O_8$–$CaAl_2Si_2O_8$–SiO_2–$NaCl$–$CaCl_2$–H_2O.

Thermodynamic Analysis of Plagioclase–Fluid Ion-Exchange Equilibria

Thermodynamic analyses of plagioclase–fluid ion-exchange equilibria in the system $NaAlSi_3O_8$–$CaAl_2Si_2O_8$–SiO_2–$NaCl$–$CaCl_2$–H_2O have been presented previously by Orville (1972) and Saxena and Ribbe (1972). In both of these investigations the main goal was to determine the thermodynamic mixing properties of $NaAlSi_3O_8$–$CaAl_2Si_2O_8$ (Ab–An) plagioclase feldspars at 2 kbars, 700°C. However, significantly, different methods were adopted to represent the concentrations of salt components in the fluids: Orville used molalities, whereas Saxena and Ribbe employed pseudobinary mole fractions. Also, accordingly, different standard states were selected for these components —an M standard state by Orville and (implicitly) an H standard state by Saxena and Ribbe. A third method, which will be examined in detail for the first time in this paper, is to specify the quantities of salts in terms of "true" mole fractions and to employ an R standard state for these components. The different approaches dictate the manner in which the chemical potentials of salt components are represented—specifically, they determine the values of standard-state chemical potentials and activity coefficients—but absolute values of the chemical potentials are unaffected. Nevertheless, rather than being arbitrary and inconsequential, it will be demonstrated here that selection of a particular method for representing the chemical potentials of aqueous salt

Table 1. Notation.

General Thermochemical Symbols

a	activity
i	a component
j	a component other than component i
K_D	an equilibrium distribution coefficient
m	molality
n	mole (gfw)
P	pressure
R	universal gas constant, 1.98726 cal/gfw-deg
T	temperature in Kelvins (K)
X	mole fraction (X_{NaCl}, X_{CaCl_2}, X_{Na}, and X_{Ca} are defined in Appendix B)
ΔG_f	molal Gibbs free energy of formation (from the elements)
ΔG_i^{am}	molal Gibbs free energy of fusion of crystalline component i
γ	activity coefficient
μ	chemical potential

Abbreviations

Ab–An	$NaAlSi_3O_8$–$CaAl_2Si_2O_8$
aq.	aqueous
f	a function of
gfw	gram formula weight (mole)
k	a constant

Superscripts

am	anhydrous melt
v	silicate(feldspar)-saturated vapor (fluid) in equilibrium with crystalline $NaAlSi_3O_8$–$CaAl_2Si_2O_8$ plagioclase (vapors may also contain, but are not saturated with, one or more salt components)
°	of or pertaining to an R standard state

Table 1 (continued).

□	of or pertaining to an M standard state
*	of or pertaining to Henry's law, the compositional range of Henry's law behavior, or an H standard state

Subscripts

i	a component
j	a component other than component i
P	constant pressure
T	constant temperature
X_i	constant mole fraction of component i
R	of or pertaining to an R standard state
M	of or pertaining to an M standard state
H	of or pertaining to an H standard state

Equilibrium Constants

$$K_{E,1} = \frac{\gamma_{an} X_{an} (\gamma^*_{NaCl,R} X_{NaCl})^2}{(\gamma_{ab} X_{ab})^2 \gamma^*_{CaCl_2,R} X_{CaCl_2}}$$

$$K_{E,2} = \frac{\gamma_{an} X_{an} (\gamma_{NaCl,M} m_{NaCl})^2}{(\gamma_{ab} X_{ab})^2 \gamma_{CaCl_2,M} m_{CaCl_2}}$$

$$K_{E,3} = \frac{\gamma_{an} X_{an} (\gamma_{NaCl,H} X_{Na})^2}{(\gamma_{ab} X_{ab})^2 \gamma_{CaCl_2,H} X_{Ca}}$$

Equilibrium Distribution Coefficients

$$K_{D,1} = \frac{X_{an} X^2_{NaCl}}{X^2_{ab} X_{CaCl_2}}, \quad K_{D,2} = \frac{X_{an} m^2_{NaCl}}{X^2_{ab} m_{CaCl_2}}, \quad K_{D,3} = \frac{X_{an} X^2_{Na}}{X^2_{ab} X_{Ca}}$$

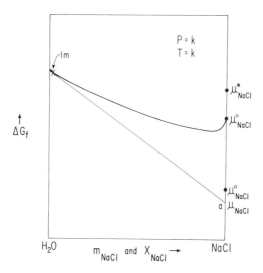

Fig. 1. Standard-state chemical potentials for the NaCl component of supercritical, "isostructural" NaCl–H$_2$O fluids at a fixed P and T. The heavy curve (curvature exaggerated for clarity) represents a single free energy–composition equation of state for these fluids; therefore, μ°_{NaCl} is the free energy of formation (ΔG_f) of pure (metastable) molten NaCl (NaCl-rich aqueous fluids are also metastable). Point X indicates a 1.0m NaCl–H$_2$O solution that is assumed to be ideally dilute, and a tangent to the free energy–composition curve at this point (the light straight line in the figure) intersects the right-hand ordinate of the diagram at point a where μ_{NaCl} represents the chemical potential of NaCl in the 1.0m solution. The small, filled circles labeled μ°_{NaCl}, $\mu^\square_{NaCl}$, and μ^*_{NaCl} are the free energies of NaCl in the R, M, and H standard states, respectively (see text and Appendix A for explanation). From the relationships between the activity coefficients for NaCl according to these standard states—$\gamma_{NaCl,R}$, $\gamma_{NaCl,M}$, and $\gamma_{NaCl,H}$—it can be shown that the corresponding standard-state chemical potentials—$\mu^\square_{NaCl}$, $\mu^\square_{NaCl}$, and μ^*_{NaCl}—are related to one another through the expressions $\mu^\square_{NaCl} = \mu^\circ_{NaCl} + RT \ln \gamma^*_{NaCl,R} - 4.017RT$ and $\mu^*_{NaCl} = \mu^\circ_{NaCl} + RT \ln \gamma^*_{NaCl,R}$. The relative positions of μ°_{NaCl} and μ^*_{NaCl} in the figure were established assuming that supercritical NaCl–H$_2$O fluids exhibit positive deviations from ideality, which requires that $\gamma^*_{NaCl,R} > 1$ and, therefore, $\mu^*_{NaCl} > \mu^\circ_{NaCl}$. Also, because it has been stipulated that the 1.0m fluid at point X is ideally dilute, it can be demonstrated that the ΔG_f gap between $\mu^\square_{NaCl}$ and μ_{NaCl} at point a is numerically equivalent to $- RT \ln X_w = 0.018RT$ (see Fig. 3). As indicated by their graphical positions relative to μ°_{NaCl}, $\mu^\square_{NaCl}$ and μ^*_{NaCl} are hypothetical in the sense that they do not correspond to any "real" stable or metastable state of pure NaCl.

(solute) components has specific and significant thermodynamic consequences.

Following Orville (1972) and Saxena and Ribbe (1972), we will adopt the key assumptions that (1) there is no significant ionization of the aqueous salt components of NaCl–CaCl$_2$–H$_2$O fluids in equilibrium with Ab–An plagioclases, and (2) these fluids exhibit a *finite compositional range* of ideally dilute behavior. As noted in numerous studies of aqueous electrolyte systems, it is reasonable to assume that even strong electrolytes (e.g., NaCl) are predominantly neutral ion-pairs in "dilute" (0.1–$2.0m$), supercritical aqueous fluids at low pressures and high temperatures. Under these circumstances, the fluids would be expected to exhibit ideally dilute behavior; that is, solute components (NaCl and $CaCl_2$) and the solvent (H_2O) would obey Henry's and Raoult's laws, respectively. Moreover, if salt concentrations are expressed as mole fractions, and with either an R or H standard state for these components, γ_{NaCl} and γ_{CaCl_2} would be constants along any compositional path.[2]

As demonstrated previously by Orville (1972), plagioclase–fluid ion-exchange equilibria in the system $NaAlSi_3O_8$–$CaAl_2Si_2O_8$–SiO_2–NaCl–$CaCl_2$–H_2O may be expressed as

$$2NaAlSi_3O_8(ab) + CaCl_2(aq.) \rightleftharpoons CaAl_2Si_2O_8(an) + 2NaCl(aq.) + 4SiO_2(qz).$$

$$(2)$$

This relation represents an energy balance among the components in the stable phases of the system—plagioclase, quartz, and fluid—and *not* a reaction involving these phases, as explained by Skippen (1974). Accordingly, at constant P and T, the equilibrium may also be written as an equality among the chemical potentials of the components

$$2\mu_{ab} + \mu_{CaCl_2} = \mu_{an} + 2\mu_{NaCl} + 4\mu_{qz} \qquad (3)$$

which, along with Eq. (1), yields

$$\ln K_E = -\frac{\Delta\mu^\circ}{RT}, \qquad (4)$$

where

$$K_E = \frac{a_{an}a_{NaCl}^2 a_{qz}^4}{a_{ab}^2 a_{CaCl_2}} = \frac{a_{an}a_{NaCl}^2}{a_{ab}^2 a_{CaCl_2}} \qquad (5)$$

and $\Delta\mu^\circ = \mu_{an}^\circ + 2\mu_{NaCl}^\circ + 4\mu_{qz}^\circ - 2\mu_{ab}^\circ - \mu_{CaCl_2}^\circ$.[3] K_E is the equilibrium constant for Eq. (2).

In order to use Eq. (5) and plagioclase–fluid ion-exchange data to derive mixing properties for Ab–An plagioclases, it is necessary to devise a means for evaluating a_{NaCl} and a_{CaCl_2} in the aqueous fluids. Herein these evaluations will be performed using an R, "molten solute" standard state for NaCl and $CaCl_2$, which yields relative activities (Guggenheim, 1967, p. 181; Blencoe, 1979) for these components. Therefore, for NaCl, μ_{NaCl}° is the free energy of formation of 1 mole of pure (metastable) molten NaCl at system P and T, and $a_{NaCl} = 1.0$ under these conditions. This state of NaCl is represented by point a in Fig. 2, which may be viewed as a starting point in a two-step "mixing" process to establish the activity of NaCl in an ideally dilute NaCl–$CaCl_2$–H_2O

---

[2] Although this circumstance is often referred to as "ideal" mixing of the components (in this case, aqueous salt components), it is more accurate to say that, in the ideally dilute range, *nonideal* solute–H_2O fluids mix ideally.

[3] Because quartz is essentially a pure crystalline phase (SiO_2), we have $a_{qz} = 1.0$ in Eq. (5).

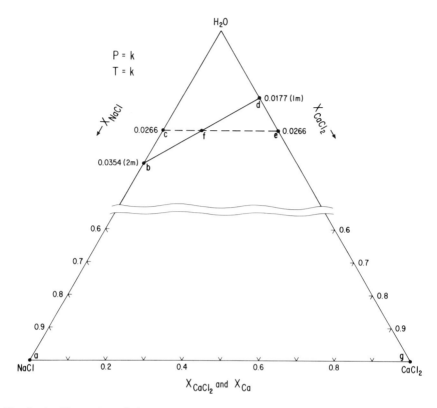

Fig. 2. An illustration of the two-step "mixing" method for deriving the activities of
NaCl and $CaCl_2$ in an aqueous fluid f. Line b-d shows the compositions of ideally
dilute, supercritical 2.0N $NaCl-CaCl_2-H_2O$ fluids (including fluid f) that are in
equilibrium with $NaAlSi_3O_8-CaAl_2Si_2O_8$ plagioclases at a fixed P and T (the small
amounts of silicate material dissolved in the fluids are not represented). For fluid f: (1)
mixing first from a to c, and then from c to f yields $a_{NaCl} = \gamma^*_{NaCl,R}(1 - X_w)X_{Na}$ and,
similarly, (2) mixing from g to e, and then from e to f yields $a_{CaCl_2} = \gamma^*_{CaCl_2,R}(1 - X_w)$
X_{Ca}, where $X_w = 0.9734$, $X_{NaCl} = 0.0181$, and $X_{CaCl_2} = 0.0085$. See text for a complete
explanation.

fluid f. To evaluate a_{NaCl} in this fluid, we first add a sufficient amount of
water to molten salt a to reach point c where $X_{NaCl} = 0.0266$. In this step X_w
increases from 0.0 at point a to 0.9734 at point c, thereby accounting for the
effect of water on the activity of NaCl in fluid f. Consequently, for NaCl
dissolved in the $NaCl-H_2O$ solution at point c, we may write $\mu_{NaCl} = \mu^\circ_{NaCl} +$
$RT \ln a_{NaCl}$ or, because $a_{NaCl} = \gamma_{NaCl}X_{NaCl}$ and $X_{NaCl} = 1 - X_w$,

$$\mu_{NaCl} = \mu^\circ_{NaCl} + RT \ln \gamma^*_{NaCl,R} + RT\ln(1 - X_w), \tag{6}$$

where $\gamma^*_{NaCl,R}$ is the Henry's law activity coefficient for NaCl with an R
standard state.

Because it has been stipulated that the fluid f is ideally dilute, $\gamma^*_{NaCl,R}$ is a
fixed value (independent of X_w) at any given P and T, and Eq. (6) could be

simplified to

$$\mu_{NaCl} = \mu_{NaCl}^* + RT\ln(1 - X_w),\qquad(7)$$

where $\mu_{NaCl}^* = \mu_{NaCl}^\circ + RT\ln\gamma_{NaCl,R}^*$ for an H standard state (Fig. 1 and Appendix A). However, other than simplifying Eq. (6), there is nothing to be gained by adopting this approach, so we will retain an R standard state for aqueous salt components. It should be kept clearly in mind that, in Eq. (6), μ_{NaCl}° refers to the free energy of formation of 1 mole of *pure molten* NaCl at system P and T, whereas $\gamma_{NaCl,R}^*$ is a function of the *mixing properties of "molten"* NaCl in H_2O. Therefore, it is evident that the H standard state combines two constants that are completely unrelated to one another; in other words, μ_{NaCl}° and $\gamma_{NaCl,R}^*$ have their own unique significance, and their individual characteristics and thermodynamic effects are obscured conceptually (if not numerically) by adopting an H standard state.

With the chemical potential of NaCl in fluid c established by Eq. (6), we may perform the second and final step of the "mixing" process to obtain μ_{NaCl} for fluid f. This step involves mixing $CaCl_2$ with fluid c at constant X_w, thereby accounting for the effect of $CaCl_2$ on the activity of NaCl in fluid f. Because X_w is constant in this step, the addition of $CaCl_2$ to fluid c is a pseudobinary mixing process, and a thermodynamic treatment of this mixing is simplified accordingly; for example, because $d\ln\gamma_w = 0$ and $d\ln X_w = 0$ along the mixing path from c to e in Fig. 2, we have $d\mu_w = 0$ and

$$X_{NaCl}\,d\mu_{NaCl} + X_{CaCl_2}\,d\mu_{CaCl_2} = 0\qquad(8)$$

which is a Gibbs–Duhem equation for *binary* mixing. Therefore, the effect of water on μ_{NaCl} in fluid f is completely accounted for in the first step of the mixing process (diluting molten salt a with water to obtain the dilute NaCl–H_2O fluid c), whereas the second step accounts entirely (and exclusively) for the effect of $CaCl_2$. It should also be evident now that μ_{CaCl_2} in fluid f can be obtained in a similar way: in the first step (Fig. 2) water is mixed with pure (metastable) molten $CaCl_2$ (point g) to obtain a $CaCl_2$–H_2O fluid with $X_{CaCl_2} = 0.0266$ (point e), and then NaCl is added to this dilute fluid until its composition reaches that of fluid f.

Because mixing along path c-e in Fig. 2 is tantamount to binary mixing, it is necessary to redefine the units of concentration of the salt components in the fluids in order to maintain internal consistency in our thermodynamic analysis. The restrictions on a thermodynamic representation of binary mixing along path c-e are that (1) $\mu_{NaCl} \to \mu_{NaCl}(c)$ and $\mu_{CaCl_2} \to -\infty$ as the composition of a fluid approaches that of fluid c, and (2) $\mu_{NaCl} \to -\infty$ and $\mu_{CaCl_2} \to \mu_{CaCl_2}(e)$ as the composition of a fluid approaches that of fluid e. These restrictions can be satisfied by (1) adopting $\mu_{NaCl}(c)$ and $\mu_{CaCl_2}(e)$ as *reference-state* chemical potentials, (2) expressing the compositions of fluids along path c-e in terms of the mole ratios (pseudobinary mole fractions) $n_{NaCl}/(n_{NaCl} + n_{CaCl_2})$ and $n_{CaCl_2}/(n_{NaCl} + n_{CaCl_2})$, hereinafter referred to as X_{Na} and X_{Ca}, respectively, and (3) defining γ_{Na} and γ_{Ca} so that, for fluid f, we now have

$$\mu_{NaCl} = \mu_{NaCl}(c) + RT\ln\gamma_{Na} + RT\ln X_{Na}\qquad(9a)$$

and

$$\mu_{CaCl_2} = \mu_{CaCl_2}(e) + RT \ln \gamma_{Ca} + RT \ln X_{Ca}. \tag{9b}$$

In Eqs. (9), γ_{Na} and γ_{Ca} are *pseudobinary* relative activity coefficients that can be derived directly from the *ternary* relative activity coefficients γ_{NaCl} and γ_{CaCl_2}. Although γ_{NaCl} and γ_{CaCl_2} cannot be evaluated from ion-exchange experimental data, at any given P and T it can be confidently predicted that both will have values $\gg 1.0$. Furthermore, in the present situation they must be constants because it has been stipulated that NaCl and $CaCl_2$ exhibit Henrian behavior in the fluids. Therefore, and because it has been established that NaCl and $CaCl_2$ exhibit binary mixing behavior along path c-e in Fig. 2 [Eq. (8)], it is evident that γ_{Na} and γ_{Ca} must both be equal to 1.0 in order to satisfy the restrictions cited above. This result translates to $RT \ln \gamma_{Na} = 0$ and $RT \ln \gamma_{Ca} = 0$ in Eqs. (9a) and (9b), respectively, so we may now write

$$\mu_{NaCl} = \mu_{NaCl}^\circ + RT \ln \gamma_{NaCl,R}^* + RT \ln(1 - X_w) + RT \ln X_{Na} \tag{10a}$$

and

$$\mu_{CaCl_2} = \mu_{CaCl_2}^\circ + RT \ln \gamma_{CaCl_2,R}^* + RT \ln(1 - X_w) + RT \ln X_{Ca} \tag{10b}$$

from which we obtain the relative activities

$$a_{NaCl} = \gamma_{NaCl,R}^*(1 - X_w)X_{Na} \tag{11a}$$

and

$$a_{CaCl_2} = \gamma_{CaCl_2,R}^*(1 - X_w)X_{Ca} \tag{11b}$$

for any ideally dilute $NaCl-CaCl_2-H_2O$ fluid such as fluid f in Fig. 2.

   If the concentrations of aqueous salt components are expressed as pseudo-binary mole fractions in a thermodynamic analysis of plagioclase–fluid ion-exchange data, and if it is assumed that the $NaCl-CaCl_2-H_2O$ fluids are ideally dilute (Saxena and Ribbe, 1972), then with an R standard state Eqs. (11) are the correct expressions for a_{NaCl} and a_{CaCl_2} to substitute into Eq. (5). However, with the definitions of X_{Na}, X_{Ca}, X_{NaCl}, and X_{CaCl_2} listed in Appendix B [see Eqs. (B3), (B4), (B9), and (B10)] it can be shown that $(1 - X_w)X_{Na} = X_{NaCl}$ and $(1 - X_w)X_{Ca} = X_{CaCl_2}$, so Eqs. (11a) and (11b) may be rewritten as

$$a_{NaCl} = \gamma_{NaCl,R}^* X_{NaCl} \tag{12a}$$

and

$$a_{CaCl_2} = \gamma_{CaCl_2,R}^* X_{CaCl_2}. \tag{12b}$$

Equations (12) demonstrate that the manner in which the preceding thermodynamic analysis deals with aqueous salt components is exactly equivalent to a treatment in which salt concentrations are expressed as ternary mole fractions. Therefore, with Eqs. (12) and the definition $a_i = \gamma_i X_i$ for crystalline components, it is now evident that Eq. (5) may be rewritten as

$$\ln K_E = \ln K_{E,1}, \tag{13}$$

where

$$K_{E,1} = \frac{\gamma_{an} X_{an} (\gamma^*_{NaCl,R} X_{NaCl})^2}{(\gamma_{ab} X_{ab})^2 \gamma^*_{CaCl_2,R} X_{CaCl_2}}$$

(Table 1).

With the equilibrium constant for plagioclase–fluid ion-exchange defined as $K_{E,1}$, we may now write

$$\ln K_{E,1} = \ln K_{\gamma,1} + \ln K_{D,1}, \tag{14}$$

where

$$K_{\gamma,1} = \frac{\gamma_{an} (\gamma^*_{NaCl,R})^2}{\gamma^2_{ab} \gamma^*_{CaCl_2,R}} \quad \text{and} \quad K_{D,1} = \frac{X_{an} X^2_{NaCl}}{X^2_{ab} X_{CaCl_2}}.$$

$K_{D,1}$, the equilibrium distribution coefficient, is obtained directly from plagioclase–fluid ion-exchange data, and by substituting the right-hand side of Eq. (14) into Eq. (4), separating the γ terms for the feldspar and salt components and rearranging, we obtain

$$\ln K_{D,1} = -\frac{\Delta\mu°}{RT} + \ln \gamma^2_{ab} - \ln \gamma_{an} - \ln \left[\frac{(\gamma^*_{NaCl,R})^2}{\gamma^*_{CaCl_2,R}} \right]. \tag{15}$$

Equation (15) shows that, at a constant P and T, Na–Ca partitioning between $NaAlSi_3O_8$–$CaAl_2Si_2O_8$ plagioclase and coexisting ideally dilute $NaCl$–$CaCl_2$–H_2O fluid is determined by the standard-state chemical potentials and activity coefficients for the Na–Ca components in the feldspar and fluid. Furthermore, because $\ln[(\gamma^*_{NaCl,R})^2/\gamma^*_{CaCl_2,R}]$ is a constant in Eq. (15), we may now differentiate this expression with respect to X_{an} at a constant P and T to obtain [with the abbreviations $d \ln K_{D,1} = (\partial \ln K_{D,1}/\partial X_{an})_{P,T}$ and $d \ln \gamma_i = (\partial \ln \gamma_i/\partial X_{an})_{P,T}$]

$$d \ln K_{D,1} = 2d \ln \gamma_{ab} - d \ln \gamma_{an}. \tag{16}$$

[The "directions" of the $d \ln K_{D,1}$ and $d \ln \gamma_i$ derivatives in Eq. (16) are parallel to path b-d and toward point d in Fig. 2.] Significantly, from the Gibbs–Duhem relation for binary solutions at a constant P and T, we also have

$$0 = X_{ab} d \ln \gamma_{ab} + X_{an} d \ln \gamma_{an}. \tag{17}$$

Equations (16) and (17) are *independent* equations that may be solved simultaneously to derive expressions for $d \ln \gamma_{ab}$ and $d \ln \gamma_{an}$, and subsequent integration of these expressions yields

$$\ln \gamma_{ab} = \int_0^{X_{an}} \left(\frac{X_{an}}{1 + X_{an}} \right) d \ln K_{D,1} \tag{18a}$$

and

$$\ln \gamma_{an} = \int_1^{X_{an}} \left(\frac{X_{an} - 1}{1 + X_{an}} \right) d \ln K_{D,1} \tag{18b}$$

[cf. Orville, 1972, Eqs. (29A) and (29B)].

Analyses Presented by Earlier Investigators

The foregoing thermodynamic analysis of plagioclase–fluid ion-exchange equilibria and treatments offered previously by Orville (1972) and Saxena and Ribbe (1972) are generally similar but different in detail. Many of the differences are superficial and attributable to variations in technique and notation, but there are also discrepancies in the presentations of thermodynamic theory which warrant explanation.

Orville (1972)

The most significant difference between the preceding thermodynamic analysis and theory presented by Orville (1972) is that the alternative approaches encompass different definitions of the ion-exchange equilibrium constant—designated $K_{T,P}$ by Orville. In accordance with an M standard state for aqueous salt components, the definition of $K_{T,P}$ listed by Orville [1972, Eqs. (11) and (14)]—hereinafter denoted $K_{E,2}$—is

$$K_{E,2} = \frac{\gamma_{an} X_{an} (\gamma_{NaCl,M} m_{NaCl})^2}{(\gamma_{ab} X_{ab})^2 \gamma_{CaCl_2, M} m_{CaCl_2}} \qquad (19a)$$

or

$$K_{E,2} = K_{\gamma,plag} K_{D,plag} K_{\gamma,salts} K_{D,salts}, \qquad (19b)$$

where $K_{\gamma,plag} = \gamma_{an}/\gamma_{ab}^2$, $K_{D,plag} = X_{an}/X_{ab}^2$, $K_{\gamma,salts} = \gamma_{NaCl,M}^2/\gamma_{CaCl_2, M}$, and $K_{D,salts} = m_{NaCl}^2/m_{CaCl_2}$. Now, in any strictly valid thermodynamic treatment of plagioclase–fluid ion-exchange data, the equilibrium constant must be *absolutely constant* at any given P and T (this is the case for $K_{E,1}$ defined previously). Therefore, following Orville [1972, Eq. (15)], we note that at constant plagioclase composition, $K_{\gamma,plag}$, $K_{D,plag}$, and the product $K_{\gamma,salts} \cdot K_{D,salts}$ must all be constants. However, significantly, Orville (1972, p. 255) also states that, if "ideal" mixing is assumed for NaCl–CaCl$_2$–H$_2$O fluids, $K_{\gamma,salts}$ is equal to unity. It is demonstrated below that this conclusion has important consequences which expose an intrinsic, detrimental characteristic of the M standard state.

From previous discussions it is evident that for $K_{E,1}$ (Table 1) we have $K_{D,salts} = X_{NaCl}^2/X_{CaCl_2}$. A close examination of the relationship between this definition of $K_{D,salts}$ and the "molal" definition reveals an important result of employing salt molalities rather than mole fractions in the definition of the equilibrium constant. Noting that $m_{NaCl} = X_{NaCl}\Sigma$moles and $m_{CaCl_2} = X_{CaCl_2}\Sigma$moles where Σmoles $= m_{NaCl} + m_{CaCl_2} + 55.51$ [Orville, 1972, Eqs. (16) and (17)], we have

$$\frac{m_{NaCl}^2}{m_{CaCl_2}} = \frac{(X_{NaCl}\Sigma moles)^2}{X_{CaCl_2}\Sigma moles} = \Sigma moles \frac{X_{NaCl}^2}{X_{CaCl_2}}. \qquad (20)$$

This relation shows that if it were possible for $\sum$moles to remain constant during plagioclase–fluid ion-exchange, then m_{NaCl}^2/m_{CaCl_2} would be directly proportional to X_{NaCl}^2/X_{CaCl_2}. Furthermore, with $K_{\gamma,salts}$ set equal to unity, $K_{E,2}$ would be a constant at any given P and T, thereby *suggesting* complete harmony with fundamental thermodynamic principles. However, the nature of plagioclase–fluid ion-exchange precludes $\sum$moles $= k$, and in Orville's $2.0N$ experiments $\sum$moles ranged from 56.51 to 57.51. This is a narrow range of variability, but with $K_{\gamma,salts} = 1.0$ *any* variation in $\sum$moles must produce a change in $K_{E,2}$. [Denbigh (1968, pp. 298–299) has described an analogous situation in a thermodynamic treatment of the equilibrium $N_2O_4 \rightleftharpoons 2NO_2$, but he did not offer a resolution to the dilemma. Consequently, we shall do so here.] Therefore, because $K_{E,2}$ *should* be a constant at any given P and T, it is evident that the identity $K_{\gamma,salts} = 1.0$ cannot be precisely correct.

The difficulty with $K_{\gamma,salts}$ can be resolved by recognizing that the M standard state is not strictly consistent with Henrian behavior of solute components, except at *infinite dilution* (Denbigh, 1968, p. 278). For solute components that exhibit a finite compositional range of Henrian behavior, it can be shown that the M standard state gives $\gamma_{i,M} = X_w$ (Appendix A and Fig. 3) and, consequently, for ideally dilute $NaCl$–$CaCl_2$–H_2O fluids,

$$K_{\gamma,salts} = \frac{\gamma_{NaCl,M}^2}{\gamma_{CaCl_2,M}} = \frac{X_w^2}{X_w} = X_w. \qquad (21)$$

Therefore, for $2.0N$ $NaCl$–$CaCl_2$–H_2O solutions as in Orville's experiments, because $X_w = 55.51/\sum$moles, $K_{\gamma,salts}$ must vary from $55.51/57.51 = 0.965$ (at $X_{CaCl_2} = 0$) to $55.51/56.51 = 0.982$ (at $X_{NaCl} = 0$) if it is assumed that these fluids are ideally dilute.

The validity of applying Eq. (21) to ideally dilute $NaCl$–$CaCl_2$–H_2O fluids can be demonstrated by deriving the relationship between $K_{E,1}$ and $K_{E,2}$. From Eqs. (20) and (21), and noting that $X_w \sum$moles $= 55.51$, it is evident that $K_{E,2}$ may now be redefined as

$$K_{E,2} = \frac{55.51\gamma_{an}X_{an}X_{NaCl}^2}{(\gamma_{ab}X_{ab})^2X_{CaCl_2}}. \qquad (22)$$

Therefore, the relationship between $K_{E,1}$ and $K_{E,2}$ becomes

$$K_{E,2} = K_{E,1} \cdot \frac{55.51\,\gamma_{CaCl_2,R}^*}{(\gamma_{NaCl,R}^*)^2}. \qquad (23)$$

Equation (23) shows that $K_{E,2}$ is now directly proportional to $K_{E,1}$, which means that, with $K_{\gamma,salts} = X_w$, $K_{E,2}$ is a valid equilibrium constant for plagioclase–fluid ion-exchange.

As noted previously, Orville analyzed his plagioclase–fluid ion-exchange data assuming that $K_{\gamma,salts} = 1.0$. However, because an M standard state gives $K_{\gamma,salts} = X_w$ for ideally dilute $NaCl$–$CaCl_2$–H_2O fluids, it is evident that assuming $K_{\gamma,salts} = 1.0$ must lead to numerical errors in calculated values of γ_{ab} and γ_{an} if $d\ln K_{D,2}$ is substituted for $d\ln K_{D,1}$ in Eqs. (18a) and (18b).

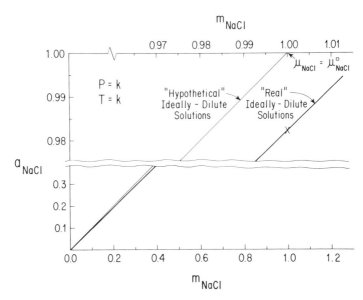

Fig. 3. Activity–composition relations for NaCl in "hypothetical" and "real" ideally dilute NaCl–H$_2$O fluids according to an M standard state (P–T conditions, the "real" fluids, and point X are the same as in Fig. 1). With this standard state, by definition, $\gamma_{NaCl,M} \to 1$ as $m_{NaCl} \to 0$ and $a_{NaCl,M} = \gamma_{NaCl,M} m_{NaCl} / \gamma^\circ_{NaCl,M} m^\circ_{NaCl}$, where $\gamma^\circ_{NaCl,M} = 1$ and $m^\circ_{NaCl} = 1$. Historically applied (mainly) to dilute solutions of aqueous electrolyte systems at "low" temperatures, and adopted routinely by investigators who prefer to express the concentrations of solute components in terms of molalities, the M standard state is increasingly being applied to undissociated solute components in "high-temperature," ideally dilute aqueous fluids. However, a detrimental characteristic of this standard state is that it does not allow a solute component to have a *finite compositional range* of Henrian behavior. Instead, as a consequence of the convention $\gamma_{i,M} \to 1$ as $m_i \to 0$, for all "real" ideally dilute fluids we have $\gamma_{i,M} = X_w$ (Appendix A) and, therefore, $a_{i,M} = X_w m_i$. Consequently, at point X, $a_{NaCl,M} = X_w = 0.982$ and $\mu_{NaCl} = \mu^\square_{NaCl} + RT \ln X_w = \mu^\square_{NaCl} - 0.018RT$ (see Fig. 1).

Alternative forms of these equations (with $d \ln K_D = d \ln K_{D,2}$) which properly account for $K_{\gamma,salts} = X_w$ are

$$\ln \gamma_{ab} = \int_0^{X_{an}} \left(\frac{X_{an}}{1 + X_{an}} \right) \left[d \ln K_{D,2} + d \ln \left(\frac{X_w}{55.51} \right) \right] \tag{24a}$$

and

$$\ln \gamma_{an} = \int_1^{X_{an}} \left(\frac{X_{an} - 1}{1 + X_{an}} \right) \left[d \ln K_{D,2} + d \ln \left(\frac{X_w}{55.51} \right) \right], \tag{24b}$$

where $d \ln(X_w / 55.51) = 1/X_w$.

Saxena and Ribbe (1972)

Shortly after Orville (1972) presented his experimental data and thermodynamic theory for plagioclase–fluid ion-exchange equilibria, Saxena and Ribbe (1972) reevaluated Orville's data in accordance with alternative interpretations of the crystallographic variations induced by $NaSi \rightleftharpoons CaAl$ substitution in high ("disordered") $NaAlSi_3O_8$–$CaAl_2Si_2O_8$ plagioclases (see discussion below). In the course of their analysis, and apparently as a matter of personal preference, Saxena and Ribbe also used pseudobinary mole fractions rather than molalities to represent the concentrations of aqueous salt components. With an H standard state for these components (which is implicit in Saxena and Ribbe's thermodynamic treatment), this approach leads to a third definition of the equilibrium constant for plagioclase–fluid ion-exchange

$$K_{E,3} = \frac{\gamma_{an}X_{an}(\gamma_{NaCl,H}X_{Na})^2}{(\gamma_{ab}X_{ab})^2\gamma_{CaCl_2,H}X_{Ca}}$$

(Table 1) in which $\gamma_{NaCl,H}$ and $\gamma_{CaCl_2,H} = 1.0$ for ideally dilute $NaCl$–$CaCl_2$–H_2O fluids.[4] Therefore, written in terms of notation used in this paper, Saxena and Ribbe's equation (11) is

$$\ln K_{E,3} = \ln K_{D,3} + \ln K_{\gamma,plag}, \tag{25}$$

where $\ln K_{D,3} = X_{an}X_{Na}^2/X_{ab}^2X_{Ca}$. However, from Eqs. (11a) and (11b), and with the definition $a_i = \gamma_i X_i$ for crystalline components, it is evident that Eq. (13) may also be written as

$$\ln K_{E,1} = \ln K_{E,3} + \ln(1 - X_w). \tag{26}$$

It has already been demonstrated that $\ln K_{E,1}$ is a constant at any given P and T, so Eq. (26) proves that $\ln K_{E,3} = f(X_w)$ under these conditions. This result is significant because, as noted previously, (1) the nature of plagioclase–fluid ion-exchange precludes $X_w = k$, and (2) in any strictly valid thermodynamic treatment of plagioclase–fluid ion-exchange data, the equilibrium constant must be absolutely constant at any given P and T. Consequently, like $K_{E,2}$, $K_{E,3}$ is not a completely legitimate equilibrium constant for plagioclase–fluid ion-exchange, and substituting $d \ln K_{D,3}$ for $d \ln K_{D,1}$ in Eqs. (18a) and (18b) must lead to numerical errors in calculated values of γ_{ab} and γ_{an}. Alternative forms of these equations (with $d \ln K_D = d \ln K_{D,3}$) that properly account for the thermodynamic effects of variable X_w during plagioclase–fluid ion-exchange are

$$\ln \gamma_{ab} = \int_0^{X_{an}} \left(\frac{X_{an}}{1 + X_{an}} \right) [d \ln K_{D,3} + d \ln(1 - X_w)] \tag{27a}$$

[4] X_{Na} (this paper) = X_{NaCl} (Saxena and Ribbe, 1972) = Na/Na + Ca (atomic ratio) in the fluid phase. Therefore, also, $X_{Ca} = 1 - X_{Na} = X_{CaCl_2}$ (Saxena and Ribbe, 1972) = Ca/Na + Ca (atomic ratio) in the fluid phase.

and

$$\ln \gamma_{an} = \int_1^{X_{an}} \left(\frac{X_{an} - 1}{1 + X_{an}} \right) \left[d\ln K_{D,3} + d\ln(1 - X_w) \right], \qquad (27b)$$

where $d\ln(1 - X_w) = -1/(1 - X_w)$.

A Direct Comparison between the Different Methods for Analyzing Plagioclase–Fluid Ion-Exchange Data

In view of the preceding discussions, it is instructive to make a direct comparison between the procedures for analyzing plagioclase–fluid ion-exchange data described by Orville (1972), Saxena and Ribbe (1972), and the present investigators. In deriving equations for calculating $\ln \gamma_{ab}$ and $\ln \gamma_{an}$ from such data [Eqs. (18a) and (18b)], it has already been demonstrated that expressing the concentrations of aqueous salt components as "true" mole fractions leads to a definition of $\ln K_D$, here denoted $\ln K_{D,1}$, which completely accounts for the thermodynamic effects of variable X_w during ion-exchange.[5] On the other hand, if the amounts of salt components are specified in terms of molalities or pseudobinary mole fractions—i.e., if $\ln K_D$ is defined as either $\ln K_{D,2}$ or $\ln K_{D,3}$, then a proper accounting for variable X_w requires an additional term in the resulting $\ln \gamma_{ab}$ and $\ln \gamma_{an}$ expressions; viz., $d\ln(X_w/55.51)$ in Eqs. (24a) and (24b) where $d\ln K_D = d\ln K_{D,2}$, and $d\ln(1 - X_w)$ in Eqs. (27a) and (27b) where $d\ln K_D = d\ln K_{D,3}$. Therefore, the task at hand is to examine the thermodynamic consequences of assuming that (1) $d\ln(X_w/55.51) = 0$ in Eqs. (24a) and (24b), and (2) $d\ln(1 - X_w) = 0$ in Eqs. (27a) and (27b). These are tacit assumptions in the thermodynamic methods presented by Orville (1972) and Saxena and Ribbe (1972), respectively.

A valid and meaningful comparison between the different approaches can be devised by using a single set of plagioclase–fluid ion-exchange data to obtain least-squares fit $\ln K_{D,1}$, $\ln K_{D,2}$, and $\ln K_{D,3}$ equations of *identical* form. The ion-exchange data listed by Orville (1972, Table 4) can be employed for this purpose, and values of $\ln K_{D,1}$, $\ln K_{D,2}$, and $\ln K_{D,3}$ derived from Orville's data are listed in Table 2 and illustrated in Fig. 4. We have used these $\ln K_D$ values to obtain least-squares fit equations of the form $\ln K_D = a + bX_{an}^3$, and Fig. 4 also shows the three different sets of $\ln K_D$ versus X_{an} relations (curves)

---

[5] Another advantage of employing "true" mole fractions to represent salt concentrations is that, as long as the aqueous fluids remain ideally dilute, $\ln K_{D,1}$ ion-exchange data are not affected by varying the normality (total chloride molality) of the fluids. Consequently, it is unnecessary to derive $(Ca/Na + Ca)_{plag}$ versus $(Ca/Na + Ca)_{salt}$ relations for different normalities of (hypothetical) ideally dilute $NaCl$–$CaCl_2$–H_2O fluids in order to determine whether experimental fluids are ideally dilute over their range of compositions [cf. Orville, 1972, Table 3 and Fig. 10].

Table 2. Selected plagioclase–fluid ion-exchange experimental data obtained at 2 kbars, 700°C with 2.0N NaCl–CaCl$_2$–H$_2$O solutions (Orville, 1972, Table 4).[a]

Run No.	X_{an}	m_{CaCl_2}	$\ln K_{D,1}$[b]	$\ln K_{D,2}$	$\ln K_{D,3}$
90-66	0.032	0.0096	− 1.4158	2.6360	1.9477
89-66	0.035	0.0089	− 1.2428	2.8090	2.1203
88-66	0.058	0.0190	− 1.4682	2.5834	1.8998
47-66	0.108	0.0300	− 1.2165	2.8349	2.1569
44-63	0.139	0.0580	− 1.6108	2.4401	1.7764
40-63	0.183	0.0690	− 1.4278	2.6229	1.9649
18-69	0.183	0.0800	− 1.5993	2.4512	1.7989
20-69	0.230	0.0800	− 1.2523	2.7983	2.1460
24-66	0.298	0.1420	− 1.5206	2.5289	1.9094
64-63	0.313	0.1510	− 1.5107	2.5387	1.9240
79-63	0.321	0.1670	− 1.6005	2.4486	1.8426
31-63	0.349	0.2010	− 1.7007	2.3478	1.7605
63-63	0.372	0.1820	− 1.4190	2.6298	2.0321
10-68	0.526	0.3170	− 1.4232	2.6233	2.1027
61-63	0.530	0.2910	− 1.2388	2.8081	2.2722
46-66	0.564	0.3360	− 1.3006	2.7455	2.2363
58-63	0.650	0.3860	− 1.0137	3.0316	2.5528
11-63	0.691	0.4170	− 0.8836	3.1610	2.7017
27-63	0.768	0.4880	− 0.6205	3.4229	3.0095
60-63	0.796	0.4850	− 0.3097	3.7338	3.3184
18-63	0.883	0.6200	0.0548	4.0959	3.7738
145-63	0.932	0.7330	0.3228	4.3620	4.1253
28-63	0.960	0.8610	− 0.0506	3.9863	3.8562

[a] Results obtained from run No. 43-63 are aberrant, so data pertaining to this run are not reproduced here. There are also minor discrepancies between the data listed in Tables 1 and 4 of Orville (1972), as well as some typographical errors in each of these tables (Orville, personal communication, 1977). However, due to their negligible numerical effects, these errors have been ignored in preparing this table.
[b] The definitions of $\ln K_{D,1}$, $\ln K_{D,2}$, and $\ln K_{D,3}$ are listed in Table 1. Values for m_{NaCl} (to calculate $\ln K_{D,2}$) are obtained from the tabulated values of m_{CaCl_2} according to the relation (for 2.0N NaCl–CaCl$_2$–H$_2$O solutions): $m_{NaCl} = 2(1 − m_{CaCl_2})$. X_{NaCl} and X_{CaCl_2} (to calculate $\ln K_{D,1}$), and X_{Na} and X_{Ca} (to calculate $\ln K_{D,3}$) may then be derived from the values of m_{NaCl} and m_{CaCl_2} using Eqs. (B3), (B4), (B9), and (B10), respectively, in Appendix B.

derived from these equations. One of the principal reasons for selecting the form $\ln K_D = a + bX_{an}^3$ is that differentiation yields the simple result $d\ln K_D = 3bX_{an}^2$ and, consequently, from Eqs. (18a) and (18b) we obtain

$$\ln \gamma_{ab} = 3b\left[\frac{X_{an}^3}{3} - \frac{X_{an}^2}{2} + X_{an} - \ln(X_{an} + 1) \right] \qquad (28a)$$

and

$$\ln \gamma_{an} = 3b\left[\frac{X_{an}^3 - 1}{3} - X_{an}^2 + 2X_{an} - 1 - 2\ln\left(\frac{X_{an} + 1}{2} \right) \right]. \qquad (28b)$$

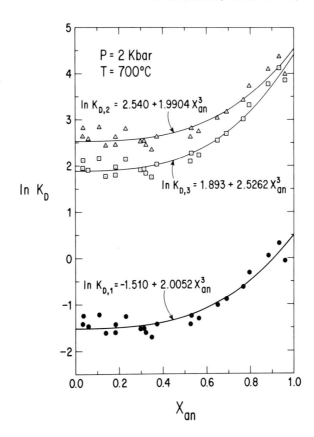

Fig. 4. Selected plagioclase–fluid ion-exchange experimental data (Orville, 1972, Table 4) illustrated as values of $\ln K_{D,1}$ (filled circles), $\ln K_{D,2}$ (open triangles), and $\ln K_{D,3}$ (open squares). All experimental data were obtained at 2 kbars, 700°C with 2.0N NaCl–CaCl$_2$–H$_2$O solutions (Table 2). The concave-upward curves represent values of $\ln K_{D,1}$, $\ln K_{D,2}$, and $\ln K_{D,3}$ calculated from least-squares fit equations of the form $\ln K_D = a + b X_{an}^3$. The curves for $\ln K_{D,1}$ and $\ln K_{D,2}$ are nearly parallel, but note that the curve for $\ln K_{D,3}$ has distinctly steeper positive slopes—especially in the range $0.5 \leqslant X_{an} \leqslant 1.0$. The thermodynamic consequences of these relationships are explained in the text.

Therefore, numerical discrepancies between calculated values of γ_{ab} and γ_{an} that result from employing the different definitions of $\ln K_D$ can be determined at any given value of X_{an} from the equalities

$$\frac{\ln \gamma_{i,1}}{\ln \gamma_{i,2}} = \frac{b_1}{b_2} \tag{29}$$

and

$$\frac{\ln \gamma_{i,1}}{\ln \gamma_{i,3}} = \frac{b_1}{b_3}, \tag{30}$$

where the subscripts 1, 2, and 3 designate quantities pertaining to the $\ln K_{D,1}$, $\ln K_{D,2}$, and $\ln K_{D,3}$ equations, respectively.

From Fig. 4 we have for Eqs. (29) and (30): $b_1 = 2.0052$, $b_2 = 1.9904$, $b_3 = 2.5262$, and, therefore, $b_1/b_2 = 1.0074$ and $b_1/b_3 = 0.7938$. Because b_1/b_2 in Eq. (29) is very nearly equal to one (this is manifested graphically by the near parallelism of the $\ln K_{D,1}$ and $\ln K_{D,2}$ curves in Fig. 4), it is evident that $\gamma_{i,1}$ and $\gamma_{i,2}$ must be very similar at all values of X_{an}. This result indicates that expressing salt concentrations as molalities almost completely accounts for the thermodynamic effects of variable X_w during ion-exchange, and that there are inconsequential numerical effects of assuming that $d\ln(X_w/55.51) = 0$ in Eqs. (24a) and (24b). On the other hand, b_1/b_3 in Eq. (30) is substantially less than unity, so $\gamma_{i,1}$ and $\gamma_{i,3}$ can be significantly different at particular values of X_{an}; for example, at $X_{an} = 0.0$ Eqs. (28b) and (30) yield $\gamma_{an,1} = 1.3752$ and $\gamma_{an,3} = 1.4939$, and at $X_{an} = 1.0$ Eqs. (28a) and (30) give $\gamma_{ab,1} = 2.3240$ and $\gamma_{ab,3} = 2.8933$. These differences demonstrate that significant errors in calculated values of γ_{ab} and γ_{an} can result from the assumption that $d\ln(1 - X_w) = 0$ in Eqs. (27a) and (27b), and this can be attributed to the fact that specifying the amounts of aqueous salt components in terms of pseudobinary mole fractions fails to account for *any* of the thermodynamic effects of variable X_w during plagioclase–fluid ion-exchange.

Reevaluation of Orville's Data

Figure 5 compares γ_{ab} versus X_{an} and γ_{an} versus X_{an} relations for Ab–An plagioclases at 2 kbars, 700°C obtained from the results presented by Orville (1972, Table 5), Saxena and Ribbe [1972, Eqs. (2) and (3), with $A_0 = 967$ cal/gfw, $A_1 = 715$ cal/gfw, $A_2 = 0$, component A = $CaAl_2Si_2O_8$, and component B = $NaAlSi_3O_8$], and the current investigators [Eqs. (28a) and (28b), with $b = 2.0052$]. The dashed curves, which represent the relations obtained *graphically* by Orville, include segments that are horizontal (indicating ideally dilute behavior) in the ranges: (1) $0.00 \leqslant X_{an} \leqslant 0.55$, where $\gamma_{ab} = 1.000$ and $\gamma_{an} = 1.276$; and (2) $0.89 \leqslant X_{an} \leqslant 1.00$, where $\gamma_{ab} = 1.893$ and $\gamma_{an} = 1.000$. According to Orville, these two compositional ranges comprise two separate isostructural feldspar series: a "high albite" series in the first range, bounded at its Ca-rich limit ($X_{an} = 0.55$) by the transition from a 7-Å to a 14-Å unit cell; and an "ordered anorthite" series in the second range. Furthermore, between these two distinct isostructural series—i.e., in the "transitional" range $0.55 < X_{an} < 0.89$ where $1.000 < \gamma_{ab} < 1.893$ and $1.276 > \gamma_{an} > 1.000$, the feldspars have a 14-Å unit cell and exhibit "intermediate" Al–Si ordering. These crystallographic and thermodynamic interpretations prompted Orville to propose that the separate "high albite" and "ordered anorthite" series could be treated as ideal binary crystalline solution series, each with one "real" and one "fictive" end-member component. Thus, the γ_{ab} versus X_{an} and γ_{an} versus X_{an} relations

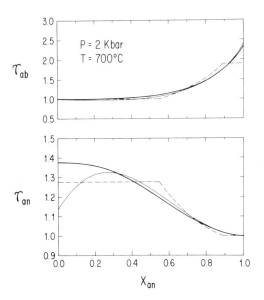

Fig. 5. Activity coefficients (γ_{ab} and γ_{an}) for $NaAlSi_3O_8-CaAl_2Si_2O_8$ plagioclase crystalline solutions at 2 kbars, 700°C: dashed curves—Orville (1972); light continuous curves—Saxena and Ribbe (1972); heavy continuous curves—this study. See text for explanations of the discrepancies between the three sets of γ_{ab} versus X_{an} and γ_{an} versus X_{an} relations.

presented by Orville are based on a partly inferred scheme of structural variations in the high plagioclase series.

Orville's approach was subsequently criticized by Saxena and Ribbe (1972) who argued that the high plagioclase series is continuous structurally. Accordingly, they reevaluated Orville's data by fitting a single least-squares $\ln K_D$ equation to all of the data [Saxena and Ribbe, 1972, Eq. (13) and Fig. 1b].[6] Significantly, and due to the disposition of Orville's ion-exchange data for Na-rich plagioclases, this equation indicates a minimum $[(\partial \ln K_D/\partial X_{an})_{P,T} = 0]$ in plagioclase–fluid $\ln K_D$ versus X_{an} relations near $X_{an} = 0.3$. The thermodynamic consequences of this minimum for deriving mixing properties of Ab–An feldspars are evident in the calculated γ_{ab} versus X_{an} and γ_{an} versus X_{an} relations represented by the light, continuous curves in Fig. 5. Although not readily discernible in Fig. 5, γ_{ab} values calculated from Eqs. (2) and (3) of Saxena and Ribbe (1972) are slightly less than unity in the range $0.00 < X_{an} \leqslant 0.44$, whereas in this same compositional range Fig. 5 also shows that γ_{an} values are continuously greater than unity. Such behavior of mixing properties of solutions has been described as "unsymmetrical" by Guggenheim (1967,

---

[6] Because $\ln K_D$ (Saxena and Ribbe) = $\ln K_{D,3}$ (this paper), we cannot explain the discrepancies between the $\ln K_D$ values plotted in Fig. 1b of Saxena and Ribbe (1972) and the $\ln K_{D,3}$ values listed in Table 2 and illustrated in Fig. 4 in this paper. Nevertheless, the trends of the two sets of data are only slightly different, and subsequent discussion would not be affected by resolution of this inconsistency.

pp. 203–205). However, in view of the results of more recent plagioclase–fluid ion-exchange experiments (Seil and Blencoe, 1979, Fig. 6) it is highly unlikely that γ_{ab} and γ_{an} actually exhibit this behavior at 2 kbars, 700°C. It should be recognized that this is not a trivial matter because, in contrast to the minimal effects on calculated values of γ_{ab}, the unsymmetrical behavior of the $\ln K_D$ equation derived by Saxena and Ribbe leads to a pronounced maximum in calculated γ_{an} versus X_{an} relations near $X_{an} = 0.3$, accompanied by rapid decreases in γ_{an} as $X_{an} \to 0$ (Fig. 5). This unusual pattern of γ_{an} versus X_{an} relations is partly an artifact of the thermodynamic methods employed by Saxena and Ribbe, but the ultimate source (in our opinion) is experimental error in Orville's ion-exchange data for Na-rich plagioclases.

In our own analysis of Orville's data, we have adopted an approach which precludes unsymmetric behavior of calculated mixing properties for the feldspars; that is, our form of an $\ln K_D$ equation, $\ln K_{D,1} = a + bX_{an}^3$ (Fig. 4), prevents the occurrence of a minimum in calculated $\ln K_{D,1}$ versus X_{an} relations (note also that, with this type of $\ln K_{D,1}$ expression, $d\ln K_{D,1} = 0.0$ at $X_{an} = 0.0$). Least-squares regression analysis of the $\ln K_{D,1}$ data in Table 2 yields

$$\ln K_{D,1} = -1.510 + 2.0052X_{an}^3 \qquad (31)$$

and resulting γ_{ab} versus X_{an} and γ_{an} versus X_{an} relations derived from Eqs. (28a) and (28b) with $b = 2.0052$ are represented by the heavy, continuous curves in Fig. 5. Figure 4 shows that Eq. (31) provides an excellent fit to the $\ln K_{D,1}$ values obtained from Orville's experimental data; furthermore, the *trend* of calculated $\ln K_{D,1}$ versus X_{an} relations is in good agreement with new 2 kbars, 700°C plagioclase–fluid ion-exchange data obtained recently by the writers (Fig. 6; Seil *et al.*, in preparation).

The R, "Molten–Solute" Standard State

The foregoing discussions of plagioclase–fluid ion-exchange equilibria provide a foundation for assessing the relative merits of different solute standard states as they apply to undissociated solute components in thermodynamic representations of crystal–fluid (heterogeneous) phase equilibria. This assessment is offered here bearing in mind that, in certain instances, thermodynamic modeling of crystal–fluid equilibria is prompted by a need to quantitatively represent the distribution of components that are *shared* between crystals and coexisting ideally dilute fluid (cf. Frantz *et al.*, 1981). To achieve this goal it is necessary to evaluate not only γ_i for each "shared" component, but also $\Delta\mu_i^\circ$ —the difference between the standard-state chemical potentials (free energies) of the pure component in the crystalline and "fluid" states. Accordingly, it is desirable to select a standard state for solute components which facilitates quantitative representations of both γ_i and $\Delta\mu_i^\circ$.

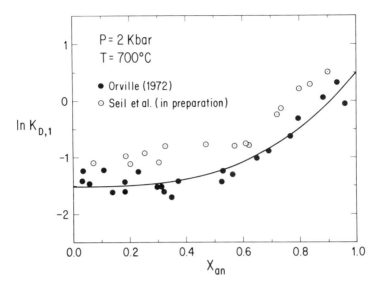

Fig. 6. Plagioclase–fluid ion-exchange ($\ln K_{D,1}$) data obtained at 2 kbars, 700°C with 2.0N NaCl–CaCl$_2$–H$_2$O solutions: filled circles—Orville (1972); open circles—Seil and Blencoe (1979), and Seil *et al.* (in preparation). Note that scatter in Orville's data for Na-rich plagioclases permits a minimum $[(\partial \ln K_{D,1}/\partial X_{an})_{P,T} = 0]$ in the trend of $\ln K_{D,1}$ versus X_{an} near $X_{an} = 0.3$, but that no such minimum is evident from the data obtained by Seil *et al.* Therefore, the somewhat more precise ion-exchange data of Seil *et al.* do not support the conclusion that Ab–An plagioclases exhibit "unsymmetrical" behavior at 2 kbars, 700°C (cf. Saxena and Ribbe, 1972). On the other hand, the trends of the Orville and Seil *et al.* data are obviously parallel within experimental error, so activity coefficients derived separately (but in a similar way, graphically or numerically) from each set of data would be nearly identical. The offset but parallel trends of the two sets of data may be due to small, systematic differences between the pressures (ostensibly, 2 kbars) at which the experiments were conducted.

Of course, even for the somewhat special application presently under consideration, any of the three principal types of standard states—herein denoted the *R*, *M*, and *H* standard states—can legitimately be applied to the solute components. Nevertheless, it should be recognized that the *M* and *H* standard states are not well suited for this purpose. The conventions $\gamma_i \to 1$ as $m_i \to 0$ (*M* standard state) and $\gamma_i \to 1$ as $X_i \to 0$ (*H* standard state) ensure that, no matter what the degree of nonideality of a solute component *i*, γ_i will not deviate greatly from unity in dilute solutions. However, as a consequence of this "simplified" numerical behavior of activity coefficients, in their standard states (Fig. 1) solute components possess *hypothetical* thermochemical properties. This result is evident from the relations

$$\mu_i^{\square} = \mu_i^{\circ} + RT \ln \gamma_{i,R}^* - 4.017RT \tag{32}$$

and

$$\mu_i^* = \mu_i^{\circ} + RT \ln \gamma_{i,R}^* \tag{33}$$

which show that the M and H standard states combine two constants—μ_i° and $\gamma_{i,R}^*$—that are completely unrelated to one another (the first, μ_i°, refers to a state of pure component i, whereas the second, $\gamma_{i,R}^*$, pertains to the mixing properties of component i). In other words, in these standard states an important numerical consequence of nonideal mixing of a solute component i ($\gamma_{i,R}^*$) is "superimposed" upon the value of an R standard state chemical potential (μ_i°).

Although it is only infrequently utilized for such a purpose, the R standard state is readily applied to undissociated solute components in thermodynamic representations of crystal–fluid equilibria. It should be recognized, first of all, that μ_i° is fundamentally different from the corresponding standard-state chemical potentials of the M and H standard states—$\mu_i^\square$ and μ_i^*, respectively (Fig. 1): μ_i° is a nonhypothetical definition of the standard-state chemical potential of component i because, with an R standard state, the chemical potential of component i approaches that of pure i as $X_i \to 1$. Therefore, for example, in Fig. 1 where $\mu_i^\circ = \mu_{NaCl}^\circ$, we note that μ_{NaCl}° is the free energy of pure NaCl in a physical state that can be envisioned as an end member of a continuum of "structures" (and compositions) that extends to pure H_2O. Furthermore, in this situation it is both reasonable and compelling to specify that μ_{NaCl}° represents the free energy of pure *molten* NaCl, despite the fact that this is a metastable state for NaCl at the P–T conditions under consideration (2 kbars, 700°C).[7] Therefore, in this example, the free energy–composition relations of the molten NaCl–H_2O mixtures can be represented by a single, "isostructural" ΔG_f–X_{NaCl} equation of state. It should now be evident that this approach is particularly advantageous for silicate($\pm$salt)–H_2O systems because, at hypersolidus temperatures where solute components can also be major components of stable melts, only a single equation of state is required to represent the mixing properties of the components in the melt and vapor phases.

Still another advantageous feature of the R standard state is that the definition of $\gamma_{i,R}$ provides a conceptually realistic representation of the mixing properties of a solute component. In the ideally dilute range, values of $\gamma_{i,R}$ are fixed (Henrian) and significantly less than or greater than unity depending on whether the solute exhibits positive or negative deviations from Raoult's law behavior. Therefore, the values of $\gamma_{i,R}$ distinctly reflect both the "direction" and magnitude of nonideal behavior of solute components. This characteristic is in marked contrast to the M and H standard states, both of which tend to obscure the differences between the mixing properties of solute components because $\gamma_{i,M} \to 1$ and $\gamma_{i,H} \to 1$ as $X_w \to 1$.

We conclude that there is no intrinsic advantage to be gained by applying either the M or H standard states to undissociated solute components in thermodynamic representations of crystal–fluid equilibria. Furthermore, if

---

[7]We assume that there are no first-order phase transitions in mixtures of (metastable) molten NaCl and supercritical H_2O. Consequently, undissociated NaCl in a dilute NaCl–H_2O fluid is treated as a trace amount of NaCl melt dissolved in the aqueous fluid.

instead an R, "molten solute" standard state is applied to these components, then it is possible to take advantage of a rapidly growing body of information on the free energies of fusion of components that comprise the compositions of many common rock-forming minerals (e.g., Burnham, 1979). We will demonstrate this point by showing how values for the free energy of fusion (ΔG_i^{am}) of $NaAlSi_3O_8$ and $CaAl_2Si_2O_8$ can be employed to calculate the amounts of these components in dilute aqueous fluids equilibrated with Ab–An plagioclases.

Crystal–Fluid Equilibria in the System $NaAlSi_3O_8$–$CaAl_2Si_2O_8$–H_2O

The general applicability of thermodynamic procedures described in this paper can be demonstrated by calculating the compositions of aqueous fluids in equilibrium with $NaAlSi_3O_8$–$CaAl_2Si_2O_8$ feldspars at 2 kbars, 700°C. This exercise will exemplify the practicality of employing pseudobinary mole fractions and a "molten solute" standard state to develop equations for the chemical potentials of undissociated solute components in ideally dilute aqueous fluids. In treating the plagioclase–fluid equilibria, we will adopt the ad hoc assumption that the compositions of the fluids can be described completely in terms of the components $NaAlSi_3O_8$, $CaAl_2Si_2O_8$, and H_2O. Thus, it is assumed that feldspar components are "shared" by the feldspars and fluids.

Under the conditions stipulated above, equilibrium between coexisting plagioclase and fluid is defined thermodynamically by the chemical potential equivalence equations $\mu_{ab} = \mu_{ab}^v$ and $\mu_{an} = \mu_{an}^v$, or

$$\mu_{ab}^\circ + RT \ln a_{ab} = \mu_{ab}^{\circ,v} + RT \ln a_{ab}^v \qquad (34)$$

and

$$\mu_{an}^\circ + RT \ln a_{an} = \mu_{an}^{\circ,v} + RT \ln a_{an}^v. \qquad (35)$$

In order to determine the conditions that satisfy Eqs. (34) and (35) for a given bulk composition in the ternary system, it is advantageous to express the compositions of fluids in terms of two mole fractions: X_w and a pseudobinary mole fraction X_{Ca}^v where

$$X_{Ca}^v = \frac{X_{an}^v}{X_{ab}^v + X_{an}^v} \,.$$

Therefore, written in a form similar to that of Eqs. (11a) and (11b) derived previously, equations for the activities of aqueous feldspar components in *ideally dilute* $NaAlSi_3O_8$–$CaAl_2Si_2O_8$–H_2O fluids are

$$a_{ab}^v = \gamma_{ab}^v (1 - X_w)(1 - X_{Ca}^v) \qquad (36)$$

and

$$a_{an}^v = \gamma_{an}^v (1 - X_w) X_{Ca}^v. \qquad (37)$$

Substituting Eqs. (36) and (37) into Eqs. (34) and (35), respectively, and rearranging, we obtain

$$\frac{\mu_{ab}^{o,v} - \mu_{ab}^{o}}{RT} = \ln a_{ab} - \ln \gamma_{ab}^{v} - \ln(1 - X_w) - \ln(1 - X_{Ca}^{v}) \tag{38}$$

and

$$\frac{\mu_{an}^{o,v} - \mu_{an}^{o}}{RT} = \ln a_{an} - \ln \gamma_{an}^{v} - \ln(1 - X_w) - \ln X_{Ca}^{v}. \tag{39}$$

Note also that, because a "molten solute" standard state has been adopted for aqueous feldspar components, we have $\mu_{ab}^{o,v} - \mu_{ab}^{o} = \Delta G_{ab}^{am}$ and $\mu_{an}^{o,v} - \mu_{an}^{o} = \Delta G_{an}^{am}$, where ΔG_{ab}^{am} and ΔG_{an}^{am} are the free energies of fusion for the $NaAlSi_3O_8$ (analbite) and $CaAl_2Si_2O_8$ (high anorthite) components, respectively.

An explicit equation for calculating X_{Ca}^{v} of fluids in equilibrium with Ab–An plagioclases can now be obtained by subtracting Eq. (39) from Eq. (38), which after rearranging terms yields

$$X_{Ca}^{v} = \frac{\exp(F_1)}{1 + \exp(F_1)}, \tag{40}$$

where

$$F_1 = \frac{\Delta G_{ab}^{am} - \Delta G_{an}^{am}}{RT} + \ln\left(\frac{a_{an}}{a_{ab}}\right) + \ln\left(\frac{\gamma_{ab}^{v}}{\gamma_{an}^{v}}\right). \tag{41}$$

An equation for calculating X_w of the fluids is similarly derived by adding Eqs. (38) and (39), which following algebraic manipulation gives

$$X_w = 1 - \exp(F_2), \tag{42}$$

where

$$F_2 = \frac{1}{2}\left[-\frac{\Delta G_{ab}^{am} + \Delta G_{an}^{am}}{RT} + \ln(a_{ab}a_{an}) - \ln(\gamma_{ab}^{v}\gamma_{an}^{v}) - \ln\left[X_{Ca}^{v}(1 - X_{Ca}^{v})\right]\right]. \tag{43}$$

To solve Eqs. (40) and (42), it is necessary to have values for ΔG_{ab}^{am}, ΔG_{an}^{am}, a_{ab}, a_{an}, γ_{ab}^{v}, and γ_{an}^{v}. The first two quantities are derived from the $\Delta G_{ab}^{am}/RT$ and $\Delta G_{an}^{am}/RT$ curves presented by Burnham (1979, Figs. 16-5 and 16-8), which give at 2 kbars, 700°C: $\Delta G_{ab}^{am}/RT = 2.1704$ and $\Delta G_{an}^{am}/RT = 4.8661$. From the relation $a_i = \gamma_i X_i$, it is evident that the activities of the Ab–An plagioclase components—a_{ab} and a_{an}—can be obtained from the $\ln \gamma_{ab}$ and $\ln \gamma_{an}$ equations listed earlier [Eqs. (28a) and (28b) with $b = 2.0052$]. Finally, although γ_{ab}^{v} and γ_{an}^{v} cannot be evaluated from available experimental data, each must be $\gg 1.0$, and it is reasonable to assume that they are approximately equal with values somewhere in the range 10–50. Therefore, with all terms in Eqs. (40) and (42) as known quantities, calculations of X_w and X_{Ca}^{v} proceed as follows: (1) select a composition (X_{an}) for plagioclase, (2) calculate X_{Ca}^{v} from Eq. (40), and (3) substitute the value of X_{Ca}^{v} into Eq. (42) and solve for X_w. Note that

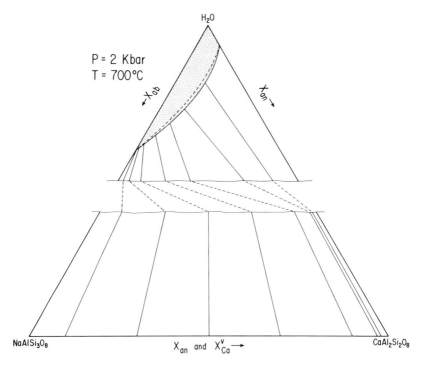

Fig. 7. A *schematic* vaporus for the $NaAlSi_3O_8$–$CaAl_2Si_2O_8$–H_2O system at 2 kbars, 700°C. The light straight lines, connected by dashed lines in the center of the diagram, represent crystal–fluid tie-lines which show that aqueous fluids preferentially dissolve $NaAlSi_3O_8$ from $NaAlSi_3O_8$–$CaAl_2Si_2O_8$ plagioclases (Adams, 1968). (For clarity, the proportions of $CaAl_2Si_2O_8$ in the fluids have been *greatly exaggerated*.) The dashed curve within the vaporus illustrates the limits of this phase field if the plagioclases are assumed to be ideal. The effects of this assumption are inconsequential numerically, thus indicating that the result $X_{ab}^v \gg X_{an}^v$ for the fluids is not due to the positive deviations from ideality exhibited by the feldspars. Instead, $X_{ab}^v \gg X_{an}^v$ is attributable to the fact that $\Delta G_{ab}^{am} < \Delta G_{an}^{am}$ in Eqs. (40) and (42) (see text for explanation).

$\gamma_{ab}^v = \gamma_{an}^v$ is a constant value in the calculations because it is assumed that the fluids are ideally dilute.

Solution of Eqs. (40) and (42) for the full range of plagioclase compositions in the $NaAlSi_3O_8$–$CaAl_2Si_2O_8$–H_2O system at 2 kbars, 700°C yields the *schematic* vaporus field illustrated in Fig. 7. For *all* estimated values of $\gamma_{ab}^v = \gamma_{an}^v$ in the range 10–50, the calculations indicate that $X_{ab}^v(ab-w) \gg X_{an}^v(an-w)$ for the fluids. This qualitative result is consistent with the experimental data of Adams (1968) who found that the solubility of albite in H_2O is much greater than that of anorthite: 0.3 wt% versus 0.09 wt%, respectively, at 2 kbars, 700°C. However, as noted previously, we have here adopted the simplifying assumption that the compositions of fluids in equilibrium with Ab–An plagioclases can be expressed entirely in terms of the components $NaAlSi_3O_8$, $CaAl_2Si_2O_8$, and H_2O. This supposition is equivalent to stipulating that Ab–An plagioclases dissolve congruently in H_2O at 2 kbars, 700°C, which is not strictly true, at least for Na-rich plagioclases [Currie (1968) has shown that aqueous fluids equilibrated with albite are persilicic, and it is likely that this would also be the case for fluids in equilibrium with Ca-rich plagioclases]. It is therefore evident that *at least* four components— $NaAlSi_3O_8$, $CaAl_2Si_2O_8$, SiO_2, and H_2O—are required to accurately express the compositions of aqueous fluids in equilibrium with Ab–An plagioclases. Nevertheless, the qualitative result $X^v_{ab}(ab-w) \gg X^v_{an}(an-w)$ obtained in the present analysis is correct and verifies the general validity of our thermodynamic approach.

Conclusions

Thermodynamic evaluations of plagioclase–fluid equilibria in the system $NaAlSi_3O_8$–$CaAl_2Si_2O_8$–SiO_2–$NaCl$–$CaCl_2$–H_2O have shown that:

(1) By taking the pure (metastable) molten solute at system P and T as the standard state for aqueous salt components, activities for these components in ideally dilute $NaCl$–$CaCl_2$–H_2O fluids can be derived by a two-step thermodynamic "mixing" process in which (1) water is combined with a pure molten salt component, yielding a dilute saline solution, followed by (2) addition of the second molten salt component. Accordingly, formulated as generalized equations applicable to any ideally dilute solute(i)–solute(j)–H_2O(w) fluid, activities of the solute components can be expressed as arithmetic products of a binary activity and a pseudo-binary (constant X_w) mole fraction—viz.,

$$a_i = (a_{i-w})_{j=0} \left[\frac{X_i}{(X_i + X_j)} \right]_{X_w}$$

and

$$a_j = (a_{j-w})_{i=0} \left[\frac{X_j}{(X_i + X_j)} \right]_{X_w},$$

where $(a_{i-w})_{j=0} = \gamma_i(1 - X_w)$ and $(a_{j-w})_{i=0} = \gamma_j(1 - X_w)$. However, written in this way, and because $X_i + X_j + X_w = 1$, it is apparent that a_i and a_j must be exactly equivalent *numerically* to the ternary activities $a_i = \gamma_i X_i$ and $a_j = \gamma_j X_j$.

(2) The equilibrium constant for plagioclase–fluid ion-exchange should be defined with the proportions of aqueous salt components represented by mole fractions. Expressing the concentrations of these components in terms of either molalities or pseudobinary mole fractions yields an (apparent) equilibrium constant that varies with X_w for the fluids.

(3) Due to variable X_w in plagioclase–fluid ion-exchange, equations of the form

$$\ln \gamma_{ab} = \int_0^{X_{an}} \left(\frac{X_{an}}{1 + X_{an}} \right) d \ln K_D$$

and

$$\ln \gamma_{an} = \int_1^{X_{an}} \left(\frac{X_{an} - 1}{1 + X_{an}} \right) d \ln K_D$$

for deriving activity coefficients of Ab–An plagioclases are only strictly correct when K_D is defined with the concentrations of aqueous salt components expressed as mole fractions. Calculated values of γ_{ab} and γ_{an} are affected only slightly if salt molalities are used in the definition of K_D, but employing pseudobinary mole fractions for this purpose leads to significant errors in γ_{ab} and γ_{an} because, in effect, this approach *completely* ignores the thermodynamic consequences of variable X_w in the fluids.

(4) In their analysis of plagioclase–fluid ion-exchange equilibria, previous investigators have used molalities or pseudobinary mole fractions to represent the concentrations of salt components in $NaCl–CaCl_2–H_2O$ fluids. However, for the reasons presented in (2) and (3) above, we suggest that it is more correct and straightforward to employ ternary mole fractions to specify the proportions of aqueous salt components.

(5) An R (Raoultian), "molten solute" standard state allows thermodynamic data on free energies of fusion to be incorporated into "isostructural" equations of state for representing the thermodynamic properties of solute components in dilute aqueous fluids. Activity coefficients for these components can then be estimated with reasonable accuracy from experimental data on the solubilities of the components in aqueous fluids.

(6) Expressing the concentrations of undissociated solute components as pseudobinary mole fractions is advantageous for developing equations of state to calculate the distribution of components that are "shared" between coexisting crystals and aqueous fluid. This approach, along with an R, "molten solute" standard state for solute components, has been adopted to obtain a calculated vaporus for the $NaAlSi_3O_8–CaAl_2Si_2O_8–H_2O$ system at 2 kbars, 700°C. The calculations indicate $X_{ab}^v \gg X_{an}^v$ for fluids in equilibrium with Ab–An plagioclases, and this result is in accord with experimental data obtained by Adams (1968).

The applicability of thermodynamic theory presented in this paper extends well beyond analyses of subsolidus plagioclase–fluid equilibria in the system $NaAlSi_3O_8–CaAl_2Si_2O_8–SiO_2–NaCl–CaCl_2–H_2O$. In subsequent papers it will be demonstrated that similar approaches can be taken to analyze hypersolidus equilibria of feldspar($\pm$salt)–water and feldspar–quartz($\pm$salt)–water systems.

Appendix A: The Relationships between the Activity Coefficients $\gamma_{i,R}$ (R standard state), $\gamma_{i,M}$ (M standard state), and $\gamma_{i,H}$ (H standard state)

Following Denbigh (1968, pp. 270–278), we recognize three "conventional" methods for representing the chemical potential of a component i at a constant P and T:

(I) According to an R (Raoultian) standard state,

$$\mu_i = \mu_i^\circ + RT\ln(\gamma_{i,R}X_i), \qquad (A1)$$

where $\gamma_{i,R} \to 1$ as $X_i \to 1$.

(II) According to an M (1 molal) standard state,

$$\mu_i = \mu_i^\square + RT\ln(\gamma_{i,M}m_i), \qquad (A2)$$

where $\gamma_{i,M} \to 1$ as $m_i \to 0$.

(III) According to an H (Henrian) standard state,

$$\mu_i = \mu_i^* + RT\ln(\gamma_{i,H}X_i), \qquad (A3)$$

where $\gamma_{i,H} \to 1$ as $X_i \to 0$.

The R and H standard states can be applied to components in any physical state—crystalline, supercritical fluid, liquid, or gaseous, but the M standard state is employed solely for fluid or liquid components. For each of these standard states, the standard-state chemical potential [μ_i° (R standard state), $\mu_i^\square$ (M standard state), and μ_i^* (H standard state)] is a function *only* of P and T.

Regardless of the standard state adopted for a component i, μ_i is a fixed value at a given P, T, and bulk composition of the system. Therefore, the relationship between $\gamma_{i,R}$ and $\gamma_{i,H}$ can be obtained by (1) noting that $\gamma_{i,H} = 1$ in the range of ideally dilute (Henrian) behavior of component i, and (2) equating the right-hand sides of Eqs. (A1) and (A3), which gives

$$\mu_i^* - \mu_i^\circ = RT\ln(\gamma_{i,R}^*X_i) - RT\ln X_i = RT\ln\gamma_{i,R}^*, \qquad (A4)$$

where $\gamma_{i,R}^*$ is the Henry's law activity coefficient (R standard state) for a component i as $X_i \to 0$. From Eqs. (A1), (A3), and (A4), it is now evident that

$$\ln(\gamma_{i,R}X_i) = \ln\gamma_{i,R}^* + \ln(\gamma_{i,H}X_i) \qquad (A5)$$

or

$$\frac{\gamma_{i,R}}{\gamma_{i,H}} = \gamma_{i,R}^*. \qquad (A6)$$

The relationship between $\gamma_{i,M}$ and $\gamma_{i,H}$ has been shown by Denbigh [1968, Eq. (9 · 20)] to be

$$\frac{\gamma_{i,M}}{\gamma_{i,H}} = \frac{1000X_i}{M_0m_i} \qquad (A7)$$

where M_0 is the molecular weight of the solvent. Consequently, when water is the solvent we have $M_0 = 18.015$ and

$$\frac{\gamma_{i,M}}{\gamma_{i,H}} = \frac{55.51 X_i}{m_i} = \frac{55.51}{\sum \text{moles}} = X_w. \tag{A8}$$

Therefore, in the range of ideally dilute behavior where $\gamma_{i,H} = 1$, Eq. (A8) simplifies to $\gamma_{i,M} = X_w$. This result serves to emphasize that, even *within* the range of ideally dilute behavior of component i, $\gamma_{i,M} \neq 1$ except at *infinite dilution*.

Finally, from Eqs. (A6) and (A8) it is evident that

$$\frac{\gamma_{i,M}}{\gamma_{i,R}} = \frac{X_w}{\gamma_{i,R}^*}. \tag{A9}$$

Appendix B: The Relationships between $\ln K_{D,1}$, $\ln K_{D,2}$, and $\ln K_{D,3}$ for Plagioclase–Fluid Equilibria in the System $NaAlSi_3O_8–CaAl_2Si_2O_8–SiO_2–NaCl–CaCl_2–H_2O$

From Table 1 we have

$$\ln K_{D,1} = \ln\left(\frac{X_{an} X_{NaCl}^2}{X_{ab}^2 X_{CaCl_2}}\right) \tag{B1}$$

and

$$\ln K_{D,2} = \ln\left(\frac{X_{an} m_{NaCl}^2}{X_{ab}^2 m_{CaCl_2}}\right). \tag{B2}$$

To obtain the relationship between $\ln K_{D,1}$ and $\ln K_{D,2}$ we substitute the expressions

$$X_{NaCl} = \frac{n_{NaCl}}{n_{NaCl} + n_{CaCl_2} + 55.51} = \frac{m_{NaCl}}{m_{NaCl} + m_{CaCl_2} + 55.51} \tag{B3}$$

and

$$X_{CaCl_2} = \frac{n_{CaCl_2}}{n_{NaCl} + n_{CaCl_2} + 55.51} = \frac{m_{CaCl_2}}{m_{NaCl} + m_{CaCl_2} + 55.51} \tag{B4}$$

into Eq. (B1), which gives

$$\ln K_{D,1} = \ln\left(\frac{X_{an} m_{NaCl}^2}{X_{ab}^2 m_{CaCl_2}}\right) - \ln\left(m_{NaCl} + m_{CaCl_2} + 55.51\right). \tag{B5}$$

Therefore

$$\ln K_{D,1} = \ln K_{D,2} - \ln\left(m_{NaCl} + m_{CaCl_2} + 55.51\right) \tag{B6}$$

or, because $m_{NaCl} + m_{CaCl_2} + 55.51 = \sum moles = 55.51/X_w$,

$$\ln K_{D,1} = \ln K_{D,2} + \ln\left(\frac{X_w}{55.51}\right). \tag{B7}$$

From Table 1 we also have

$$\ln K_{D,3} = \ln\left(\frac{X_{an}X_{Na}^2}{X_{ab}^2 X_{Ca}}\right). \tag{B8}$$

To obtain the relationship between $\ln K_{D,3}$ and $\ln K_{D,2}$ we substitute the expressions

$$X_{Na} = \frac{n_{NaCl}}{n_{NaCl} + n_{CaCl_2}} = \frac{m_{NaCl}}{m_{NaCl} + m_{CaCl_2}} \tag{B9}$$

and

$$X_{Ca} = \frac{n_{CaCl_2}}{n_{NaCl} + n_{CaCl_2}} = \frac{m_{CaCl_2}}{m_{NaCl} + m_{CaCl_2}} \tag{B10}$$

into Eq. (B8), which yields

$$\ln K_{D,3} = \ln\left(\frac{X_{an}m_{NaCl}^2}{X_{ab}^2 m_{CaCl_2}}\right) - \ln(m_{NaCl} + m_{CaCl_2}) \tag{B11}$$

or

$$\ln K_{D,3} = \ln K_{D,2} - \ln(m_{NaCl} + m_{CaCl_2}). \tag{B12}$$

The relationship between $\ln K_{D,1}$ and $\ln K_{D,3}$ can be derived by subtracting Eq. (B12) from Eq. (B6), which gives

$$\ln K_{D,1} - \ln K_{D,3} = \ln\left(\frac{m_{NaCl} + m_{CaCl_2}}{m_{NaCl} + m_{CaCl_2} + 55.51}\right). \tag{B13}$$

However,

$$\frac{m_{NaCl} + m_{CaCl_2}}{m_{NaCl} + m_{CaCl_2} + 55.51} = X_{NaCl} + X_{CaCl_2} = 1 - X_w. \tag{B14}$$

Consequently, we have

$$\ln K_{D,1} = \ln K_{D,3} + \ln(1 - X_w). \tag{B15}$$

Finally, by subtracting Eq. (B15) from Eq. (B7) it is evident that the relationship between $\ln K_{D,3}$ and $\ln K_{D,2}$ [Eq. (B12)] can also be expressed as

$$\ln K_{D,3} = \ln K_{D,2} + \ln\left[\frac{X_w}{55.51(1 - X_w)}\right]. \tag{B16}$$

Acknowledgments

We thank Dr. A. C. Lasaga for numerous enlightening discussions of standard states for aqueous solute components. (However, commentary and opinions on this subject offered herein are entirely attributable to the authors, and, therefore, we must accept full responsibility for any errors of omission or commission.) Special thanks are extended to Dr. A. Navrotsky for suggesting ways to improve the original manuscript, and to Dr. S. K. Saxena for rendering several Solomon-like decisions regarding how best to change that manuscript. We would also like to acknowledge the cheerful assistance of Mrs. D. E. Detwiler who typed the manuscripts, and Mr. R. J. Texter who expertly drafted the figures. Financial support for our research was provided mainly by NSF Grant EAR-7822443 (awarded to J. G. Blencoe), but most of the costs of computer calculations and related activities were defrayed by The Pennsylvania State University.

References

Adams, J. B. (1968) Differential solution of plagioclase in supercritical water, *Amer. Mineral.* **53**, 1603–1613.

Blencoe, J. G. (1979) The use of thermodynamic excess functions in the Nernst distribution law: discussion, *Amer. Mineral.* **64**, 1122–1128.

Burnham, C. W. (1979) The importance of volatile constituents, in *The Evolution of the Igneous Rocks: Fiftieth Anniversary Perspectives*, edited by H. S. Yoder, Jr., pp. 439–482. Princeton University Press, Princeton, N.J.

Denbigh, K. (1968) *The Principles of Chemical Equilibrium*. Cambridge University Press, Cambridge.

Frantz, J. D., Popp, R. K. and Boctor, N. Z. (1981) Mineral-solution equilibria—V. Solubilities of rock-forming minerals in supercritical fluids, *Geochim. Cosmochim. Acta* **45**, 69–77.

Guggenheim, E. A. (1967) *Thermodynamics*. North-Holland, Amsterdam.

Orville, P. M. (1972) Plagioclase cation exchange equilibria with aqueous chloride solution: Results at 700°C and 2000 bars in the presence of quartz, *Amer. J. Sci.* **272**, 234–272.

Saxena, S. K. and Ribbe, P. H. (1972) Activity–composition relations in feldspars. *Contrib. Mineral. Petrol.* **37**, 131–138.

Seil, M. K. and Blencoe, J. G. (1979) Activity–composition relations of $NaAlSi_3O_8$–$CaAl_2Si_2O_8$ feldspars at 2 kb, 600–800°C. *Geol. Soc. Amer. Abstr. Progs.* **11**, 513.

Skippen, G. B. (1974) Thermodynamics of experimental sub-solidus silicate systems including mixed volatiles, *Fortschr. Mineral.* **52**, 75–99.

III. Thermodynamic Methods and Data

Chapter 7
Computation of Multicomponent Phase Equilibria

S. K. Saxena

Introduction

With the publication of thermodynamic data by Helgeson *et al.* (1978) and Robie *et al.* (1978), we have data on the Gibbs free energy of formation of many geologically important substances. It is, therefore, appropriate to consider suitable techniques of calculating multicomponent phase equilibria. Geochemists and petrologists are familiar with the method of equilibrium constants used in the calculation of equilibrium pressure and temperature for single or multiple reactions. This is the method which, for example, has been used by Helgeson *et al.* in deriving the free-energy data on many minerals. Once there is an accumulation of a sizable amount of such data, the method of minimization of total Gibbs free energy as described here may be used advantageously to calculate equilibrium assemblages. With this method, it is not necessary to write down any specific reactions and little chemical intuition or experience is needed to predict the course of complex equilibria. This method, although used by some geochemists (e.g., Grossman, 1972) in nebular condensation calculations, has not been given due attention by petrologists (but see Brown and Skinner (1974) and Nicholls (1977) for alternative methods).

The method employs linear algebraic techniques and is therefore particularly suitable for a computer. It was originally described by White *et al.* (1958) and reviewed recently by Van Zeggeren and Storey (1970) and Eriksson and Rosen (1973). The thermodynamics presented below follows largely the discussion presented by the latter authors.

It is appropriate to mention that Brown and Skinner (1974) have developed an alternative method which also serves the purpose of calculating multiple component equilibria. One reason these authors seemed to have been discouraged with the free-energy minimization method was the computation time required. With improved programming techniques, this problem could be overcome.

The thermodynamic relations which are used in computing the equilibrium assemblage with the minimum free energy are presented in the first part of the paper. This is followed by an estimation of the Gibbs free energy of formation of orthoenstatite and a discussion of Fe–Mg mixing in olivine. The Gibbs free energy of formation of some other minerals at high temperatures (> 1000 K), e.g., pyrope and the Al_2O_3–component in orthopyroxene are next determined from the experimental data of Lane and Ganguly (1980) and Perkins et al. (1981). It is possible then to present some examples of calculations in multicomponent systems using the method of free-energy minimization.

Symbols and Abbreviations

G	total Gibbs free energy of the system
n_i	number of moles of i
$\bar{G}_i$	$= \mu_i$, the chemical potential of i
T	absolute temperature
R	gas constant, $\gamma =$ activity coefficient
P	pressure
K	equilibrium constant
K_D	distribution coefficient
G_f°	Gibbs free energy of formation at 1 bar and T from elements in their standard state
G^g	total free energy of the gas phase at any T and P
G^s	total free energy of the solid pure phase at any T and P
X_i	mole fraction of component i
m_p	total number of components in the pth phase
Ann	annite
Phl	phlogopite
Ol	olivine
Opx	orthopyroxene
Sill	sillimanite
Bi	biotite
Q	quartz
Sp	spinel
Mt	magnetite
Py	pyrope

Thermodynamics

The total Gibbs free energy of a chemical system can be expressed as

$$G = \sum n_i G_i = \sum n_i \mu_i. \tag{1}$$

Let us consider a system with one ideal gas phase, q ideal liquid and solid solutions, and s pure condensed species. The total free energy is

$$G = G^g + G^q + G^s, \tag{2}$$

where

$$G^g = \sum n_i G_{fi}^{\circ} + \sum n_i (RT \ln P + RT \ln X_i) \qquad (3)$$

for the gas phase,

$$G^q = \sum_{p=1}^{q} \sum_{i=1}^{m_p} n_{pi} \left[(G_{fi}^{\circ})_{pi} + RT \ln X_i \right] \qquad (4)$$

for the solutions (n_{pi} is n_i in the pth phase, m_p is the total number of components in the pth phase), and

$$G^s = \sum_{p=1}^{s} n_{pi} (G_{fi}^{\circ})_{pi} \qquad (5)$$

for "pure" solids.

Equations (3)–(5) may be combined by counting the phases consecutively beginning with the gas phase ($p = 1$) to the solids with fixed compositions ($p = q + s + 1$):

$$G = \sum_{p=1}^{1} \sum_{p=1}^{m_p} n_{pi} \left[G_{fpi}^{\circ} + RT \ln P + \ln \left(\frac{n_{pi}}{N_p} \right) \right]$$
$$+ \sum_{p=2}^{q+1} \sum_{i=1}^{m_p} n_{pi} \left[G_{fpi}^{\circ} + \ln X_i \right] + \sum_{p=q+2}^{q+s+1} \sum_{i=1}^{m_p} n_{pi} G_{fpi}^{\circ}, \qquad (6)$$

where N_p is the total amount of substances in the pth phase.

The minimization of free energy is achieved with the constraints imposed by the mass balance relationships represented as

$$\sum_{p=1}^{q+s+1} \sum_{i=1}^{m_p} n_{pi} A_{pij} = bj \qquad (j = 1, 2, \ldots, l), \qquad (7)$$

where A_{pij} represents the number of atoms of the jth element in a molecule of the ith component in the pth phase, l is the total number of elements, and b the initial amount of the element. For each element j in a pth phase we have

$$\sum_{i=1}^{m_p} n_{pi} A_{pij} - b_{pj} = 0. \qquad (8)$$

This is multiplied by an undetermined constant λ (Lagrangian multiplier) and the expression is summed up over all j, giving

$$\sum_{p=1}^{q+s+1} \left[\lambda_{pj} \left(\sum_{i=1}^{m_p} n_{pi} A_{pij} - b_{pj} \right) \right] = 0. \qquad (9)$$

Equation (9) is added to the total free energy (6). Partial differentiation (P, T, n_j) yields several linear equations (10)–(13) as shown below. For the gas phase

$$G_{fi}^{\circ} + RT \ln P + RT \ln X_i + \sum_{i=1}^{j} \lambda_j A_{ij} = 0. \qquad (10)$$

For the liquid and solid solutions

$$G_{fi}^{\circ} + RT \ln X_i + \sum_{i=1}^{j} \lambda_j A_{ij} = 0. \tag{11}$$

For the pure phases

$$G_{fi}^{\circ} + \sum_{i=1}^{j} \lambda_j A_{ij} = 0. \tag{12}$$

There will be as many Eqs. (10) and (11) as there are components (m_p). In addition there are s Eqs. (12) for pure phases, l material balances, and other equations resulting from

$$\sum X_i = 1.0. \tag{13}$$

Further discussion may best be presented with the help of an actual system, e.g., Mg–Al–Si–O–Ar. Let there be a gas phase composed of oxygen and argon, a solid solution orthopyroxene with $MgSiO_3$ and Al_2O_3, and two pure solids pyrope $Mg_3Al_2Si_3O_{12}$ and spinel $MgAl_2O_4$. The composition matrix is shown in Table 1. For this example, Eqs. (10)–(12) at $P = 1$ bar (after neglecting the vapor pressure of solids in the gas phase) are as follows:

O: $\qquad\qquad 0 + 2\lambda_O = 0;$

$MgSiO_3$: $\qquad G_{En}^{\circ} + RT \ln X_{En} + \lambda_{Mg} + \lambda_{Si} + 3\lambda_O = 0;$

Al_2O_3: $\qquad G_{Al_2O_3}^{\circ} + RT \ln X_{Al_2O_3} + 2\lambda_{Al} + 3\lambda_O = 0;$ $\qquad$ (14)

$Mg_3Al_2Si_3O_{12}$: $\quad G_{Py}^{\circ} + 3\lambda_{Mg} + 2\lambda_{Al} + 3\lambda_{Si} + 12\lambda_O = 0;$

$MgAl_2O_4$: $\qquad G_{Sp}^{\circ} + \lambda_{Mg} + 2\lambda_{Al} + 4\lambda_O = 0.$

From Eq. (8) there are five material balances:

Mg: $\quad n_{En}X_{En} + 3n_{Py} + n_{Sp} = b_{Mg};$

Si: $\quad n_{En}X_{En} + 3n_{Py} = b_{Si};$

Al: $\quad 2n_{Py} + 2n_{Sp} + 2n_{Al_2O_3}X_{Al_2O_3} = b_{Al};$ $\qquad$ (15)

O: $\quad 3n_{En}X_{En} + 3n_{Al_2O_3}X_{Al_2O_3} + 12n_{Py} + 4n_{Sp} + 2n_{O_2} = b_O.$

Table 1. The composition matrix.

Gas phase	Mg	Al	Si	O
O_2	0	0	0	2
Pyroxene				
$MgSiO_3$	1	0	1	3
$AlAlO_3$	0	2	0	3
Pyrope				
$Mg_3Al_2Si_3O_{12}$	3	2	3	12
Spinel				
$MgAl_2O_4$	1	2	0	4

In addition, we have

$$X_{En} + X_{Al_2O_3} = 1.0. \tag{16}$$

The unknown quantities n_i and λ_i can be determined by solving the equations simultaneously. For nonideal solutions activity coefficients must be used in Eqs. (10) and (11). A complete listing of phases in the Mg–Al–Si–O system would include several other minerals, e.g., MgO, Al_2SiO_5, Al_2O_3, SiO_2, Mg_2SiO_4, etc. The assemblage which would result in minimizing the total free energy of the given system has to be found by determining the total free energy of all possible assemblages. Eriksson's (1975) computer program uses an iteration procedure (Gaussian elimination) for solving the system of linear equations starting from an initial assemblage of phases with estimated non-negative values of n.

A search for the minimum energy assemblage is continued by storing relevent information and comparison with the next estimated composition. During these iterations, the activities of the components (expressions written by the users) also approach equilibrium values. The search for the lowest energy assemblage is also constrained by the Gibbs phase rule.

Eriksson's original program has been modified to permit calculations at high pressure as well as at high temperature. By using appropriate thermal expansion and compressibility data when available precise calculations can be done. For the gas phase, an ideal solution behavior is assumed at 298 K and 1 bar. High-pressure fugacities for water can be introduced from Burnham *et al.* (1969) and by the use of the modified Redlich–Kwong equation for pure gases and mixtures (Holloway, 1977; Kerrick and Jacobs, 1981).

Free Energy of Orthoenstatite

The chemical species $MgSiO_3$ occurs in three polymorphs—clinoenstatite (up to 580°C), orthoenstatite (580–990°C), and protoenstatite (990°C–melting). The free energy of formation of enstatite in all the polymorphic transitions has been estimated by Helgeson *et al.* (1978) from the phase equilibrium data of Boyd and England (1965), Atlas (1952), and Boyd, England, and Davis (1964). The recently acquired data on the enstatite–diopside solvus (see the review by Lindsley *et al.* (1981)) are very well constrained by reversals and should be used in determining the free energy of formation of orthoenstatite.

Holland *et al.* (1979) and Lindsley *et al.* (1981) consider the following transfer reaction:

$$\underset{\text{Opx}}{MgSiO_3} \rightleftharpoons \underset{\text{Cpx}}{MgSiO_3} \tag{a}$$

across the enstatite-diopside solvus. Each of the solutions Opx and Cpx are binary solutions of two chemical species $MgSiO_3$ and $Ca_{0.5}Mg_{0.5}SiO_3$. In orthopyroxene these are crystal-structurally orthopyroxene and fictive

Table 2. Calculation of G_f° of orthoenstatite using regular solution model.

T (K)	Clinoenstatite $-G_f^\circ$ (KJ/M)	Orthoenstatite $-G_f^\circ$ (KJ/M) (1)	Orthoenstatite $-G_f^\circ$ (KJ/M) (2)
1000	1255.42	1259.47	1257.07
1100	1225.78	1229.56	1227.24
1200	1196.22	1199.72	1197.49
1300	1166.78	1170.00	1167.85
1400	1133.82	1136.77	1134.71
1500	1095.28	1097.96	1095.98
1600	1056.92	1059.32	1057.40
1700	1018.31	1020.44	1018.62

Note: Clinoenstatite data from Helgeson *et al.* (1978); orthoenstatite (1) by the use of data from Holland *et al.* (1979) and orthoenstatite (2) by the use of data from Lindsley *et al.* (1981).

orthodiopside, respectively. In clinopyroxene, they are clinoenstatite and diopside. Saxena (1973) considered each solution to be nonideal regular. Recently, Holland *et al.* (1979) proposed the symmetric model and Lindsley *et al.* (1981) proposed the asymmetric model for clinopyroxene. The exchange free energy for reaction (a) according to these models are

$$\Delta G_a^\circ = 3.561-0.00191\,T \quad (\mathrm{KJ}/M)\,(\text{Lindsley } et\,al.\ 1981),$$

$$\Delta G_a^\circ = 6.800-0.00275\,T \quad (\mathrm{KJ}/M)\,(\text{Holland } et\,al.\ 1979).$$

Using G_f° for clinoentatite from Helgeson *et al.* (1978), the free energy of formation of orthoenstatite can be determined by using the two equations. The G_f° data for orthoenstatite as calculated from the two equations are listed in Table 2. The difference of 2 to 3 KJ/M in each value could be important in considering exchange equilibrium reactions. Both sets of data could be used at present until one proves to be more consistent with other experimental or natural observations. It should be emphasized, however, that the Lindsley *et al.* model is the only one at present that successfully predicts the occurrence of pigeonite.

Iron–Magnesium Solid Solution in Olivine and Orthopyroxene

Wood and Kleppa (1980) measured the heats of solution in Fe–Mg olivines and they found that the excess enthalpy of the solution can be adequately represented by an asymmetric regular solution:

$$H^{XS} = 2\left[W_{Mg}X_{Mg}X_{Fe}^2 + W_{Fe}X_{Fe}X_{Mg}^2 \right].$$

The constants are the Margules parameters for enthalpy (W_H) and are related to the Margules parameter for free energy by the relation

$$W_G = W_H - TW_S,$$

where W_S is the excess entropy parameter. If the excess entropy is zero, W_H equals W_G.

A number of investigators (Nafziger and Muan, 1967; Medaris, 1969; Williams, 1971; Larimer, 1968; Kitayama and Katsura; 1968; Matsui and Nishizawa, 1974) have experimentally determined the equilibrium

$$0.5Mg_2SiO_4 + FeSiO_3 \rightleftharpoons 0.5Fe_2SiO_4 + MgSiO_3. \qquad \text{(b)}$$

Orthopyroxene is closely ideal above 900°C irrespective of the models used (Saxena, 1973; Sack, 1980). For reaction (b) we have

$$K_b \exp(- P\Delta V_b / RT) = K_D K_\gamma,$$

where

$$K_D = \frac{X_{En}X_{Fa}}{X_{Fs}X_{Fo}}, \qquad (17)$$

$$K_\gamma = \frac{\gamma_{Fa}}{\gamma_{Fo}}, \qquad (18)$$

Equation (17) may be written as

$$\frac{X_{En}}{(1 - X_{En})} = K_b\left(\frac{X_{Fo}}{X_{Fa}} \right)\left(\frac{\gamma_{Fo}}{\gamma_{Fa}} \right). \qquad (19)$$

In Eq. (19) for a chosen value of X_{Fo} or X_{Fa} and temperature, the right-hand side is completely determined by the following substitutions:

$$K_b = \exp\left(- \frac{\Delta G_b^\circ}{RT} \right), \qquad (20)$$

$$RT \ln\left(\frac{Fa}{\gamma_{Fo}} \right) = W_{Mg}X_{Mg}(X_{Mg} - 2X_{Fe}) + W_{Fe}X_{Fe}(2X_{Mg} - X_{Fe}), \qquad (21)$$

where ΔG_b° is from Table 3, W_{Mg} and W_{Fe} are the excess enthalpy parameters of Wood and Kleppa (1980). It is assumed that excess entropy is zero—an assumption shown to be correct by application by Wood and Kleppa and later in this paper. Although the subscripts Fo and Fa are used, the mixing is considered on a one-cation basis. Directly or by an iterative procedure, it is possible to calculate X_{En} in an orthopyroxene coexisting with olivine. Table 4 shows the results of such calculations which are then compared to the experimentally determined values of X_{En}. The almost perfect match shows that (a) excess entropy in olivine solid solution is zero and (b) the ΔG_b° and, therefore, all the thermochemical data used in Table 3 are consistent internally

Table 3. Free energy of formation of olivines and pyroxenes and Fe–Mg exchange energy (all values $-$ KJ$/M$).

T (K)	Mg_2SiO_4	Fe_2SiO_4	$MgSiO_3$	$FeSiO_3$	ΔG_b°
1000	1777.29	1152.83	1259.47	939.69	-7.54
1100	1736.33	1120.50	1229.56	914.61	-7.03
1200	1695.49	1088.21	1199.72	889.53	-6.55
1300	1654.80	1056.04	1170.00	864.56	-6.06
1400	1607.05	1024.10	1136.77	839.72	-5.57
1500	1548.08	992.34	1097.95	815.01	-5.08

Note: Forsterite and fayalite from Helgeson *et al.* (1978), enstatite from Table 2, and ferrosilite from Bohlen *et al.* (personal communication, 1981).

Table 4. Comparison of calculated and experimentally determined compositions of coexisting olivine and orthopyroxene.

Sample	Larimer (1968)		Calculated
No.	X_{Mg}^{Ol}	X_{Mg}^{Opx}	X_{Mg}^{Opx}
$T = 1373$ K			
1	0.979	0.981	0.981
2	0.943	0.948	0.949
3	0.921	0.929	0.930
4	0.924	0.932	0.933
5	0.887	0.900	0.900
6	0.864	0.880	0.881
7	0.805	0.829	0.831
$T = 1473$ K			
1	0.982	0.983	0.983
2	9.950	0.955	0.953
3	0.940	0.944	0.944
4	0.917	0.927	0.922
5	0.837	0.856	0.849
6	0.652	0.694	0.694
$T = 1573$ K			
1	0.972	0.975	0.973
2	0.974	0.976	0.975
3	0.927	0.935	0.930
4	0.925	0.934	0.928
5	0.919	0.927	0.923
6	0.833	0.853	0.842

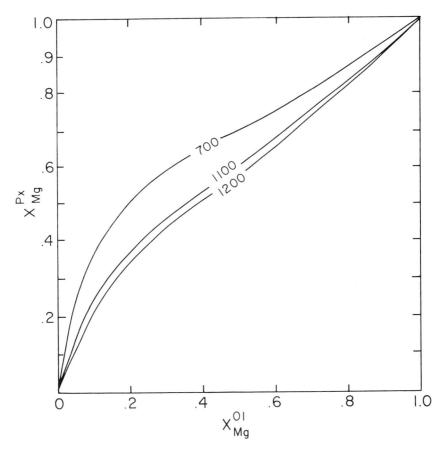

Fig. 1. Composition of coexisting olivine and orthopyroxene at 700, 1100, and 1200°C. Olivine is nonideal solution (Wood and Kleppa, 1980), $\Delta G°$ for the ion-exchange is from the free-energy values discussed in the text.

and with the experiments on olivine–pyroxene equilibrium. Another assumption, that orthopyroxene is closely ideal above 900°C, is also borne out by this analysis. Figure 1 shows several calculated isotherms for reaction (b).

Aluminous Orthopyroxene and Pyrope

In computations that follow, orthopyroxene is considered as an ideal solution of $MgSiO_3$–$FeSiO_3$–$AlAlO_3$. The component Al_2O_3 is a fictive component with the same structure as orthoenstatite and it is designed to mix ideally with $MgSiO_3$. Saxena (1981) estimated the Gibbs free energy of formation of orthoenstatite in the temperature range 1000–1600 K from experimental data on phase equilibrium in the system $MgO-Al_2O_3-SiO_2$. Instead of a fictive Al_2O_3, Ganguly and Ghose (1979) recommend the use of the Opy $(Mg_3Al_2Si_3O_{12})$ component which mixes ideally with enstatite. Because of the Al-avoidance principle (the instability of Al–O–Al linkage) (Lowenstein, 1953), the substitution of Al in the tetrahedral sites is limited to one-fourth of the total Si atoms. This, in turn, requires that only one-fourth of the Mg atoms can be replaced by Al. Therefore, the Opy component is crystal-chemically more feasible than the Al_2O_3 component.

In the ternary solution, a fictive Al_2O_3 or an Opy component may not mix ideally with $FeSiO_3$. However, since $MgSiO_3$ and $FeSiO_3$ from a nearly ideal solution at high temperature, we may expect that Al_2O_3 and $FeSiO_3$ also form an ideal solution. Unfortunately, there are no experimental data to confirm this.

Recently, Perkins et al. (1981) have presented new data on the $MgO-Al_2O_3-SiO_2$ system. In view of these data, it is necessary to revise Saxena's (1981) estimated data on Al_2O_3–Opx and pyrope. Perkins et al. have listed over ten data points on the composition of orthopyroxene coexisting with garnet at $\sim 1090°C$ and in the pressure range of 20–40 kbars. A regression analysis using the expression

$$\Delta G_C^\circ(1, T) + P\Delta V_C = RT \ln X_{Al_2O_3}X_{En}^3, \tag{22}$$

where ΔG_C° is the standard free-energy change for the reaction

$$\underset{Opx}{Al_2O_3} + \underset{Opx}{3MgSiO_3} = \underset{pyrope}{Mg_3Al_2Si_3O_{12}} \tag{c}$$

yields $\Delta G_C^\circ = -16699\ (\pm 974)\ J/M$ and $\Delta V_C^\circ = -0.783\ (\pm 0.032)\ J/M/bar$. The effect of pressure and temperature on the molar volume has been ignored. Lane and Ganguly's (1980) calculations show a small dependence of ΔV_C° on P and T. The ΔG_C° calculated at other temperatures for compositions given by Perkins et al. (1981) does not change significantly and mostly lies within $-16699\ (\pm 974)\ J$.

The ΔV_C° of -7.83 cm results in a molar volume of 27.03 cm^3 for the Al_2O_3 component in orthopyroxene. The reaction for the boundary between the garnet and spinel fields is given by

$$\underset{Opx}{MgSiO_3} + \underset{Sp}{MgAl_2O_4} \rightleftharpoons \underset{Opx}{Al_2O_3} + \underset{Fo}{Mg_2SiO_4}. \tag{d}$$

The molar volume of 27.03 cm^3 for Al_2O_3 component when used to calculate ΔV_d° results in a small ΔV of -0.2 cm^3. This ΔV_d° combined with the experimental data of Danckwerth and Newton (1979) and Perkins et al. (1981) results in the following ΔG_d° equation:

$$\Delta G_d^\circ(1, T) = 24409 + 3.8(\pm 0.05)T\ K$$

in the temperature range 1200–1700 K. Using these data on ΔG_C°, ΔG_d°, ΔV_C°, and ΔV_d°, the isopleth diagram shown in Fig. 2 has been constructed. Because of the small value of ΔV_d° the Al_2O_3 isopleths in the spinel field are pressure insensitive.

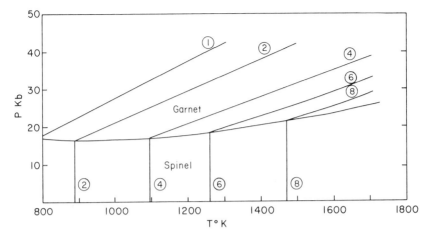

Fig. 2. Al_2O_3 in orthopyroxene coexisting with garnet or spinel. The isopleths are based on thermodynamic data on fictive Al_2O_3–orthopyroxene and pyrope calculated using experimental phase equilibria data of Perkins *et al.* (1981). For data on other phases see text.

The ΔG_C° and ΔG_d° can be used to determine the G_f° of pyrope and Al_2O_3–orthopyroxene (fictive). These free energies will depend on the set of thermodynamic values used for other phases in the system, i.e., forsterite, orthoenstatite, and spinel. Table 5 shows the G_f° of Al_2O_3–orthopyroxene and pyrope. The G_f° of forsterite and spinel are from Helgeson *et al.* (1978) and G_f° of orthoenstatite from Table 2. No error estimates have been made but an error of more than 2000 J/M in any one of the free-energy data would result in a significant change in the $X_{Al_2O_3}$ as plotted in Fig. 2. R. C. Newton (personal communication, 1981) calculated the G_f° of pyrope at 1100 K from ΔH_f° of Charlu *et al.* (1975), ΔS_f°, and free-energy function of Haselton and Westrum (1980) and found it to be -4997.32 ± 4.57 KJ/M as compared to -4993.34 calculated here. The G_f° of pyrope in Table 5 increases systematically up to 1500 K but not at 1600 and 1700 K.

Examples of Computations

Thermochemical Data

Phases in the computed assemblages are listed in Table 6. The thermochemical data on pyrope and Al_2O_3–Opx are from Table 4. All other thermochemical data are from Helgeson *et al.* (1978) except for orthoenstatite (Table 3). Biotite and garnet are considered as ideal solutions (see Mueller (1972) for biotite and Newton and Haselton (1981) for discussion of garnet). Gas is considered as an ideal mixture.

Table 5. $-G_f^{\circ}$ Al$_2$O$_3$–Opx and pyrope.

T (K)	Al$_2$O$_3$–Opx	Pyrope
1100	1287997	4993367
1200	1254207	4870856
1300	1220476	4747175
1400	1186880	4613895
1500	1153363	4463927
1600	1119962	4314609
1700	1086691	4164698

Note: Error estimates may be based on changes produced in the experimentally determined $X_{Al_2O_3}$ which becomes significant if an individual G_f° is changed by more than 2000 J/M. The G_f° are consistent with G_f° of other minerals as determined by Helgeson *et al.* (1978) and as in Table 2 but may not be consistent with data from other sources.

Table 6. Phases used in various calculations.

Elements	Al, Fe, H, K, Mg, O, Si, Ar
Gas Phase:	Fe, FeO, OH, SiH, H$_2$, O$_2$, H, O, SiO, Si, K, Mg, MgO, Ar
Solid Solutions:	olivine (Mg$_2$SiO$_4$–Fe$_2$SiO$_4$), orthopyroxene (Al$_2$O$_3$–MgSiO$_3$–FeSiO$_3$), biotite (KFe$_3$AlSi$_3$O$_{10}$(OH)$_2$–KMg$_3$AlSi$_3$O$_{10}$(OH)$_2$), garnet (Mg$_3$Al$_2$Si$_3$O$_{12}$–Fe$_3$Al$_2$Si$_3$O$_{12}$)
Pure Phases:	Orthoclase, sanidine, spinel (MgAl$_2$O$_4$), kyanite, andalusite sillimanite, corundum, cordierite, quartz, wustite, melilites, muscovite, magnetite

The System: Ar–Al–Fe–H–K–Mg–O–Si

Table 7 shows the result of computed assemblage in this system at 1000 K and 1 kbar. The gas phase will be nonideal at this pressure and, therefore, the gas composition and to some extent the composition of solid solutions with Fe : Mg variation would not be accurate. Table 7 shows the assemblage with a gas phase, olivine, orthopyroxene, biotite, orthoclase, magnetite, and corundum. It is rather an unusual assemblage resulting from an arbitrarily chosen composition. Argon is included mainly to ascertain that a gas phase is always present. Table 8 shows the summary of results obtained by changing the Fe : Mg ratio in the initial composition. The following changes may be noted:

(A) Fe–Mg exchange between olivine and pyroxene;
(B) presence of biotite only in the most magnesium-rich composition;
(C) a correlation between hydrogen pressure and the modal amount of Fe$_3$O$_4$;

Table 7. An example of a computed assemblage at 1000 K and 1 Kbar. Initial composition (atoms): Al 16, Ar 100, Fe 20, H 30, K 10, Mg 60, O 305, Si 90.

Gas Phase	Moles	P/bar	Activity	Mode %
$Fe(OH)_2$	87E − 7	0.775E − 6	0.775E − 6	—
H_2O	12.32	109.42	109.42	—
H	0.39E − 9	0.35E − 8	0.35E − 8	—
Ar	1000.0	888.20	888.20	—
H_2	0.27	2.42	2.42	—
Solid Solutions	Moles	X_i	Activity	Mode %
Olivine: Mg_2SiO_4	11.69	0.647	0.733	14.04
Fe_2SiO_4	6.38	0.353	0.481	7.66
Pyroxene: Al_2O_3	2.32	0.044	0.044	2.78
$MgSiO_3$	41.22	0.773	0.773	49.48
$FeSiO_3$	9.76	0.183	0.183	11.73
Biotite: Fe–Bi	0.04	0.017	0.017	0.05
Mg–Bi	2.36	0.983	0.983	2.84
Pure Phases				
K–feldspar	7.59	—	—	9.11
Fe_3O_4	1.24	—	—	1.49
Al_2O_3	0.67	—	—	0.81

Note: E format used, e.g. 1.E − 2 = 0.01.

Table 8. Summary of several computed assemblages with a variable Fe : Mg at 1000 K and 1 Kbar. Initial composition: as in Table 7.

	Gas Pressure		Olivine		Opx			% Modes				
Fe : Mg	H_2O	H_2	Mg	Fe	Al	Mg	Fe	Ol	Opx	K–F	Fe_3O_4	Sill
1. 50 : 30	126	4.27	0.23	0.77	0.047	0.501	0.452	36.50	49.80	11.11	1.65	0.98
2. 40 : 40	126	4.61	0.360	0.640	0.044	0.622	0.334	34.30	51.82	11.12	1.69	1.08
3. 30 : 50	127	3.74	0.520	0.480	0.043	0.727	0.230	34.93	51.31	11.09	1.57	1.10
4. 20 : 60	109	2.41	0.647	0.353	0.044	0.773	0.183	21.70	63.99	9.11	1.49	—

other phases in this sample biotite $X_{Mg} = 0.98$, mole % 2.89; Al_2O_3 mole % 0.81

(D) a negative correlation between $X_{Al_2O_3}$ in orthopyroxene and the modal amount of sillimanite;

(E) a decrease in modal olivine and increase in modal orthopyroxene with increasing Mg : Fe in the initial composition.

Table 9 shows the stable coexistence of quartz and fayalitic olivine in the iron-rich composition. Comparisons of the quartz-present and quartz-absent assemblages show that Al_2O_3 in orthopyroxene remains fixed with changing Fe/Mg ratio in quartz-present compositions but not in others indicating the reaction:

$$Al_2O_3 + SiO_2 = Al_2SiO_5 . \qquad (e)$$

Opx   Quartz   Sill

Table 9. Equilibrium assemblages in quartz-saturated composition at 1000 K and 1 Kbar. Initial composition (atoms): Al 25, Ar 100, H 30, K 10, O 305, Si 90.

		Gas Pressure Bars			Pyroxene			Olivine		% Modes			
Fe	Mg	H_2O	O_2	H_2	Al	Mg	Fe	Mg	Fe	K–F	Sill	Mt	Quartz
1. 55	5	128	—	2.2	—	—	—	0.136	0.834	10.0	7.50	7.76	34.13
2. 50	10	128	—	2.9	0.026	0.267	0.710	—	—	10.0	7.84	6.51	17.01
3. 30	204	129	—	0.8	0.0264	0.523	0.451	—	—	10.0	7.59	6.49	16.27
4. 20	40	130	3.7	—	0.026	0.973	—	—	—	10.0	6.42	6.65	13.54
5. 10	50	128	18	—	0.0264	0.973	—	—	—	10.0	6.15	3.32	3.82

This result, which could be deduced from theory (e.g., Kretz, 1964), shows quantitatively the danger of using the composition of individual minerals in geothermometry–geobarometry. However, when suitably buffered, the composition of orthopyroxene coexisting with Al_2SiO_5 may be used for P–T estimates. ΔV_e° for this reaction is small, but perhaps in the kyanite field and andalusite field, ΔV_e° may be large enough for the reaction to be pressure sensitive.[1]

Finally, an attempt is made to introduce Fe in the MgO–Al_2O_3–SiO_2 system. The G_f° for the $Fe_3Al_2Si_3O_{12}$ and $FeAl_2O_4$ components in garnet and spinel solid solutions respectively has been calculated as follows. For almandine, the enthalpy of formation at 298 K is from Zen (1972) and the C_p function from Ganguly (1978). For spinel $FeAl_2O_4$, the G_f° is approximately calculated using the data from Fabries (1979). Spinel is assumed to be ideal solution of Fe_3O_4, $FeAl_2O_4$, and $MgAl_2O_4$. This is definitely not true at metamorphic temperatures where there is a solvus relationship between coexisting magnetite and Al–spinel. The result of computation as shown in Table 10 is likely to change substantially with a new G_f° value for $FeAl_2O_4$. Note that with oxidation, there is a small change in Al_2O_3 in orthopyroxene but a significant change in $FeSiO_3$.

Table 10. Calculation of Fe–Mg orthopyroxene composition coexisting with spinel at 1200 K and 1 bar. Initial composition (atoms): Al 16, Ar 100, Fe 20, Mg 40, O variable, Si 60.

(bars)	Spinel			Opx			Mode %		
P_{O_2}	X_{Mt}	X_{Mg}	X_{Fe}	Al_2O_3	En	Fs	Sp	Opx	Quartz
7E-7	0.515	0.457	0.027	0.061	0.894	0.044	15.71	52.41	31.88
0.309	0.531	0.466	0.002	0.059	0.937	0.004	16.68	48.90	34.42

[1]Note that if the ternary solution Al_2O_3–$MgSiO_3$–$FeSiO_3$ is not ideal, $X_{Al_2O_3}$ would vary as a function of X_{MgSiO_3} and X_{FeSiO_3} in Opx. Because of disequilibrium, the compositional data on coexisting Opx and Sill as presented by Grew (1980) do not provide information in this regard.

Conclusions

The method of free-energy minimization is a very powerful tool for computing phase equilibria in assemblages of minerals. The method is particularly important in assemblages where multicomponent solid solutions are involved. However, the full advantage of this method will be realized only when the basic work of collecting thermochemical data on the end-member components and their mixing behavior has progressed significantly. In the meantime the method can be used in providing us with significant insights into the thermo-dynamics of complex phase equilibria with the help of real or hypothetical models of solutions. The method can be used to great advantage in checking the consistency of multicomponent solution models which are constructed from data on binary and ternary systems. An example of the ternary solution of orthopyroxene was used in this paper where the binary data on fictive Al_2O_3–$MgSiO_3$ was combined with the binary data on $MgSiO_3$–$FeSiO_3$. Estimation of Gibbs free energy of formation of the mineral phases has been usually done by the equilibrium constant method. Such free energies, because their derivation is only relative to the participating phases in a specific reaction, are relative free energies (see Helgeson *et al.* (1978)). With the present method it is possible to estimate the free energies of formation which may be consistent with all phases in a multireaction system. This could be accomplished by iterating for the unknown free energy until a solution consistent with the experimental data can be found. The method also provides us with the possibility to calculate the total composition of coexisting minerals and their modal proportions and, therefore, it can be eventually used to simulate crustal and mantle material.

Acknowledgments

Thanks are due to R. Kretz and R. C. Newton for reviews. Dr. G. Eriksson's program and advice is gratefully acknowledged.

References

Atlas, L. (1965) The polymorphism of $MgSiO_3$ and solid state equilibria in the system $MgSiO_3$–$CaMgSi_2O_6$, *J. Geology* **60**, 125–147.

Bohlen, S. R., Essene, E. J. and Boettcher, A. L. (1980) Reinvestigation and application of olivine–quartz–orthopyroxene barometry, *Earth Planet. Sci. Lett.* **47**, 1–10.

Boyd, F. R., and England, J. L. (1965) The rhombic enstatite–clinoenstatite inversion, *Carnegie Inst. Washington Yearbook* **64**, 117–120.

Boyd, F. R., England, J. L., and Davis, B. T. C. (1964) The effect of pressure on the melting and polymorphism of enstatite, MgSiO₃, *J. Geophys. Res.* **69**, 2101–2109.

Brown, T. H., and Skinner, B. J. (1974) Theoretical prediction of equilibrium phase assemblages in multicomponent systems, *Amer. J. Sci.* **274**, 961–986.

Charlu, T. V., Newton, R. C., and Kleppa, O. J. (1975) Enthalpies of formation at 970°K of compounds in the system $MgO–Al_2O_3–SiO_2$ from high temperature solution calorimetry, *Geochim. Cosmochim. Acta* **39**, 1487–1497.

Danckwerth, P. A., and Newton, R. C. (1978) Experimental determination of the spinel peridotite to garnet peridotite reaction in the system $MgO–Al_2O_3–SiO_2$ in the range 900–1100 C and Al_2O_3 isopleths of enstatite in the spinel field, *Contrib. Mineral. Petrol.* **66**, 189–201.

Eriksson, G. (1976) Quantitative equilibrium calculations in multiphase systems at high temperatures with special reference to the roasting of chalcopyrite $(CuFeS_2)$, *Akedemisk Avhandling*, Umea University, Sweden.

Eriksson, G., and Rosen, E. (1973) Thermodynamic studies of high temperature equilibria, *Chem. Scr.* **4**, 193–194.

Fabries, J. (1979) Spinel–olivine geothermometry in peridotites from ultramafic complexes, *Contrib. Mineral. Petrol.* **69**, 329–336.

Ganguly, J., and Ghose, S. (1979) Aluminous orthopyroxene: order–disorder, thermodynamic properties, and petrologic implications, *Contrib. Mineral. Petrol.* **69**, 375–382.

Grew, E. S. (1980) Sapphirine + quartz association from Archean rocks in Enderby Land, Antarctica, *Amer. Mineral.* **65**, 821–836.

Grossman, L. (1972) Condensation in the primitive solar nebula, *Geochim. Cosmochim. Acta* **36**, 109–128.

Guggenheim, E. A. (1967) *Thermodynamics*. North-Holland, Amsterdam.

Haselton, H. T., Jr., and Westrum, E. F., Jr. (1980) Low temperature heat capacities of synthetic pyrope, grossular, and pyrope₆₀, grossular₄₀, *Geochim. Cosmochim. Acta* **44**, 701–709.

Helgeson, H. C., Delany, J. M., Nesbitt, H. W., and Bird, D. K. (1978) Summary and critique of the thermodynamic properties of rock-forming minerals, *Amer. J. Sci.* **278A**, 1–250.

Holloway, J. R. (1977) Fugacity and activity of molecular species in supercritical fluids, in *Thermodynamics in Geology*, edited by D. G. Fraser, pp. 161–181, Reidel, Dordrecht, The Netherlands.

Kerrick, D. M., and Jacobs, G. K. (1982) A modified Redlich–Kwong equation for H_2O, CO_2, and $H_2O–CO_2$ mixtures at elevated pressures and temperatures, *Amer. J. Sci.* 281–290.

Kitayama, K., and Katsura, T. (1968) Activity measurements in orthosilicate and metasilicate solid solutions. I. $Mg_2SiO_4–Fe_2SiO_4$ and $MgSiO_3–FeSiO_3$ at 1204 C, *Chem. Soc. Japan J.* **41**, 1146–1151.

Kretz, R. (1964) Analysis of equilibrium in garnet–biotite–sillimanite gneisses from Quebec, *J. Petrology* **5**, 1–20.

Lane, D. L., and Ganguly, J. (1980) Al_2O_3 solubility in orthopyroxene in the system $MgO–Al_2O_3–SiO_2$: A reevaluation and mantle geotherm, *J. Geophys. Res.* **85**, 6963–6972.

Larimer, J. W. (1968) Experimental studies on the system $Fe-MgO-SiO_2-O_2$ and their bearing in the petrology of chondritic meteorites, *Geochim. Cosmochim. Acta* **32**, 1187–1207.

Lowenstein, W. (1953) The distribution of aluminum in the tetrahedra of silicates, *Amer. Mineral.* **38**, 92–96.

Matsui, Y., and Nishizawa, O. (1974) Iron(II)–magnesium exchange equilibrium between olivine and calcium-free pyroxene over a temperature range 800°C to 1300°C, *Bull. Soc. Fr. Mineral. Cristallogr.* **97**, 122–130.

Medaris, L. G., Jr. (1972) Partitioning of Fe and Mg between coexisting synthetic olivine and orthopyroxene, *Amer. J. Sci.* **267**, 945–968.

Mueller, R. F. (1972) Stability of biotite: A discussion, *Amer. Mineral.* **57**, 300–316.

Nafziger, R. H., and Muan, A. (1967) Equilibrium phase compositions and thermodynamic properties of olivine and pyroxenes in the system $MgO-FeO-SiO_2$, *Amer. Mineral.* **52**, 1364–1385.

Nicholls, J. (1977) The calculation of mineral compositions and modes of olivine–two pyroxene–spinel assemblages, *Contrib. Mineral. Petrol.* **60**, 119–142.

Perkins, D., Holland, T. J. B., and Newton, R. C. (1981) The Al_2O_3 contents of enstatite in equilibrium with garnet in the system $MgO-Al_2O_3-SiO_2$ at 15–40 kbar and 900–1600°C, *Contrib. Mineral. Petrol.* **76**, 530–541.

Robie, R., Hemingway, B. S., and Fisher, J. R. (1978) Thermodynamic properties of minerals and related substances at 298°K and 1 bar (10 pascals) pressure and at higher temperatures, *U. S. Geol. Survey Bull.* **1452** (1978).

Sack, R. (1980) Some constraints on the thermodynamic mixing properties of Fe–Mg orthopyroxenes and olivines, *Contrib. Mineral. Petrol.* **71**, 257–269.

Saxena, S. K. (1973) *Thermodynamics of Rock-Forming Crystalline Solutions*, Springer-Verlag, New York.

Saxena, S. K. (1981) The $MgO-Al_2O_3-SiO_2$ System: Free energy of pyrope and Al_2O_3–enstatite, *Geochim. Cosmochim. Acta*.

Van Zeggeren, F. and Storey, S. H. (1970) *The Computation of Chemical Equilibria*, Cambridge University Press, New York.

White, W. B., Johnson, S. M. and Dantzig, G. B. (1958) Chemical equilibrium in complex mixtures, *J. Chem. Phys.* **28**, 751–755.

Williams, R. J. (1971) Reaction constants in the system $Fe-MgO-SiO_2-O_2$ at 1 atm between 900°C and 1300°C: Experimental result, *Amer. J. Sci.* **270**, 334–360.

Wood, B. J. and Kleppa, O. J. (1981) Thermochemistry of forsterite–fayalite olivine solutions, *Geochim. Cosmochim. Acta* **45**, 529–534.

Chapter 8

Thermodynamic Procedures for Treating the Monoclinic/Triclinic Inversion as a High-Order Phase Transition in Equations of State for Binary Analbite–Sanidine Feldspars

G. A. Merkel and J. G. Blencoe

Introduction

Many solid-state phase transitions (transformations) can be classified thermo-dynamically according to their *order* (Ehrenfest, 1933): an nth-order phase transition is characterized by discontinuities in the nth and all higher-order derivatives of G with respect to an intensive variable—e.g., pressure, tempera-ture, or composition—while lower-order derivatives are continuous.[1] There-fore, a *first-order phase transition* produces discontinuities in V, H, and μ_i because these properties are related to the first derivatives $\partial G / \partial P$, $\partial G / \partial T$, and $\partial G / \partial X_i$, respectively, and properties related to higher-order derivatives of G—e.g., $C_p = -T(\partial^2 G / \partial T^2)$, $\alpha = (1/V)(\partial^2 G / \partial P \, \partial T)$, and $\beta = -(1/V)(\partial^2 G / \partial P^2)$—also exhibit discontinuities. On the other hand, for a *second-order phase transition*, V, H, and μ_i are all continuous at the point of phase change, so ΔV and ΔH of transition are zero, but C_p, α, and β are discontinuous. Finally, at a *third-order phase transition* all properties related to the first and second derivatives of G are continuous, and only third- and higher-order derivatives of G are discontinuous. In this chapter, transitions that fit into this classification scheme and whose order is greater than one will be referred to as *high-order transitions.*

However, there is yet another type of transition, one that cannot be categorized according to the scheme devised by Ehrenfest—the lambda (λ)-transition. For such transitions, V, H, and μ_i are continuous at the point of phase change, but properties such as C_p, α, and β approach infinity, thus distinguishing λ-transitions from any "classical" high-order transition. [De-tailed discussions of the differences between high-order and λ-transitions have been presented by Pippard (1966), Rao and Rao (1978), and Thompson and Perkins (1981).] Herein, λ- and high-order transitions will be referred to collectively as *non-first-order* (NFO) *transitions.*

NFO transitions occur in many binary and multicomponent systems of metallurgical and petrological interest, and numerous studies have shown that

---

[1] Notation used in this chapter is defined in Table 1.

Table 1. Notation.

General Thermodynamic Symbols

a	relative activity
C_p	isobaric heat capacity
G	molal Gibbs free energy
G^{mm}	molal Gibbs free energy of *mechanical* mixing
G^{mix}	molal Gibbs free energy of mixing
G^{id}	molal Gibbs free energy of ideal mixing
G^{ex}	excess molal Gibbs free energy
H	molal enthalpy
P	pressure
R	universal gas constant, 1.98726 cal/gfw deg
T	temperature in Kelvins (K)
V	molal volume
X	mole fraction
α	isobaric thermal expansivity
β	isothermal compressibility
γ	activity coefficient
μ	chemical potential

Abbreviations

CMG	conditional miscibility gap
MIS	monoclinic isostructural solvus
NFO	non-first-order
TIS	triclinic isostructural solvus

Superscripts

A	pertaining to a phase with structure A
B	pertaining to a phase with structure B
inv	evaluated at the inversion
m	monoclinic
t	triclinic
v	vapor
ϕ	a phase
$^{\circ}$	pertaining to a standard state

Subscripts

ab	$NaAlSi_3O_8$ component
i	a component
iso	isostructural
or	$KAlSi_3O_8$ component
ss	solid solution
1	component 1
2	component 2
2nd	second order
3rd	third order

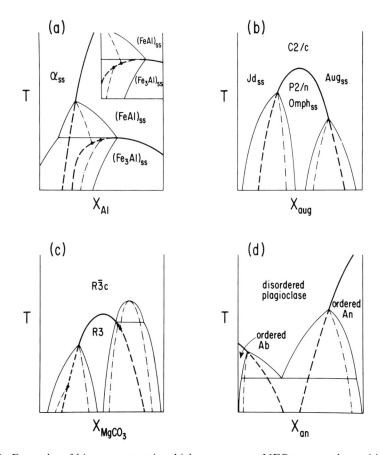

Fig. 1. Examples of binary systems in which one or more NFO structural transitions of the order–disorder type are associated with subsolidus two-phase regions. Heavy curves illustrate the T–X positions of structural transitions (dashed where metastable), light curves are boundaries of stable two-phase regions, and dash–dot curves are spinodals. (a) Schematic T–X relations (incoherent phase boundaries) for a portion of the system Fe–Al after Allen and Cahn (1975, 1976a). Inset shows alternative spinodal relations proposed in this chapter. (b) Schematic T–X relations for the system jadeite–augite (Carpenter, 1980). (c) Schematic T–X relations proposed by the writers for the systems $CaCO_3$–$MgCO_3$ and $CdCO_3$–$MgCO_3$. This phase equilibrium topology is partly based on the experimental data of Goldsmith and Heard (1961) and Goldsmith (1972). (d) Schematic T–X phase relations for the system $NaAlSi_3O_8$–$CaAl_2Si_2O_8$ (Carpenter, 1981).

these transitions can have an important effect on the subsolidus phase equilibria of these systems. For example, Allen and Cahn (1976a, 1976b) discovered that a two-phase field in the system Fe–Al (Fig. 1(a)) is related to the occurrence of two NFO transitions: between α_{ss} and $(FeAl)_{ss}$ and between $(FeAl)_{ss}$ and $(Fe_3Al)_{ss}$. Additionally, Carpenter (1980) has presented evidence indicating that, at low temperatures in the system jadeite–augite (Fig. 1(b)), the two-phase regions separating the omphacite phase field from adjacent jadeite and augite phase fields are associated with an NFO transition that exhibits a thermal maximum at or near the pure omphacite composition. In light of the thermodynamic analysis of Carpenter for the system jadeite–augite, phase relations proposed previously for the systems $CaCO_3$–$MgCO_3$ (Goldsmith and Heard, 1961) and $CdCO_3$–$MgCO_3$ (Goldsmith, 1972) are reinterpreted here and presented in Fig. 1(c). Carpenter (1981) has also suggested that NFO transitions could account for the peristerite and Hutten-locher two-phase regions in the system albite–anorthite (Fig. 1(d)).

In the phase diagrams displayed in Fig. 1, the transitions could also be classified *structurally* as NFO transformations of the order–disorder type, with the structure of the low-temperature phase being of lower symmetry due to increased cationic ordering. Another type of structural transition that can be NFO is the displacive transformation: an abrupt but minor structural read-justment in which bonds of secondary coordination are broken while bonds of primary coordination remain intact. Displacive transformations include the so-called "polyhedral tilt transitions" which occur in a number of minerals, including several important framework silicates (Hazen and Finger, 1979). Among the transformations believed to be polyhedral tilt transitions are the OC/OS transition in tridymite (Nukui *et al.*, 1978), the orthorhombic/hexa-gonal inversion in subpotassic nepheline (Henderson and Roux, 1977), and the monoclinic/triclinic (m/t) inversion in alkali feldspars (Laves, 1952). Heat-capacity measurements have shown that the inversions in tridymite (Thompson and Wennemer, 1979) and nepheline (Henderson and Thompson, 1980) are of the λ-type, but the nature of the inversion in alkali feldspars is less certain (Thompson and Hovis, 1979b).

In this chapter we first review the available evidence suggesting that the m/t inversion in analbite–sanidine feldspars can be treated as a "classical" high-order phase transition. Due to the presence of this transition, the anal-bite–sanidine system must be treated as nonisostructural in thermodynamic modeling of G^{mix}–X_{or} relations for the crystalline solutions. Consequently, G^{mix}–X relations in hypothetical binary systems exhibiting first-, second-, and third-order transitions are examined, and the principles developed are then applied in an analysis of hydrothermal ion-exchange data for analbite–sanidine feldspars to derive expressions for G_{2nd}^{mix} and G_{3rd}^{mix}. To generate a wide spectrum of theoretically valid phase relations for these feldspars, a $(\partial G^{mix}/\partial X_{or})_{P,T}$ equation for monoclinic feldspars was fixed while the coeffi-cients of $(\partial G^{mix}/\partial X_{or})_{P,T}$ expressions for triclinic feldspars were varied sys-tematically within broad limits established by regression analysis of 2 kbar feldspar–fluid ion-exchange data. The resulting calculated subsolidus phase relations include most of those likely to be encountered in other binary systems that exhibit NFO transitions, and they provide a basis for thermody-namic interpretations of subsolidus, two-phase equilibria associated with such transitions in these systems. Finally, solvi calculated from isostructural and "preferred" nonisostructural G^{mix} equations are evaluated in light of solvus experimental data for the analbite–sanidine system.

Nature and Significance of the
Monoclinic/Triclinic Inversion

From X-ray diffraction data indicating a continuous variation in unit cell parameters of binary analbite–sanidine feldspars at 1 atm, 25°C, Donnay and Donnay (1952) concluded that there was no two-phase region associated with the displacive m/t transformation, and that the inversion was a high-order transition. Results of subsequent X-ray work at these conditions (Orville, 1967; Wright and Stewart, 1968; Luth and Querol-Suñé, 1970; Hovis, 1980), as well as at higher temperatures (MacKenzie, 1952; Laves, 1952; Kroll, 1971; Okamura and Ghose, 1975; Henderson, 1979; Winter *et al.*, 1979; Kroll *et al.*, 1980) and higher pressures (Hazen, 1976), are consistent with these conclusions, and they show that the composition of the inversion, X_{or}^{inv}, is nearly a linear function of pressure and temperature (Fig. 2). Regression analysis of the available P–T–X data for the inversion yields the following least-squares fit equation:

$$X_{or}^{inv} = 0.474 - 0.361 \left[\frac{T(K)}{1000} \right] + 0.0252 P \,(\text{kbars}). \qquad (1)$$

Additionally, Willaime *et al.* (1974) measured lattice parameters for high albite between 25 and 1000°C, and they found that the volume expansion coefficient showed no discontinuity or anomaly at the inversion. Thus, collectively, the available X-ray diffraction data indicate that ΔV^{inv} and $\Delta \alpha^{inv}$ are zero.

MacKenzie (1952) investigated four natural anorthoclases by DTA and could find "no definite evidence of a discontinuity in the differential thermal curves of the natural feldspars." Furthermore, in DSC measurements on natural anorthoclases, Thompson and Perkins (1981) found no evidence for an anomaly in heat capacity at the inversion. These results imply, respectively, that ΔH^{inv} and ΔC_P^{inv} are zero.

Because ΔV and ΔH can be defined in terms of the first derivative of ΔG with respect to P and T, respectively, the results $\Delta V^{inv} = 0$ and $\Delta H^{inv} = 0$ suggest that $(\partial G / \partial P)^{inv}$ and $(\partial G / \partial T)^{inv}$ are continuous and, therefore, that the inversion is a second- or higher-order transition in the classical approach of Ehrenfest (1933). Furthermore, ΔC_p^{inv} and $\Delta \alpha^{inv}$, properties that can be defined in terms of second derivatives of G, are also apparently zero, thus suggesting that the inversion is perhaps third order. However, Thompson and Hovis (1979b) have noted that two of the principal coefficients of the thermal expansion tensor for high albite reported by Willaime *et al.* (1974) appear to exhibit a λ-type anomaly near the transition temperature. Therefore, while it is obvious that the m/t inversion is not a first-order transition, it is still uncertain whether, in a strict sense, it is either second order, third order, or of the λ-type. Considering the resolution of the bulk-property physicochemical measurements that have been performed to date, we conclude that it is reasonable to treat the inversion as a classical high-order transition.

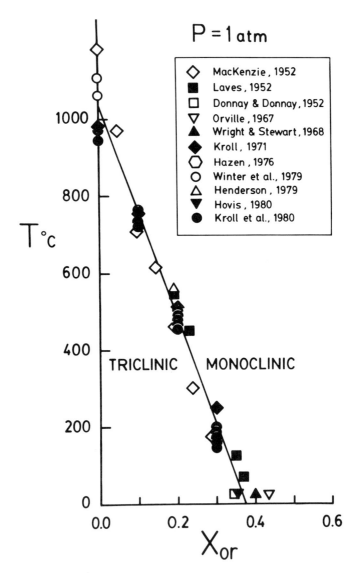

Fig. 2. Composition (X_{or}^{inv}) of $(Na, K)AlSi_3O_8$ analbite–sanidine feldspar at the mono-clinic/triclinic (m/t) inversion as a function of (1) temperature at 1 atm and (2) pressure at 25°C. The sloping lines that show $T-X_{or}$ (1 atm) and $P-X_{or}$ (25°C) relations for the inversion were obtained from Eq. (1), which is a least-squares fit equation derived from the data illustrated in this figure.

In describing the thermodynamic significance of the m/t inversion, Thompson and Waldbaum (1968, p. 1975) noted that this "symmetry change . . . must be accompanied by a change in the configuration of the oxygen atoms about the alkali site and should thereby have some effect on the Na–K mixing properties." They further speculated that "it seems likely . . . that the transi-

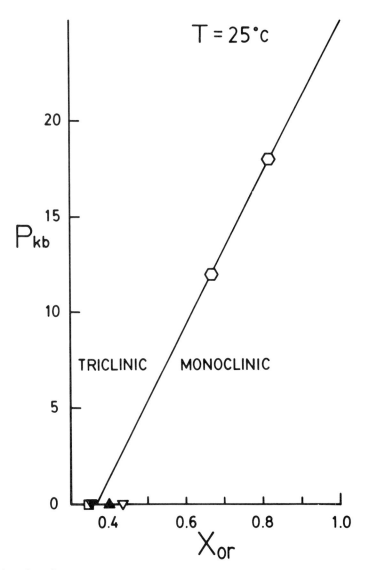

Fig. 2. (continued)

tion is of second or higher order. If it is in fact second order, a plot of . . . $(\partial \bar{G}/\partial N_{or})_{P,T}$ or of $(\partial \bar{G}^{ex}/\partial N_{or})_{P,T}$ against composition at constant P and T should show a break in slope at the symmetry change. If such a break is well marked, a fitted curve that neglected it could lead to serious error on both sides of the transition. The two sides would better be fitted separately."[2] However, Thompson and Waldbaum did not specify a procedure for deriving

---

[2]Thompson and Waldbaum used bars above symbols that designate molal quantities, a practice which we forego here to avoid confusing these properties with *partial* molal properties.

separate polynomial $G^{mix}-X_{or}$ expressions for the two structures such that the equations intersect one another in a manner that is thermodynamically consistent with the stipulation that the inversion is a high-order phase transition. Consequently, in their thermodynamic treatment of the alkali feldspar–aqueous chloride solution ion-exchange data obtained by Orville (1963), Thompson and Waldbaum (1968), and later Thompson and Hovis (1979a), circumvented the difficulties associated with a quantitative treatment of the m/t inversion by deleting most of the data for triclinic feldspars. Therefore, the G^{ex} equations of state derived in these two previous studies pertain to an *isostructural monalbite–sanidine* feldspar series. In similar attempts to derive G^{ex} expressions from ion-exchange data, other workers have simply ignored the inversion and combined the data for monoclinic and triclinic feldspars (Traetteberg and Flood, 1972; Zyrianov *et al.*, 1978).

Thermodynamic Characteristics and Treatments of High-Order Phase Transitions in Binary Systems

Blencoe (1977) has described the free energy–composition relations of a hypothetical binary system exhibiting a first-order phase transition (Fig. 3). In this situation a discontinuity in $(\partial G/\partial X_2)_{P,T}$, herein abbreviated ∂G, occurs at the transition point i, thereby producing a break in slope in the minimum $G-X_2$ curve.[3] Consequently, any phase with a composition in the range $a < X_2 < b$ will tend to exsolve by nucleation and growth. In developing equations of state to represent $G-X$ relations for the phases, it is theoretically justifiable to treat the system as consisting of two independent solution series, one with the structure of phase A, the other with the structure of phase B. The $G-X$ relations can thus be modeled thermodynamically by developing expressions for $\Delta\mu_1^\circ$, $\Delta\mu_2^\circ$, $G^{ex,A}$, and $G^{ex,B}$. Blencoe (1979a, 1979b) and Lindsley *et al.* (1980, 1981) have taken this approach in their treatments of the non-isostructural orthoenstatite–diopside series which exhibits a first-order (ortho-enstatite–pigeonite) transition.

In a system exhibiting a classical second-order transition, ∂G is continuous at the transition composition X_2^{inv}. Therefore, the $G-X$ curves for the two structures must intersect one another with the same slope ($\mu_1^A = \mu_1^B$ and $\mu_2^A = \mu_2^B$) at X_2^{inv}, and no miscibility gap is generated. However, for a second-order transition $\partial^2 G$ is discontinuous at X_2^{inv}, and either (1) $\partial^2 G^A < \partial^2 G^B$ (Fig. 4) or (2) $\partial^2 G^A > \partial^2 G^B$ (Fig. 5). (In both Figs. 4 and 5, structure A is of lower symmetry than structure B.) In Fig. 4, the relation $\partial^2 G^A < \partial^2 G^B$ at X_2^{inv} requires that the "metastable extension" of curve G^A on the right-hand side of

---

[3]For brevity in notation, hereinafter, unless indicated otherwise, all partial derivatives are understood to be taken with respect to mole fraction (X_2 or X_{or}) at a fixed P and T.

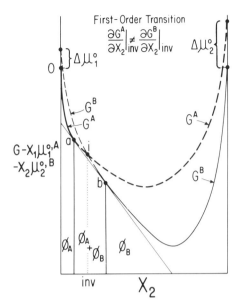

Fig. 3. Generalized free energy–composition relations for a system exhibiting a first-order phase transition. Points *a* and *b* mark the compositional limits (X_2^a and X_2^b) of a stable two-phase region associated with the transition. $\Delta\mu_1^\circ = \mu_1^{\circ,B} - \mu_1^{\circ,A}$, and $\Delta\mu_2^\circ = \mu_2^{\circ,A} - \mu_2^{\circ,B}$. (See text and references cited there for further explanation.) Additional notation for Figs. 3–6: *i*—point of phase transition; ϕ_A—phase(s) with structure A; ϕ_B—phase(s) with structure B; heavy curves (G^A)—free energy–composition relations for phases with structure A (lower-symmetry structure); light curves (G^B)—free energy–composition relations for phases with structure B (higher-symmetry structure); dashed curves—physically realistic metastable extensions of free energy–composition curves; dotted curves—physically meaningless extensions of free energy–composition curves; sloping straight lines—tangents to G–X_2 curves (for clarity, these tangents have not been extended to the $X_2 = 1$ sideline).

the inversion lie *below* curve G^B for some finite range of $X_2 > X_2^{\text{inv}}$. This compositional range is within the stability field for phases with structure B, so if a second-order transition were treated simply as a degenerate analogue of a first-order transition, then it is apparent that phase relations calculated from the resulting $\Delta\mu_1^\circ$, $\Delta\mu_2^\circ$, $G^{\text{ex,A}}$, and $G^{\text{ex,B}}$ equations of state would be inconsistent with the observed phase relations. These equations would fail to predict a change in phase at X_2^{inv} because $G^A < G^B$ on both sides of this composition; instead, a miscibility gap would be generated somewhere to the right of X_2^{inv}, as can be seen in Fig. 4. Increasing the magnitude of $\mu_2^{\circ,B} - \mu_2^{\circ,A}$ (with subsequent readjustments of the $G^{\text{ex,A}}$ and $G^{\text{ex,B}}$ expressions to restore the condition $\partial G^A = \partial G^B$ at X_2^{inv}) would reduce the range of X_2 to the right of X_2^{inv} where $G^A < G^B$, but a finite two-phase region would always remain. Accordingly, for the G–X relations of Fig. 4, we conclude that a metastable extension of curve G^B is theoretically valid, whereas curve G^A must terminate at X_2^{inv} without a metastable extension. An analogous situation is depicted in

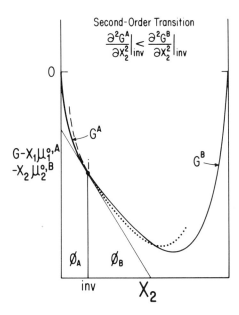

Fig. 4. Generalized free energy–composition relations for a binary system exhibiting a second-order phase transition at point i where $\partial^2 G^A < \partial^2 G^B$.

Fig. 5 where, as a consequence of $\partial^2 G^A > \partial^2 G^B$ at X_2^{inv}, the "metastable extension" of curve G^B lies below curve G^A within the stability field of phases with structure A. In this case only curve G^A is allowed to have a metastable extension beyond the inversion. Similar thermodynamic restrictions on metastable extensions of G functions have been recognized by earlier workers with regard to variation of G with temperature or pressure across a second-order transition in a one-component system (Epstein, 1937; Pippard, 1966; Rao and Rao, 1978).

In view of the thermodynamic principles presented above, we have devised procedures for developing empirical equations of state which are consistent with the constraints imposed by high-order (second- or third-order) phase transitions. Because they are empirical, these equations are of general applicability in thermodynamic modeling of systems that exhibit NFO transitions (e.g., the systems illustrated in Fig. 1), but particular algebraic forms of the relations are necessarily specific to the inferred order of the transition. Thus, for a second-order phase transition, and in contrast to the treatment of a first-order transition, the standard states of components 1 and 2 along the *minimum G–X path* are $\mu_1^{\circ,A}$ and $\mu_2^{\circ,B}$, respectively, across the *entire join*, regardless of the actual structure observed for any given phase. This approach to modeling G–X relations has the following consequences:

(1) The function for the free energy of mechanical mixing, $G^{\text{mm}} = X_1\mu_1^{\circ,A} + X_2\mu_2^{\circ,B}$, is the same for each stable G–X segment. This is in contrast to thermodynamic modeling of first-order transitions for which $G^{\text{mm,A}}$ and $G^{\text{mm,B}}$ must be represented by different functions owing to the different standard states for the two structures

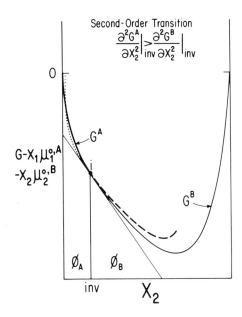

Fig. 5. Same as Fig. 4, but with $\partial^2 G^A > \partial^2 G^B$ at the transition point i.

(2) The ideal free energy of mixing, $G^{id} = G^{mm} + RT(X_1 \ln X_1 + X_2 \ln X_2)$, is a single function in the range $0 \leqslant X_2 \leqslant 1$, and *all* compositional derivatives of G^{id} are continuous across the transition.

(3) The discontinuity in $\partial^2 G$ at X_2^{inv} is attributable entirely to a discontinuity in $\partial^2 G^{ex}$ at that point. Therefore, separate equations are required for $G^{ex,A}$ and $G^{ex,B}$ which, at the inversion, obey the restrictions $G^{ex,A} = G^{ex,B}$, $\partial G^{ex,A} = \partial G^{ex,B}$, and $\partial^2 G^{ex,A} \neq \partial^2 G^{ex,B}$. These G^{ex} relations can be obtained by deriving expressions for the *activity coefficients* of components 1 and 2 in the two structures according to the following procedure. First, an empirical equation for $\ln \gamma_2^B$ of a form such as $\ln \gamma_2^B = \sum_{j=1}^{n} a_j (X_2^j - 1)$ is established, and then an expression for $\ln \gamma_2^A$ is formulated to be consistent with the constraints that, at X_2^{inv}, $\ln \gamma_2^A = \ln \gamma_2^B$ but $\partial \ln \gamma_2^A \neq \partial \ln \gamma_2^B$. [An example of such an expression is $\ln \gamma_2^A = (\ln \gamma_2^B)^{inv} + \sum_{k=1}^{n} a_k (X_2^k - (X_2^{inv})^k)$.] Subsequently, equations for $\ln \gamma_1^A$ and $\ln \gamma_1^B$ are obtained from the expressions for $\ln \gamma_2^A$ and $\ln \gamma_2^B$ via the Gibbs–Duhem relation and the boundary conditions $\ln \gamma_1^A = 0$ at $X_2 = 0$ and $\ln \gamma_1^B = \ln \gamma_1^A$ at X_2^{inv}.[4]

Figure 6 shows $G^{mix}-X$ relations for a system exhibiting a classical third-order transition where, by definition, both ∂G and $\partial^2 G$ are continuous. With the selection of standard states as for a second-order transition, there is again a single G^{mm} function for the stable phases, but now both ∂G^{ex} *and* $\partial^2 G^{ex}$ are continuous at X_2^{inv}. Activity–coefficient equations would be formulated in a manner similar to that for a second-order transition, with the additional

[4] An explicit equation for X_2^{inv} as a function of P and T is required in the expressions for $\ln \gamma_2^A$ and $\ln \gamma_1^B$. Also, to ensure that the appropriate $\ln \gamma_i^\phi$ equations are used, X_2^{inv} must be known in order to determine which of the structures A or B is stable for a given X_2.

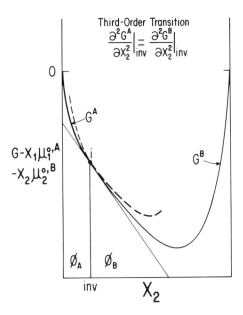

Fig. 6. Generalized free energy–composition relations for a binary system exhibiting a third-order phase transition.

constraint that, at X_2^{inv}, $\partial \ln \gamma_i^A = \partial \ln \gamma_i^B$ for both components. Accordingly, with an equation for $\ln \gamma_2^B$, an empirical expression of the form $\ln \gamma_2^A = (\ln \gamma_2^B)^{inv} + (X_2 - X_2^{inv})(\partial \ln \gamma_2^B/\partial X_2)^{inv} + \sum_{m=2}^{n} a_m(X_2 - X_2^{inv})^m$ is derived. Note that this equation satisfies all constraints on $\ln \gamma_i^\phi$ and $\partial \ln \gamma_i^\phi$ at the inversion. Expressions for $\ln \gamma_1^A$ and $\ln \gamma_1^B$ are then derived from the $\ln \gamma_2^A$ and $\ln \gamma_2^B$ equations as described previously for a second-order transition.

Finally, it should be recognized that the analysis presented above can be extended to include thermodynamic treatments of metastability associated with high-order phase transitions in binary systems. Thus, in Appendix A, we present derivations of generalized expressions for μ_1, μ_2, and G along the metastable extensions of curves G^A and G^B in Figs. 4–6.

Thermodynamic Analysis of Ion-Exchange Data for Analbite–Sanidine Feldspars

Methods for deriving the thermodynamic mixing properties of high albite–sanidine feldspars from hydrothermal ion-exchange data have been discussed previously by Thompson and Waldbaum (1968), Perchuk and Ryabchikov (1968), Wood (1977), and Merkel and Blencoe (1978). These investigators have shown that, with the definition

$$K_D \equiv \frac{X_{or}X_{NaCl}^v}{X_{b}X_{KCl}^v}, \tag{2}$$

equations for $\ln \gamma_{ab}$ and $\ln \gamma_{or}$ can be derived from the relations[5]

$$\ln \gamma_{ab} = \int_0^{X_{or}} X_{or} \partial \ln K_D, \tag{3a}$$

$$\ln \gamma_{or} = \int_1^{X_{or}} (X_{or} - 1)\partial \ln K_D. \tag{3b}$$

These equations were derived assuming that the feldspars form an isostructural crystalline solution series. However, treatment of the m/t inversion as a second- or third-order transition requires separate G^{ex} functions for the triclinic and monoclinic feldspars, and, as discussed previously, calculated G^{ex} curves must intersect at X_{or}^{inv} according to constraints imposed by the assumed order for the transition. These constraints, in turn, impose restrictions on the manner in which calculated $\ln K_D'$ and $\ln K_D^m$ curves intersect at the inversion. The following equations relate $\ln K_D$ to G^{ex} (Merkel and Blencoe, in preparation) at all X_{or}, independent of the order of the m/t inversion:

$$\ln K_D = const_{P,T} + \frac{\mu_{ab}^\circ - \mu_{or}^\circ}{RT} - \frac{\partial G^{ex}/\partial X_{or}}{RT} \tag{4}$$

and, therefore,

$$\frac{\partial \ln K_D}{\partial X_{or}} = -\frac{\partial^2 G^{ex}/\partial X_{or}^2}{RT}. \tag{5}$$

These relations apply to both triclinic and monoclinic feldspars, and the relationships between $\ln K_D'$ and $\ln K_D^m$ at the inversion can be derived from the relationships between $G^{ex,t}$ and $G^{ex,m}$ at that point.

For a nonisostructural crystalline solution series exhibiting a first-order transition, $\ln K_D$ curves must display a discontinuity at the inversion (Blencoe, 1979b). Therefore, *if* the m/t inversion were first order (which it is not), ion-exchange curves for analbite–sanidine feldspars would be similar to the schematic curves depicted at the top of Fig. 7. Points t' and m' represent $\ln K_D'$ and $\ln K_D^m$ at the inversion, and they must lie metastably within a two-phase region. Values of $\ln K_D'$ and $\ln K_D^m$ for a pair of feldspars coexisting stably on the boundaries of such a two-phase region are represented by points t'' and m'' in Fig. 7.

If the m/t transformation is second order, the quantities $\mu_{ab}^\circ - \mu_{or}^\circ$ and $\partial G^{ex}/\partial X_{or}$ are both continuous at the inversion; therefore, from Eq. (4) it is evident that $\ln K_D'$ and $\ln K_D^m$ curves must intersect at X_{or}^{inv}. However, it is also true that $\partial^2 G^{ex}$ must be discontinuous at the inversion, so from Eq. (5) we have $\partial \ln K_D' \neq \partial \ln K_D^m$ at X_{or}^{inv}, which means that the two $\ln K_D$ curves must intersect with different slopes. These relations are shown in the middle of Fig. 7.

Finally, if the m/t transformation is third order, $\mu_{ab}^\circ - \mu_{or}^\circ$, $\partial G^{ex}/\partial X_{or}$, and $\partial^2 G^{ex}/\partial X_{or}^2$ are all continuous at the inversion. From Eqs. (4) and (5) it is

---

[5]Note errors in, first of all, Eq. (6) of Perchuk and Ryabchikov (1968) in which the minus sign preceding the integral should not be present, and also in Eq. (14) of Wood (1977) wherein the term "X_{or}" within the integral should be "$(X_{or} - 1)$".

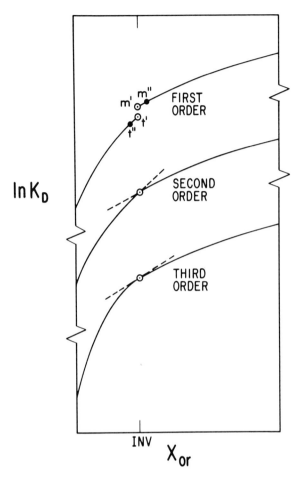

Fig. 7. Schematic ion-exchange ($\ln K_D$–X_{or}) relations for analbite–sanidine feldspars at an arbitrary but constant P and T. Regardless of the nature of the m/t inversion at X_{or}^{inv} (open circles), triclinic feldspars are more stable than monoclinic feldspars in the range $0 \leqslant X_{or} < X_{or}^{inv}$, and monoclinic feldspars are more stable than triclinic feldspars in the range $X_{or}^{inv} < X_{or} \leqslant 1$. However, the manner in which the *separate* $\ln K_D^t$ and $\ln K_D^m$ curves intersect at X_{or}^{inv} is dictated by the order of the m/t inversion (see text for explanation). For the *hypothetical* first-order m/t inversion in this figure, t' and m' represent $\ln K_D$ values for a metastable triclinic feldspar and a metastable monoclinic feldspar, respectively, each with a composition X_{or}^{inv} that lies within a stable two-phase region delimited by the solid circles labeled t'' and m''. For the cases in which the m/t inversion is assumed to be second or third order, the dashed lines represent tangents (*not* metastable extensions) to the $\ln K_D^t$ and $\ln K_D^m$ curves at X_{or}^{inv}. These tangents show that, for a second-order inversion, the $\ln K_D^t$ and $\ln K_D^m$ curves have different slopes at X_{or}^{inv}, whereas for a third-order inversion these slopes must be identical.

therefore evident that, at X_{or}^{inv}, $\ln K_D' = \ln K_D^m$ and $\partial \ln K_D' = \partial \ln K_D^m$. Consequently, the $\ln K_D$ curves intersect with the same slope at X_{or}^{inv}, a relationship illustrated by the schematic ion-exchange curves depicted at the bottom of Fig. 7.

Regardless of whether the m/t inversion is assumed to be second or third order, the ion-exchange phase relations of monoclinic feldspars can be represented by an equation of the form

$$\ln K_D^m = a_0 + b_0 X_{or} + c_0 X_{or}^2 + d_0 X_{or}^3 + \left(\frac{1000}{T} \right)(a_1 + b_1 X_{or}). \qquad (6)$$

Subsequently, and in accordance with the assumed order of the transition, an $\ln K_D'$ expression is formulated in a way that ensures proper intersection of the $\ln K_D^m$ and $\ln K_D'$ curves at the inversion. If the inversion is assumed to be second order, then the equation for triclinic feldspars may be of the form

$$\ln K_{D,2nd}' = (\ln K_D^m)^{inv} + b_0'(X_{or} - X_{or}^{inv})$$

$$+ c_0'\left(X_{or}^2 - (X_{or}^{inv})^2 \right) + b_1'\left(\frac{1000}{T} \right)(X_{or} - X_{or}^{inv}) \qquad (7)$$

with the value of X_{or}^{inv} at a given T supplied by Eq. (1). For a third-order inversion, a valid form for an $\ln K_D'$ equation is

$$\ln K_{D,3rd}' = (\ln K_D^m)^{inv} + (X_{or} - X_{or}^{inv})(\partial \ln K_D^m)^{inv} + c_0''(X_{or} - X_{or}^{inv})^2. \qquad (8)$$

Derivations of these equations are given by Merkel and Blencoe (in preparation). Values of the regression coefficients obtained from fitting hydrothermal ion-exchange data for analbite–sanidine feldspars at 2 kbars are presented in Table 2.

Substitution of Eqs. (6) and (7) or (6) and (8) into Eqs. (3a) and (3b) yields the following relations from which $\ln \gamma_{ab}$ and $\ln \gamma_{or}$ may be calculated for triclinic and monoclinic feldspars:

At $X_{or} \leqslant X_{or}^{inv}$ (triclinic feldspars)

$$\ln \gamma_{ab}^t = \int_0^{X_{or}} X_{or} \partial \ln K_D', \qquad (9a)$$

$$\ln \gamma_{or}^t = \int_1^{X_{or}^{inv}} (X_{or} - 1) \partial \ln K_D^m + \int_{X_{or}^{inv}}^{X_{or}} (X_{or} - 1) \partial \ln K_D', \qquad (9b)$$

and at $X_{or} \geqslant X_{or}^{inv}$ (monoclinic feldspars)

$$\ln \gamma_{ab}^m = \int_0^{X_{or}^{inv}} X_{or} \partial \ln K_D' + \int_{X_{or}^{inv}}^{X_{or}} X_{or} \partial \ln K_D^m, \qquad (10a)$$

$$\ln \gamma_{or}^m = \int_1^{X_{or}} (X_{or} - 1) \partial \ln K_D^m. \qquad (10b)$$

Appendix B contains generalized polynomial expressions for $\ln \gamma_{ab}^t$, $\ln \gamma_{or}^t$, $\ln \gamma_{ab}^m$, and $\ln \gamma_{or}^m$ that we have derived from Eqs. (9) and (10) assuming that the m/t inversion is either second or third order, and Table 2 lists numerical values for the coefficients in these expressions to calculate subsolidus phase relations for analbite–sanidine feldspars at 2 kbars.

Table 2. Regression coefficients for $\ln K_D$ equations derived from 2 kbar hydrothermal ion-exchange data for $(Na, K)AlSi_3O_8$ analbite–sanidine feldspars (Merkel and Blencoe, in preparation).

	Equation in Text	a_0	b_0	c_0	d_0	a_1	b_1
(a) Coefficients for "Isostructural" $\ln K_D$ Equations							
Set A[a]	(6)	−1.558	2.941	−5.941	2.357	0.0096	4.479
		(0.219)[b]	(0.893)	(1.715)	(0.968)	(0.1981)	(0.303)
Set B[c]	(6)	−1.387	4.181	−9.515	4.383	−0.413	5.090
		(0.178)	(0.446)	(0.879)	(0.583)	(0.161)	(0.264)

	Equation in Text	b_0'	c_0'	b_1'
(b) Coefficients for an $\ln K_{D,2nd}'$ Equation				
Set C	(7)	3.506	−13.356	7.114
		(2.811)	(7.361)	(3.129)

	Equation in Text	c_0''
(c) Coefficients for an $\ln K_{D,3rd}'$ Equation		
Set D	(8)	−14.3
		(8.6)

[a] These coefficients are for an $\ln K_D$ equation based on 2 kbar ion-exchange data for monoclinic high albite–sanidine feldspars.
[b] Numbers in parentheses are standard errors of the coefficients.
[c] These coefficients are for an $\ln K_D$ equation based on 2 kbar ion-exchange data for monoclinic *and* triclinic high albite–sanidine feldspars.

Calculated Solvus Relations for Analbite–Sanidine Feldspars

In this section we present a range of calculated subsolidus phase relations for analbite–sanidine feldspars assuming that the m/t inversion is either second or third order. Common to virtually all of these phase relations is the presence of an isostructural solvus at high temperatures and a nonisostructural solvus at low temperatures. The phase relations have been generated by employing a fixed equation for $\partial G^{ex,m}$, while an equation for $\partial G^{ex,t}$ was varied to determine the relationships between calculated phase equilibria and varying *relative* degrees of nonideality exhibited by the monoclinic and triclinic phases. Only one of the calculated phase diagrams can *possibly* provide a completely accurate schematic portrayal of subsolidus equilibria for analbite–sanidine feldspars at low pressures, but at the present time it is not certain which diagram this would be, or even which diagram is most nearly correct, and for this reason we present a range of possible calculated phase relations for this system. All phase equilibrium calculations were performed using methods modified from those described by Luth and Fenn (1973), and $G^{mix}-X_{or}$ relations at selected temperatures in Figs. 8(a), 8(b), 9(b), and 9(c) are presented in Appendix C.

Thermodynamic Treatment Consistent with a Second-Order m/t Inversion

We have employed both symmetric and asymmetric $G^{ex,t}$ expressions to calculate analbite–sanidine phase relations in which the m/t inversion is assumed to be second order. For a second-order transition, as noted previously, $\partial^2 G^t \neq \partial^2 G^m$ at X_{or}^{inv}; consequently, the triclinic and monoclinic spinodals cannot terminate at a common point along the $T-X_{or}^{inv}$ curve because, in this circumstance, both $\partial^2 G^t$ and $\partial^2 G^m$ would equal zero at X_{or}^{inv} (see Figs. 8–10).[6] In the examples shown in Figures 8 and 9 the $G^{ex,t}$ equations yield $\partial^2 G^t < \partial^2 G^m$ at X_{or}^{inv}. This result forces the triclinic spinodal to intersect the inversion at a temperature higher than that at which the monoclinic spinodal terminates. The opposite relationship is obtained if $\partial^2 G^t > \partial^2 G^m$ at the inversion; for example, as in Fig. 10. In Figs. 8–10 the points at which the $T-X_{or}^{inv}$ curve intersects (1) the triclinic spinodal (point b) and (2) the monoclinic spinodal (point a) are *tricritical points* according to the terminology of Allen and Cahn (1976a, 1976b).

Figures 8(a)–8(c) illustrate possible phase relations when $G^{ex,t}$ is either symmetric or only slightly asymmetric. In Fig. 8(a), $G^{ex,t}$ is low enough that the triclinic spinodal terminates at a temperature (T_b) below that at which the monoclinic isostructural solvus (MIS) intersects the inversion (T_c).[7] Due to the discontinuity in $\partial^2 G$ at point c, a change in slope of the sodic solvus limb occurs at temperature T_c, thus producing a kink.[8] However, our calculations show that the potassic solvus limb does *not* exhibit a change in slope at T_c, but instead shows only a slight change in curvature there. By analogy, these calculated results are consistent with the theoretical predictions of Allen and Cahn (1975, p. 1022) regarding the nature of the intersection of the FeAl/Fe$_3$Al λ-transition with the miscibility gap in the Fe–Al system.

The systems CaCO$_3$–MgCO$_3$ (Goldsmith and Heard, 1961) and CdCO$_3$–MgCO$_3$ (Goldsmith, 1972) may exhibit phase relations in the range $0.5 \leqslant X_{MgCO_3} \leqslant 1.0$ that are topologically similar to those illustrated in Fig. 8(a).

---

[6] The spinodal is the locus of points (spinodes) in $P-T-X$ space where $\partial^2 G/\partial X^2 = 0$ (Gordon, 1968, pp. 90–95).

[7] Here a subscript on T denotes the point in the interior of the $T-X$ diagram to which the temperature refers.

[8] Here and in subsequent discussion the term "kink" is used to indicate a discontinuity in the *first* derivative of a curve, whereas "change in curvature" refers to a discontinuity in the *second* derivative.

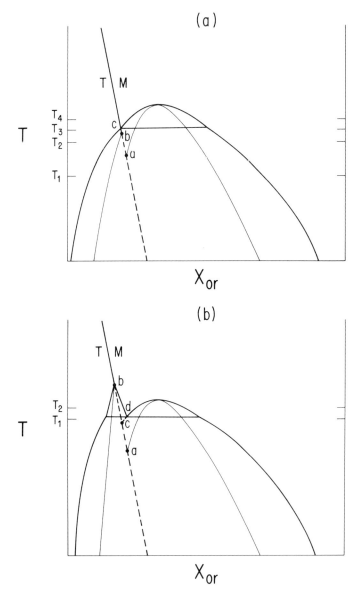

Fig. 8. Theoretically viable solvus relations for analbite–sanidine feldspars assuming that (1) the m/t inversion is second order, (2) $\partial^2 G^t < \partial^2 G^m$ at X_{or}^{inv}, and (3) there is no thermal maximum on the triclinic spinodal. Heavy curves are boundaries of two-phase regions, and light curves are spinodals. Points a, b, and c mark the terminations of the monoclinic spinodal (point a), the triclinic spinodal (point b), and the monoclinic binodal (point c) at the m/t inversion. Here and in subsequent figures, the T–X position of the m/t inversion (dashed where metastable) was calculated from Eq. (1). Free energy–composition relations for temperatures T_1–T_4 in (a) and T_1–T_2 in (b) are described in Appendix C. (a) Phase relations for a comparatively small $G^{ex,t}$ so that $T_b < T_c$. (b) Phase relations for a larger $G^{ex,t}$ so that $T_b > T_c$. This circumstance produces a conditional miscibility gap (CMG) that terminates at point b.

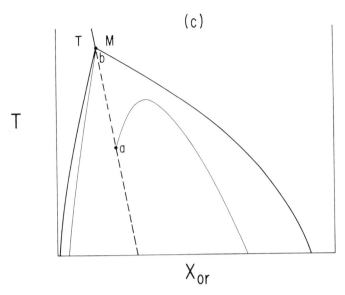

Fig. 8. (c) Phase relations when $G^{ex,t}$ is very large and the entire monoclinic isostructural solvus (MIS) lies metastably within a CMG.

In these systems the high-temperature $R\bar{3}c$ and low-temperature $R\bar{3}$ phase fields are separated by an order–disorder transformation that intersects the Mg-poor boundary of a two-phase region (Fig. 1(c)). This intersection subdivides the miscibility gap into an isostructural solvus at high temperatures and a nonisostructural solvus at low temperatures. Orientations of the spinodals within the solvi in the magnesian portions of these carbonate systems are based on analogy with phase relations depicted in Fig. 8(a).

Figure 8(b) illustrates phase relations that result when the magnitude of $G^{ex,t}$ is increased to the point that the triclinic spinodal intersects the inversion (point b) at a temperature above the metastable terminal point of the MIS (point c). In this case a small, nonisostructural two-phase region, which we will refer to here as a *conditional miscibility gap*[9] (CMG), is generated next to the MIS. This CMG terminates sharply at a tricritical point at T_b. In Fig. 8(b), X_{or}^{inv} is a "pseudoinflection point" on the G^{mix} curve at temperatures from T_a to T_b; consequently, the T–X_{or}^{inv} curve may be viewed as a "pseudospinodal" within the CMG.[10]

Figure 8(c) shows that, if $G^{ex,t}$ is very large, the CMG envelops the (now

---

[9] In their studies of the Fe–Al system, Allen and Cahn (1976a, 1976b) refer to the spinodal of the lower-symmetry phase as a "conditional spinodal" because its existence is "conditional on the presence of the lower-symmetry phase." That is, if the higher-symmetry phase persists metastably upon cooling through the transition, no conditional spinodal will exist on the (metastable) G–X curve, and there will be no free energy driving force for phase separation. Therefore, the miscibility gap itself may also be termed "conditional."

[10] The composition of the inversion is not a true inflection point on the G–X curve because, at such a point, in addition to changing sign, $\partial^2 G$ would have to be a continuous function and equal to zero. Neither of these criteria are satisfied at X_{or}^{inv} between T_a and T_b.

entirely metastable) MIS and becomes continuous with the low-temperature nonisostructural solvus. Of course, a miscibility gap generally similar to that in Fig. 8(c) could be generated even if nonideality in the monoclinic phase were low enough that no monoclinic spinodal existed. Other systems that may contain one or more such conditional miscibility gaps are illustrated in Figs. 1(a)–1(d).

Figures 9(a)–9(d) illustrate phase equilibria calculated from $G^{ex,t}$ equations that are sufficiently asymmetric to generate a thermal maximum on the triclinic spinodal. In Fig. 9(a), this type of maximum indicates the existence of a small, metastable, triclinic isostructural solvus (TIS) over a narrow temperature interval (see inset in Fig. 9(a)), but otherwise the stable phase relations are similar to those depicted in Fig. 8(a).

By increasing the magnitude of $G^{ex,t}$, a sufficiently asymmetric equation will cause the thermal maximum on the triclinic spinodal to lie at a temperature above T_c (Figs. 9(b)–9(d)). In this circumstance the TIS exists stably to the sodic side of the MIS and terminates at a consolute point (in contrast to a CMG which terminates at a tricritical point, as in Fig. 8(b)). It is to be expected that the T–X_{or}^{inv} curve will intersect one of the isostructural solvi at a higher temperature than the other, thereby creating a nonisostructural two-phase region over a narrow temperature interval. In Fig. 9(b) the inversion intersects the stable TIS giving rise to the two-phase region d-e-f-g. On the other hand, in Fig. 9(c) the inversion intersects the stable MIS, thereby producing the two-phase field d'-c-f'-g'.

Phase relations in Fig. 9(d) were calculated for a larger $G^{ex,t}$ which causes an additional increase in the temperature of both the triclinic consolute point and the tricritical point b, as well as an expansion of the two-phase field d''-e''-f''-g'' and a decrease in the thermal stability range of the overlying TIS. Further increase in $G^{ex,t}$ would cause points b, e'', and f'' to merge with the triclinic consolute point at a higher temperature on the T–X_{or}^{inv} curve, while for a still larger $G^{ex,t}$ a maximum on the triclinic spinodal would no longer exist and phase relations would become identical to those in Fig. 8(b).

Phase relations described up to this point were generated by treating the m/t inversion as second order with $\partial^2 G^t < \partial^2 G^m$ at X_{or}^{inv}. For comparison, phase relations calculated from a $G^{ex,t}$ equation that yields $\partial^2 G^t > \partial^2 G^m$ at X_{or}^{inv} are shown in Fig. 10. Note that, in contrast to the triclinic spinodals depicted in Figs. 8(a) and 9(a), the triclinic spinodal in Fig. 10 intersects the inversion at a temperature *below* that at which the monoclinic spinodal terminates. Also, the isostructural and nonisostructural solvi intersect at X_{or}^{inv} to form a kink that points toward $X_{or} = 0$ rather than toward $X_{or} = 1$ as observed for the cases where $\partial^2 G^t < \partial^2 G^m$ (Figs. 8(a) and 9(a)). Although this result would appear to violate the boundary curvature rule (Gordon, 1968, p. 188), this rule is based on orientations of metastable extensions of phase boundaries and G–X curves for *first-order* transitions, and therefore may not necessarily apply to high-order transitions.

A phase–boundary intersection similar to that at point c in Fig. 10 may occur in the system Fe–Al at the point where the T–X curve for the FeAl/Fe$_3$Al λ-transition transects the CMG (Fig. 1(a)). In their thermody-

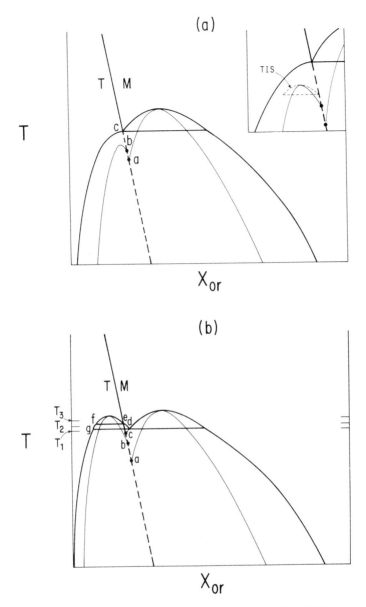

Fig. 9. Theoretically viable solvus relations for analbite–sanidine feldspars assuming that (1) the m/t inversion is second order, (2) $\partial^2 G^t < \partial^2 G^m$ at X_{or}^{inv}, and (3) $G^{ex,t}$ is sufficiently asymmetric to generate a thermal maximum on the triclinic spinodal. The heavy curves, light curves, and points *a*, *b*, and *c* in this figure are explained in the caption to Fig. 8. Free energy–composition relations for temperatures $T_1–T_3$ in (b) and $T_1–T_2$ in (c) are described in Appendix C. (a) Phase relations for a comparatively small $G^{ex,t}$ so that $T_b < T_c$. The inset shows a metastable triclinic isostructural solvus (TIS, light dashed curve) that exists over a limited temperature range due to the presence of a maximum on the triclinic spinodal. (b) Phase relations for a $G^{ex,t}$ that is sufficiently large to cause the maximum on the triclinic spinodal to lie at a temperature above T_c, thereby generating a stable TIS. Also, due to the intersection of the m/t inversion with the TIS, a nonisostructural two-phase region *d-e-f-g* is stable over a narrow temperature interval.

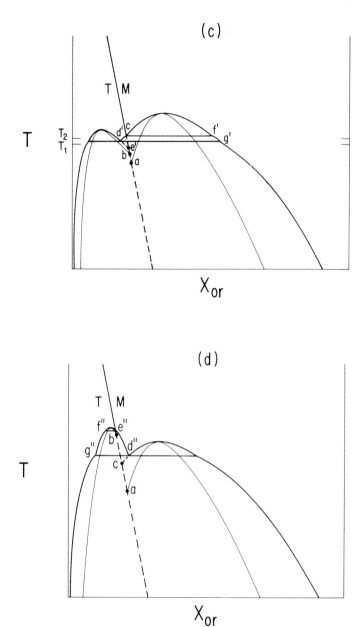

Fig. 9. (c) Similar to (b) in that $G^{ex,t}$ is sufficiently large to generate a stable TIS, but in this case it is the MIS that is intersected by the m/t inversion, thereby generating the underlying nonisostructural two-phase region d'-c-f'-g'. Point e' marks the termination of the TIS (metastable at $T < T_{d'}$) at the m/t inversion. (d) Phase relations for a very large $G^{ex,t}$ which causes T_b to lie well above T_c. The nonisostructural two-phase region d''-e''-f''-g'', generated by the intersection of the m/t inversion with the TIS, is stable over a considerably larger temperature interval than the analogous region d-e-f-g in (b).

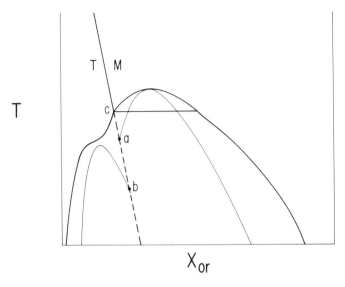

Fig. 10. An example of possible analbite–sanidine solvus relations given that the m/t inversion is second order and $\partial^2 G^{t,\mathrm{inv}} > \partial^2 G^{m,\mathrm{inv}}$. The heavy curves, light curves, and points *a*, *b*, and *c* in this figure are explained in the caption to Fig. 8.

namic analysis of this system, Allen and Cahn (1976a) postulate that the Fe_3Al spinodal terminates at the metastable extension of the $FeAl/Fe_3Al$ transition at a higher temperature than the point at which the FeAl spinodal terminates (Fig. 1(a)). However, by analogy with the phase relations depicted in Fig. 10, and as an alternative to the relations presented by Allen and Cahn, it is proposed here that the Fe_3Al spinodal could intersect this λ-transition at a temperature *lower* than that at which the FeAl spinodal terminates, as shown in the inset to Fig. 1(a)).

Thermodynamic Treatment Consistent with a Third-Order m/t Inversion

Figures 11(a)–11(c) illustrate the range of theoretically valid subsolidus phase relations for analbite–sanidine feldspars when the m/t inversion is assumed to be third order. For a third-order transition, $\partial^2 G$ is continuous; consequently, when $\partial^2 G^m = 0$ at $X_{\mathrm{or}}^{\mathrm{inv}}$, it necessarily follows that $\partial^2 G^t = 0$ at this point, so the triclinic and monoclinic spinodals must always terminate at a common point along the $T–X_{\mathrm{or}}^{\mathrm{inv}}$ curve.

Figure 11(a) shows phase relations for the case where the triclinic spinodal does not exhibit a thermal maximum. Raising $G^{\mathrm{ex},t}$ must simultaneously increase its asymmetry due to (1) the negative slope of the $T–X_{\mathrm{or}}^{\mathrm{inv}}$ curve, and (2) the constraint that $\partial^2 G^{\mathrm{ex}} = 0$ at the temperature and composition of point

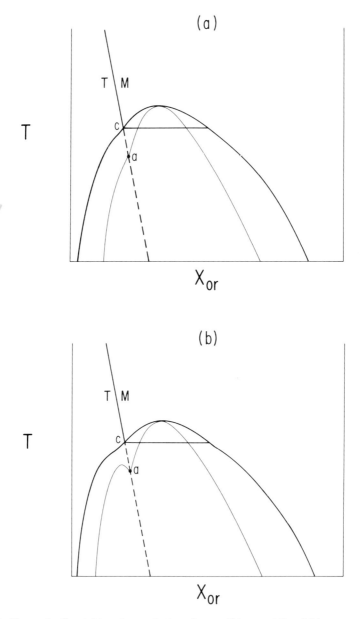

Fig. 11. Theoretically viable solvus relations for analbite–sanidine feldspars assuming that the m/t inversion is third order. The heavy and light curves in this figure are explained in the caption to Fig. 8. Points *a* and *c* mark the terminations of the triclinic and monoclinic spinodals (point *a*), and the monoclinic binodal (point *c*), at the m/t inversion. (a) Phase relations for a comparatively small $G^{ex,t}$ which does not produce a maximum on the triclinic spinodal. (b) Phase relations for a larger and more asymmetric $G^{ex,t}$ so that the triclinic spinodal exhibits a thermal maximum. A metastable TIS (not shown), similar to that illustrated in Fig. 9(a), must exist over a limited temperature range.

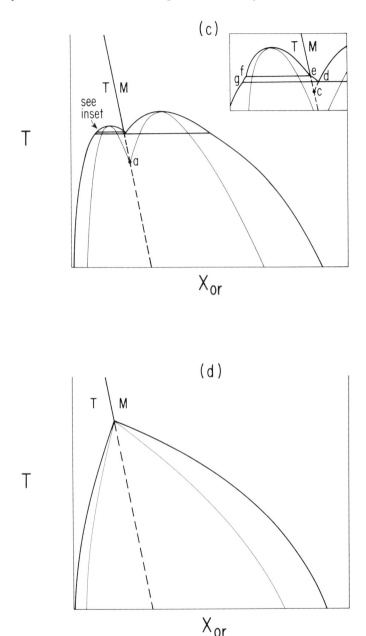

Fig. 11. (c) Phase relations for an even larger $G^{ex,t}$ so that the maximum on the triclinic spinodal lies above T_c, and a stable TIS is generated. Intersection of the m/t inversion with the TIS gives rise to a nonisostructural two-phase region *d-e-f-g* over a narrow temperature interval below the TIS (see inset). (d) Phase relations for $G^{ex,t}$ and $G^{ex,m}$ functions which do *not* produce a thermal maximum on *either* the triclinic or monoclinic spinodals. These spinodals and the limbs of the nonisostructural solvus all terminate at the same point on the T–X curve for the m/t inversion.

a. Figure 11(b) depicts phase relations for a larger $G^{ex,t}$ which produces a thermal maximum on the triclinic spinodal. In Figs. 11(a) and 11(b), the limbs of the nonisostructural solvus and those of the MIS meet with the same slope at X_{or}^{inv}; that is, there is no kink in the sodic limb at the inversion as in the case of a second-order transition. This behavior observed in our *calculated* phase relations when the m/t inversion is assumed to be third order is consistent with the qualitative predictions of these equilibria presented by Thompson and Waldbaum (1969).

Further increase in $G^{ex,t}$ causes the thermal maximum of the triclinic spinodal to rise to a sufficiently high temperature that a stable TIS is generated alongside the MIS, as shown in Fig. 11(c). These phase relations are analogous to those in Fig. 9(b), except that now the triclinic and monoclinic spinodals meet at the inversion. Although we have not performed the calculations, it is also likely that phase relations analogous to those of Fig. 9(c) could be generated for a third-order transition.

In a system exhibiting a third-order transformation, owing to the constraint that the triclinic and monoclinic spinodals must intersect at the inversion, it is impossible to generate a miscibility gap terminated by a tricritical point (such as in Figs. 8(b) and 8(c)) if one or both of the spinodals possess a thermal maximum. However, if *neither* spinodal exhibits a thermal maximum, then stable phase relations include a miscibility gap that terminates sharply at a tricritical point where the triclinic and monoclinic spinodals intersect one another at the inversion (Fig. 11(d)).

Comparison of Calculated and Experimentally Determined Solvi

In deriving their "preferred" monoclinic and triclinic activity–coefficient equations for analbite–sanidine feldspars from 2 kbar hydrothermal ion-exchange data, Merkel and Blencoe (in preparation) selected those equations that yield calculated solvi that are in closest agreement with the available solvus experimental data. Figure 12 illustrates solvus data from the literature, obtained at pressures from 1 to 15 kbars, which have been normalized to 2 kbars with a "correction factor" of 16°C/kbar (Parsons, 1978). Scatter in the data is attributable to disequilibrium in some experiments, to analytical errors, and perhaps also in part to varying degrees of Al–Si ordering which is known to affect the T–X limits of the solvus (Müller, 1971). Although the disposition of the data does not require the presence of a stable TIS or CMG, the scatter in the data is significant; therefore, the existence of a small two-phase region of either type cannot be completely ruled out.

Figure 13 illustrates various calculated solvi from Merkel and Blencoe (in preparation), as well as a solvus calculated from Eq. (10) of Thompson and Hovis (1979a). For comparison, the solvus data of Fig. 12 are represented by

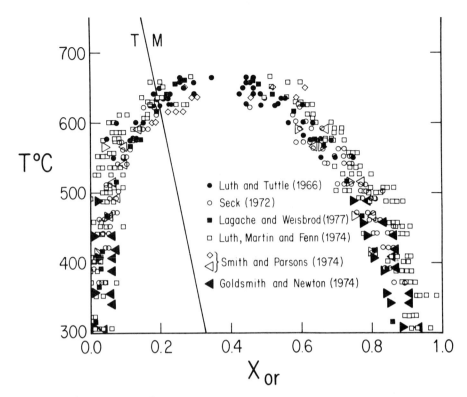

Fig. 12. High albite–sanidine solvus experimental data obtained at 1–15 kbars and normalized to 2 kbars by a "correction factor" of 16°C/kbar (Parsons, 1978).

horizontal bars drawn at selected temperatures, the width of each bar indicating approximately the compositional range spanned by the solvus data within ±10–15°C of the indicated temperature. It is noteworthy that the positions of the bars are consistent with an anomalous change in curvature of the sodic solvus limb in the temperature range 600–650°C. This feature is reasonably attributed to effects of the m/t inversion on the subsolidus phase relations of the feldspars.

Solvus A in Fig. 13 was calculated from activity–coefficient equations derived by regression analysis of 2 kbar ion-exchange data for *monoclinic* high albite–sanidine feldspars only. Activity coefficients were calculated from Eqs. (3a) and (3b) using the regression coefficients designated as set A in Table 2. The critical temperature (674°C), critical composition ($X_{or} = 0.325$), and position of the potassic limb of this isostructural solvus are in good agreement with the experimental data. However, the position of the sodic limb is considerably too potassic at temperatures below that at which the m/t inversion intersects the solvus, thus suggesting that thermodynamic mixing properties for analbite–sanidine crystalline solutions are not satisfactorily approximated by equations of state based entirely on ion-exchange data for monoclinic feldspars.

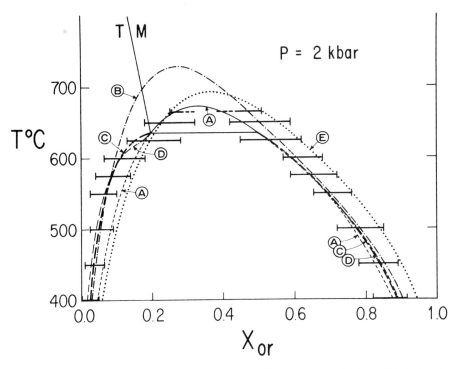

Fig. 13. Calculated solvi for the high albite–sanidine system at 2 kbars (Merkel and Blencoe, in preparation). Horizontal bars illustrate X_{or} ranges spanned by available solvus experimental data obtained within ± 10–$15°C$ of the indicated temperatures (see Fig. 12). Solvus A: a monoclinic isostructural solvus calculated from a G_{iso}^{ex} equation based on 2 kbar ion-exchange data for monoclinic high albite–sanidine feldspars. Solvus B: an isostructural solvus calculated from a G_{iso}^{ex} equation based on 2 kbar ion-exchange data for monoclinic *and* triclinic high albite–sanidine feldspars. Solvus C: a calculated nonisostructural (analbite–sanidine) solvus that is consistent with the assumption that the m/t inversion is second order. Solvus D: a calculated nonisostructural (analbite–sanidine) solvus that is consistent with the assumption that the m/t inversion is third order. Solvus E: a monoclinic isostructural solvus calculated from Eq. (10) of Thompson and Hovis (1979a). This equation is based on (1) the ion-exchange data of Orville (1963) and (2) the calorimetric (excess enthalpy) data of Hovis and Waldbaum (1977).

Solvus B was calculated from a G_{iso}^{ex} equation derived by regression analysis of 2 kbar ion-exchange data for both monoclinic *and* triclinic feldspars. The $\ln K_D$ equation fit to these data is of the form of Eq. (6), and activity–coefficient equations were derived from Eqs. (3a) and (3b). Values of the regression coefficients are those designated as set B in Table 2. This G_{iso}^{ex} equation was derived to investigate the possibility that a single equation could adequately describe the G^{mix} relations of the analbite–sanidine feldspars, and Fig. 13 shows that while calculated and experimentally determined phase relations are in good agreement below 600°C, solvus B is excessively asymmetric at higher temperatures and its critical temperature is much too high. These discrepancies indicate an inaccuracy in the activity–coefficient equations from which solvus B was calculated, and we believe that this inaccuracy is attributable to effects of the m/t inversion on the disposition of the ion-exchange data. Evidently, slight but "abrupt" changes in the slopes of ion-exchange curves near the inversion (Fig. 7) preclude accurate representation of the data with a single, "simple" polynomial $\ln K_D$ expression. Consequently, it is to be expected that a G_{iso}^{ex} equation derived from such an expression will fail to yield an accurate calculated solvus for the feldspars.

Figure 13 also shows phase relations calculated from the "preferred" G_{2nd}^{ex} equations of Merkel and Blencoe (in preparation). These equations were derived by fitting ion-exchange data for the monoclinic and triclinic feldspars separately with $\ln K_D^m$ and $\ln K'_{D,2nd}$ expressions of the forms of Eqs. (6) and (7), respectively. Activity–coefficient equations consistent with a second-order m/t inversion are presented in Appendix B, and regression coefficients for these equations are listed in Table 2 (coefficient sets A and C). In Fig. 13, the low-temperature, nonisostructural section of the solvus is indicated by the letter C, whereas the high-temperature, isostructural "cap" of the solvus is coincident with solvus A (see Appendix D).

"Preferred" G_{3rd}^{ex} equations were derived from $\ln K_D^m$ and $\ln K'_{D,3rd}$ expressions with the forms of Eqs. (6) and (8), respectively. [The "preferred" G_{2nd}^{ex} and G_{3rd}^{ex} equations of Merkel and Blencoe (in preparation) are based on the same $\ln K_D^m$ function that was used (indirectly) to calculate solvus A. However, our $\ln K'_{D,2nd}$ and $\ln K'_{D,3rd}$ equations were derived from somewhat different sets of 2 kbar hydrothermal ion-exchange data for triclinic (analbite) feldspars.] Activity–coefficient equations consistent with a third-order m/t inversion are given in Appendix B, and values for the regression coefficients are designated as sets A and D in Table 2. The low-temperature, nonisostructural section of the solvus calculated from the G_{3rd}^{ex} equations is indicated by the letter D in Fig. 13, and the high-temperature, isostructural cap of the solvus is coincident with solvus A.

G_{2nd}^{ex} and G_{3rd}^{ex} equations yield calculated solvi that are in much better agreement with available experimental data than are solvi calculated from either of the two G_{iso}^{ex} equations described previously. Figure 13 shows that, although the positions of the potassic limbs of all of the calculated solvi are consistent with the experimental data, only the G_{2nd}^{ex} and G_{3rd}^{ex} equations give solvi that are completely compatible with the experimental data for the sodic limb. Solvi calculated from the G_{2nd}^{ex} and G_{3rd}^{ex} equations also exhibit a critical temperature and composition that are completely accordant with the experimental data.

Conclusions

Non-first-order (NFO) transitions have significant effects on the subsolidus phase relations of many systems of metallurgical and petrological interest. Experimental data on the volumes, thermal expansivities, heat contents, and heat capacities of analbite–sanidine feldspars suggest that the mono-clinic/triclinic (m/t) inversion is an NFO transition. This inversion may be a λ-transition, but measurements of heat capacity and *bulk* thermal expansivity have failed to detect any λ anomalies. Consequently, at the present time it is reasonable to assume that the inversion is either a classical second- or third-order phase transition.

Equations of state for analbite–sanidine feldspars should be developed in a way that properly accounts for the thermodynamic effects of the m/t inversion. In our thermodynamic analysis of hydrothermal ion-exchange data for these feldspars, we have assumed that the inversion is either a second- or third-order transition, and G^{ex} equations derived from the ion-exchange data have been used to calculate solvi for the feldspars at 2 kbars. When the inversion is assumed to be second order, the solvus exhibits a kink in the sodic limb at X_{or}^{inv} (Fig. 9(a)); however, when the inversion is treated as third order, only a change in curvature of the sodic limb occurs at this point (Fig. 11(b)). Solvi calculated from "preferred" G_{2nd}^{ex} and G_{3rd}^{ex} equations are in very good agreement with the available solvus experimental data. Two G_{iso}^{ex} equations have also been derived for the feldspars, one of which was obtained by fitting just the ion-exchange data for monoclinic feldspars, whereas the other was derived by fitting a data set composed of ion-exchange data for both triclinic *and* monoclinic feldspars. Compared to solvi calculated from the "preferred" G_{2nd}^{ex} and G_{3rd}^{ex} equations, solvi generated from these G_{iso}^{ex} equations are in much less satisfactory agreement with the solvus experimental data.

Second- and third-order phase transitions can generate numerous subsolidus phase equilibrium topologies. The various phase relations for analbite–sanidine feldspars that we have calculated provide a basis for thermodynamic interpretations of a wide range of phase equilibria that are associated with high-order phase transitions in other mineral systems.

Appendix A: Expressions for μ_1, μ_2, and G along the Metastable Extensions of G–X_2 Curves in Figs. 4–6

In some binary systems exhibiting an NFO transition it may be possible for the higher-symmetry (high-temperature) phase to persist metastably at temperatures below the transition point, thus forestalling inversion to a more stable structure of lower symmetry. This situation is illustrated schematically in Figs. 4 and 6 in this chapter where the G–X_2 curve labeled G^B is metastable in the range $0 \leqslant X_2 < X_2^{inv}$, and also in Figs. 5 and 6 where curve G^A is metastable in the range $X_2^{inv} < X_2 \leqslant 1$. In this appendix, to show how such metastability can be treated thermodynamically, we derive generalized expressions for μ_1, μ_2, and G along the metastable extensions of curves G^A and G^B in Figs. 4–6.[11]

---

[11] In these derivations, an asterisk ("*") will be used to designate thermochemical quantities pertaining to a metastable phase. Standard states for the components are (at system P and T): for the metastable extension of curve G^B, pure crystalline components 1 and 2 with structure B; and for the metastable extension of curve G^A, pure crystalline components 1 and 2 with structure A.

First of all, to obtain expressions for μ_1^{B*}, μ_2^{B*}, and G^{B*} along the metasta-ble extensions of curve G^B in Figs. 4 and 6, we note that a relation for μ_2^{B*} is readily formulated because (1) the standard-state chemical potential for com-ponent 2 with structure B, $\mu_2^{\circ,B}$, is the same for both the stable and metastable portions of curve G^B, and (2) the $\ln \gamma_2^B$ equation for the stable segment of curve G^B can be extrapolated to $X_2 = 0$. Therefore, for the compositional range $0 \leqslant X_2 < X_2^{inv}$, we have

$$\mu_2^{B*} = \mu_2^{\circ,B} + RT \ln X_2 + RT \ln \gamma_2^B. \tag{A1}$$

To obtain an expression for μ_1^{B*}, we first use the equation for $\ln \gamma_2^B$ and the Gibbs–Duhem relation to derive an equation for $\ln \gamma_1^{B*}$:

$$\ln \gamma_1^{B*} = \int_0^{X_2} \left(\frac{X_2}{X_2 - 1} \right) \partial \ln \gamma_2^B. \tag{A2}$$

Next, to complete the definition of μ_1^{B*}, we note that the standard-state chemical potential $\mu_1^{\circ,B*}$ is established by the restriction $\mu_1^{B,inv} = \mu_1^{A,inv}$ at X_2^{inv} (this restriction holds for both second- and third-order phase transitions). Expanding this equality, we may now write

$$\mu_1^{\circ,B*} + RT \ln (\gamma_1^{B*})^{inv} + RT \ln X_1^{inv} = \mu_1^{\circ,A} + RT \ln (\gamma_1^A)^{inv} + RT \ln X_1^{inv} \tag{A3}$$

or

$$\mu_1^{\circ,B*} = \mu_1^{\circ,A} + RT \ln (\gamma_1^A / \gamma_1^{B*})^{inv}. \tag{A4}$$

Therefore,

$$\mu_1^{B*} = \mu_1^{\circ,A} + RT \ln \left(\frac{\gamma_1^A}{\gamma_1^{B*}} \right)^{inv} + RT \ln X_1 + RT \ln \gamma_1^{B*}. \tag{A5}$$

G^{B*} may then be evaluated from Eqs. (A1), (A5), and the relation

$$G^{B*} = X_1 \mu_1^{B*} + X_2 \mu_2^{B*}. \tag{A6}$$

Expressions for μ_1^{A*}, μ_2^{A*}, and G^{A*} along the metastable extensions of curves G^A in Figs. 5 and 6 can be derived in a similar fashion. Thus, we may write

$$\mu_1^{A*} = \mu_1^{\circ,A} + RT \ln X_1 + RT \ln \gamma_1^A, \tag{A7}$$

$$\ln \gamma_2^{A*} = \int_1^{X_2} \left(\frac{X_2 - 1}{X_2} \right) \partial \ln \gamma_1^A, \tag{A8}$$

$$\mu_2^{A*} = \mu_2^{\circ,B} + RT \ln \left(\frac{\gamma_2^B}{\gamma_2^{A*}} \right)^{inv} + RT \ln X_2 + RT \ln \gamma_2^{A*}, \tag{A9}$$

and

$$G^{A*} = X_1 \mu_1^{A*} + X_2 \mu_2^{A*}. \tag{A10}$$

Appendix B: Generalized Activity–Coefficient Equations for $(Na, K)AlSi_3O_8$ Analbite–Sanidine Feldspars

The generalized activity–coefficient equations listed below were developed from Eqs. (9) and (10) in the text. [For the latter equations, $\partial \ln K_D'$ was derived from Eq. (7) (m/t inversion = second-order) or Eq. (8) (m/t inversion = third order).] Values of the coefficients in Eqs. (B1)–(B7) to calculate 2 kbar analbite–sanidine phase relations are presented in Table 2, and X_{or}^{inv} at 2 kbars can be evaluated from Eq. (1) in the text.

Equations Consistent with a Second-Order m/t Inversion

$X_{or} \leqslant X_{or}^{inv}$:

$$\ln \gamma_{ab}^t = \frac{1}{2}\left[b_0' + b_1'\left(\frac{1000}{T} \right) \right] X_{or}^2 + \frac{2}{3} c_0' X_{or}^3, \tag{B1}$$

$$\ln \gamma_{or}^t = (\ln \gamma_{or}^m)^{inv} + \left[b_0' + b_1'\left(\frac{1000}{T} \right) \right]\left\{ \frac{1}{2}\left[X_{or}^2 - (X_{or}^{inv})^2 \right] - (X_{or} - X_{or}^{inv}) \right\}$$

$$+ 2c_0'\left\{ \frac{1}{3}\left[X_{or}^3 - (X_{or}^{inv})^3 \right] - \frac{1}{2}\left[X_{or}^2 - (X_{or}^{inv})^2 \right] \right\}. \tag{B2}$$

$X_{or} \geqslant X_{or}^{inv}$:

$$\ln \gamma_{ab}^m = (\ln \gamma_{ab}^t)_{2nd}^{inv} + \frac{1}{2} b_0\left[X_{or}^2 - (X_{or}^{inv})^2 \right] + \frac{2}{3} c_0\left[X_{or}^3 - (X_{or}^{inv})^3 \right]$$

$$+ \frac{3}{4} d_0\left[X_{or}^4 - (X_{or}^{inv})^4 \right] + \frac{1}{2} b_1\left(\frac{1000}{T} \right)\left[X_{or}^2 - (X_{or}^{inv})^2 \right], \tag{B3}$$

$$\ln \gamma_{or}^m = \frac{1}{2} b_0 + \frac{1}{3} c_0 + \frac{1}{4} d_0 - b_0 X_{or} + \left(\frac{1}{2} b_0 - c_0 \right) X_{or}^2 + \left(\frac{2}{3} c_0 - d_0 \right) X_{or}^3$$

$$+ \frac{3}{4} d_0 X_{or}^4 + \left(\frac{1000}{T} \right)\left(\frac{1}{2} b_1 - b_1 X_{or} + \frac{1}{2} b_1 X_{or}^2 \right). \tag{B4}$$

Equations Consistent with a Third-Order m/t Inversion

$X_{or} \leqslant X_{or}^{inv}$:

$$\ln \gamma_{ab}^t = \frac{1}{2} X_{or}^2\left[b_0 + 2c_0 X_{or}^{inv} + 3d_0(X_{or}^{inv})^2 + b_1\left(\frac{1000}{T} \right) \right]$$

$$+ c_0''\left[\frac{2}{3} X_{or}^3 - (X_{or}^{inv})X_{or}^2 \right], \tag{B5}$$

$$\ln \gamma_{or}^t = (\ln \gamma_{or}^m)^{inv} + \left\{ \frac{1}{2} \left[X_{or}^2 - (X_{or}^{inv})^2 \right] - (X_{or} - X_{or}^{inv}) \right\}$$

$$\times \left[b_0 + 2c_0 X_{or}^{inv} + 3d_0 (X_{or}^{inv})^2 + b_1 \left(\frac{1000}{T} \right) \right]$$

$$+ 2c_0'' \left\{ \frac{1}{3} \left[X_{or}^3 - (X_{or}^{inv})^3 \right] - \frac{1}{2} (1 + X_{or}^{inv}) \left[X_{or}^2 - (X_{or}^{inv})^2 \right] \right.$$

$$\left. + X_{or}^{inv} (X_{or} - X_{or}^{inv}) \right\}. \tag{B6}$$

$X_{or} \geqslant X_{or}^{inv}$:

$$\ln \gamma_{ab}^m = \left(\ln \gamma_{ab}^t \right)_{3rd}^{inv} + \frac{1}{2} b_0 \left[X_{or}^2 - (X_{or}^{inv})^2 \right] + \frac{2}{3} c_0 \left[X_{or}^3 - (X_{or}^{inv})^3 \right]$$

$$+ \frac{3}{4} d_0 \left[X_{or}^4 - (X_{or}^{inv})^4 \right] + \frac{1}{2} b_1 \left(\frac{1000}{T} \right) \left[X_{or}^2 - (X_{or}^{inv})^2 \right], \tag{B7}$$

$$\ln \gamma_{or}^m = \left[\text{right-hand side of Eq. (B4)} \right].$$

Appendix C: Free Energy–Composition Diagrams for Figs. 8(a), 8(b), 9(b), and 9(c) in the Text

In any binary or multicomponent system at a constant P and T, heterogeneous phase equilibria directly reflect the free energy-composition relations of the phases. Therefore, to further demonstrate that calculated phase relations presented in this chapter have a firm foundation in thermodynamic theory, and because we also suspect that many of the phase-equilibrium topologies described herein are unfamiliar to the general reader, we present in this appendix detailed explanations of the relationships between phase stability and free energy at various temperatures in Figs. 8(a), 8(b), 9(b), and 9(c).

Figures C1(a)–C1(d) illustrate $G^{mix}-X_{or}$ relations for temperatures T_1-T_4 in Fig. 8(a). At $T_1 < T_a$ (Fig. C1(a)) $G^{mix,t}$ and $G^{mix,m}$ each exhibit a single spinode, and a common tangent joins points on the triclinic and monoclinic segments of the G^{mix} curve, thereby defining the feldspar compositions (binodes) at the nonisostructural solvus limbs. At temperatures $T_a < T_2 < T_b$ (Fig. C1(b)) a second spinode is present along the monoclinic segment of the G^{mix} curve. This spinode originates at X_{or}^{inv} at T_a, and migrates to more potassic compositions with increasing temperature. Between T_a and T_b $\partial^2 G^{m,inv}$ is positive whereas $\partial^2 G^{t,inv}$ is negative, so $\partial^2 G$ changes sign at the inversion. However, $\partial^2 G$ is discontinuous at X_{or}^{inv} and, therefore, *senso stricto*, X_{or}^{inv} is not a true inflection point on the $G^{mix}-X_{or}$ curve in this temperature range. At higher temperatures the inversion and triclinic spinode approach one another and intersect at T_b. At $T_3 > T_b$ (Fig. C1(c)) $\partial^2 G^{t,inv}$ and $\partial^2 G^{m,inv}$ are both positive, so X_{or}^{inv} is no longer a "pseudoinflection point." Increasing temperature above T_c causes the sodic binode to shift from the triclinic

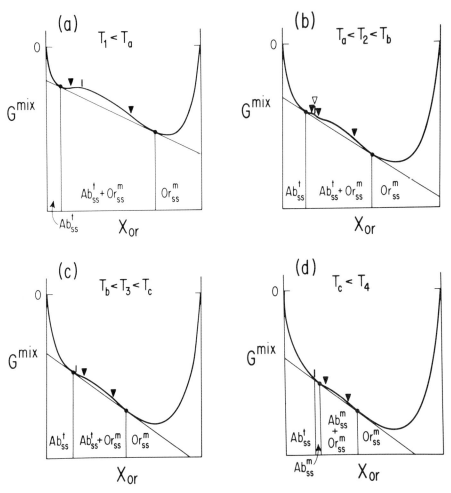

Fig. C1. $G^{mix}-X_{or}$ relations at temperatures (a) $T_1 < T_a$, (b) $T_a < T_2 < T_b$, (c) $T_b < T_3 < T_c$, and (d) $T_c < T_4$ in Fig. 8(a). Symbols for $G^{mix}-X_{or}$ curves in Figs. C1–C4: vertical bars—X_{or}^{inv}; filled triangles—spinodes; open triangles—points of *discontinuous* change in the sign of $\partial^2 G/\partial X_{or}^2$; filled circles—$X_{or}$ values at stable phase boundaries; open circles—X_{or} values at metastable phase boundaries. Compositional ranges of stable and metastable phase fields are also delimited by common-tangent lines (light lines for stable equilibria, dotted lines for metastable equilibria) that connect filled and open circles, respectively, in the figures. For clarity, extensions of common-tangent lines to $X_{or} = 1$ are not illustrated in some of the figures. See text for further explanation.

segment to the monoclinic segment of the G^{mix} curve, and at $T_4 > T_c$ the solvus becomes isostructural (Fig. C1(d)).

Figures C2(a) and C2(b) show $G^{mix}-X_{or}$ relations for temperatures T_1 and T_2 in Fig. 8(b). At $T_c < T_1 < T_d$ (Fig. C2(a)) the monoclinic isostructural solvus (MIS) is metastable due to a pronounced warping of the triclinic

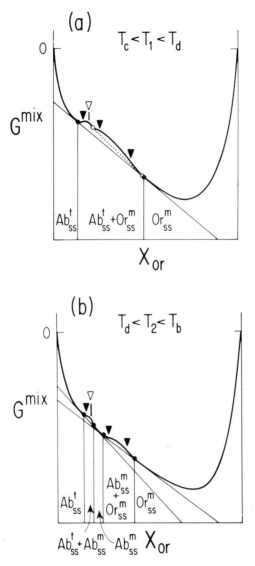

Fig. C2. G^{mix}–X_{or} relations at temperatures (a) $T_c < T_1 < T_d$ and (b) $T_d < T_2 < T_b$ in Fig. 8(b).

portion of the G^{mix}–X_{or} curve. (This major change in $\partial^2 G^{mix,t}$ near X_{or}^{inv} is attributable to strong nonideality of the triclinic phase.) With increasing temperature the warping becomes less pronounced as $G^{mix,m}$ decreases more than $G^{mix,t}$ near X_{or}^{inv}, and at T_2 (Fig. C2(b)) two stable common tangents are defined by the boundaries of the conditional miscibility gap ($Ab_{ss}^t + Ab_{ss}^m$) and the (now stable) MIS ($Ab_{ss}^m + Or_{ss}^m$). In Figs. C2(a) and C2(b), as in Fig. C1(b), X_{or}^{inv} is a "pseudoinflection point" on the G^{mix}–X_{or} curve.

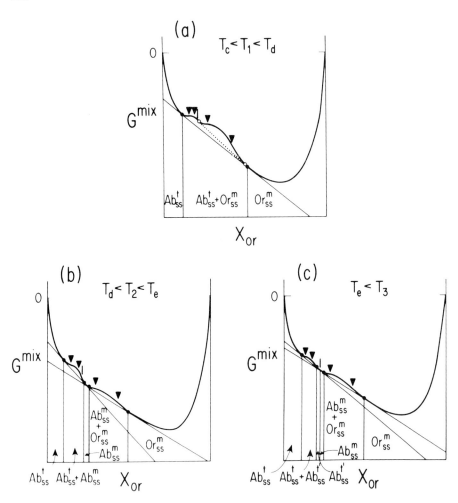

Fig. C3. G^{mix}–X_{or} relations at temperatures (a) $T_c < T_1 < T_d$, (b) $T_d < T_2 < T_e$, and (c) $T_e < T_3$ in Fig. 9(b).

Figures C3(a)–C3(c) illustrate G^{mix}–X_{or} relations for temperatures below, within, and above the thermal stability range of the two-phase region *d-e-f-g* in Fig. 9(b). At $T_c < T_1 < T_d$ (Fig. C3(a)) the MIS lies metastably within the nonisostructural solvus. With increasing temperature the free energy of the metastable, sodic, monoclinic binode decreases relative to the free energies of the nonisostructural binodes until at T_d a common tangent connects the three points. Above T_d (Fig. C3(b)) the MIS becomes stable, and the nonisostructural two-phase region *d-e-f-g* is generated simultaneously. With increasing temperature the potassic boundary (*d-e*) of this two-phase field trends toward $X_{or} = 0$ more rapidly than does X_{or}^{inv}, and at T_e the two points meet. At $T > T_e$ (Fig. C3(c)) field *d-e-f-g* is transformed to a stable triclinic isostructural solvus (TIS) because its potassic boundary now lies to the triclinic side of the inversion.

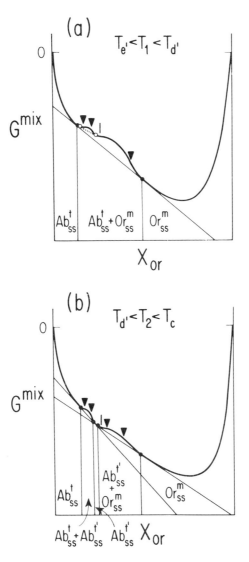

Fig. C4. G^{mix}–X_{or} relations at temperatures (a) $T_{e'} < T_1 < T_{d'}$ and (b) $T_{d'} < T_2 < T_c$ in Fig. 9(c).

G^{mix}–X_{or} relations for Fig. 9(c) at temperatures T_1 and T_2 are presented in Figs. C4(a) and C4(b). At T_1 it is the TIS, rather than the MIS as in Fig. C3(a), that exists metastably within the stable nonisostructural solvus. With increasing temperature the potassic, triclinic binode intersects the common tangent that defines the nonisostructural binodes, and at T_2 further down-warping of the G^{mix}–X_{or} curve causes splitting of the nonisostructural solvus into a stable TIS with a nonisostructural two-phase region to its potassic side. Thus, if the point of contact between the downwarping portion of the G^{mix}–X_{or} curve and the common tangent for the nonisostructural solvus lies to the potassic side of X_{or}^{inv} (Fig. C3(b)), the phase relations depicted in Fig. 9(b) are generated. Alternatively, if this contact point lies to the sodic side of X_{or}^{inv} (Fig. C4(b)), the resulting phase equilibria are as illustrated in Fig. 9(c).

Appendix D: A Proof that Isostructural and Nonisostructural G^{ex} Equations with the Same $\partial G^{ex,m}$ Function Yield Identical Monoclinic Isostructural Solvi for $(Na, K)AlSi_3O_8$ High Albite–Sanidine Feldspars

Consider a set of G_{2nd}^{ex} and G_{3rd}^{ex} equations for high albite–sanidine feldspars derived from expressions for $\ln K_D^t$ and $\ln K_D^m$, and also a G_{iso}^{ex} equation derived from the $\ln K_D^m$ expression. From Eq. (4) in the text it is evident that these G^{ex} equations must have the same $\partial G^{ex,m}$ function. Furthermore, for the stable coexistence of two monoclinic feldspars (a solvus-pair) with compositions X_{or}' and X_{or}'', respectively, the activity equivalence conditions

$$a_{ab}' = a_{ab}'' \tag{D1a}$$

and

$$a_{or}' = a_{or}'' \tag{D1b}$$

must be satisfied. Therefore, and from the definition $a_i = \gamma_i X_i$, we have

$$\gamma_{ab}' X_{ab}' = \gamma_{ab}'' X_{ab}'' \tag{D2a}$$

and

$$\gamma_{or}' X_{or}' = \gamma_{or}'' X_{or}''. \tag{D2b}$$

Rearranging terms then gives

$$X_{ab}'' / X_{ab}' = \gamma_{ab}' / \gamma_{ab}'' \tag{D3a}$$

and

$$X_{or}'' / X_{or}' = \gamma_{or}' / \gamma_{or}''. \tag{D3b}$$

Consequently, in order to obtain the same monoclinic isostructural solvus (that is, identical values for X_{ab}'' / X_{ab}' as well as X_{or}'' / X_{or}') whether calculations are performed using a G_{iso}^{ex} equation or a set of G_{2nd}^{ex} or G_{3rd}^{ex} equations, it is apparent that the ratios $\gamma_{ab}' / \gamma_{ab}''$ and $\gamma_{or}' / \gamma_{or}''$ must be the same for each case.

Considering first the ratio $\gamma_{ab}' / \gamma_{ab}''$, γ_{ab}^m for the isostructural (I) G_{iso}^{ex} equation is defined by Eq. (3a) in the text, while γ_{ab}^m for a set of nonisostructural (NI) G_{2nd}^{ex} or G_{3rd}^{ex} functions is given by Eq. (10a) in the text. By subdividing the integral in equation (3a), we obtain

$$\ln \gamma_{ab}^{m,I} = \int_0^{X_{or}^{inv}} X_{or} \partial \ln K_D^m + \int_{X_{or}^{inv}}^{X_{or}} X_{or} \partial \ln K_D^m. \tag{D4}$$

Subtracting Eq. (D4) from Eq. (10a) then yields

$$\ln \gamma_{ab}^{m,NI} - \ln \gamma_{ab}^{m,I} = \int_0^{X_{or}^{inv}} X_{or} \partial \ln K_D^t - \int_0^{X_{or}^{inv}} X_{or} \partial \ln K_D^m$$

or

$$\ln \gamma_{ab}^{m,NI} = \ln \gamma_{ab}^{m,I} + \int_0^{X_{or}^{inv}} X_{or}(\partial \ln K_D^t - \partial \ln K_D^m). \tag{D5}$$

At any fixed P and T the integral in Eq. (D5) is a constant ("C"); therefore, we may write

$$\gamma_{ab}^{m,NI} = \gamma_{ab}^{m,I} \exp(C). \tag{D6}$$

Consequently,

$$(\gamma_{ab}^{m,NI})'/(\gamma_{ab}^{m,NI})'' = \left[\gamma_{ab}^{m,I} \exp(C) \right]'/\left[\gamma_{ab}^{m,I} \exp(C) \right]''$$

or

$$(\gamma_{ab}^{m,NI})'/(\gamma_{ab}^{m,NI})'' = (\gamma_{ab}^{m,I})'/(\gamma_{ab}^{m,I})'' \tag{D7}$$

and the ratio $\gamma_{ab}'/\gamma_{ab}''$ must be identical for the G_{iso}^{ex}, G_{2nd}^{ex}, and G_{3rd}^{ex} equations.

Regarding the ratio $\gamma_{or}'/\gamma_{or}''$, Eqs. (3b) and (10b) in the text show that, at a given $X_{or} > X_{or}^{inv}$, the G_{iso}^{ex}, G_{2nd}^{ex}, and G_{3rd}^{ex} equations yield identical values for γ_{or}^m. Consequently, each of these functions would also yield the same value for $\gamma_{or}'/\gamma_{or}''$.

Thus, the activity-coefficient *ratios* in Eqs. (D3a) and (D3b) are the same for G_{iso}^{ex}, G_{2nd}^{ex}, and G_{3rd}^{ex} functions, provided that these functions are based on the same $\ln K_D^m$ expression.

Acknowledgments

We thank Drs. D. K. Bird, G. L. Hovis, A. Navrotsky, and S. K. Saxena for their constructive criticisms of the original manuscript. Thanks are also extended to Mr. R. J. Texter who drafted the figures, and to Mrs. D. E. Detwiler who typed the manuscripts. Financial support for this study was provided mainly by NSF Grant EAR-7822443, but a large proportion of computer and terminal-time costs were defrayed by The Pennsylvania State University. The first author also gratefully acknowledges a Three-Year NSF Graduate Fellowship, a Pennsylvania State University Graduate Fellowship, and a Domestic Mining (MMMFC) Fellowship from the Department of Education.

References

Allen, S. M., and Cahn, J. W. (1975) Coherent and incoherent equilibria in iron-rich iron–aluminum alloys, *Acta Metall.* **23**, 1017–1026.

Allen, S. M., and Cahn, J. W. (1976a) On tricritical points resulting from the intersection of lines of higher-order transitions with spinodals, *Scr. Metall.* **10**, 451–454.

Allen, S. M., and Cahn, J. W. (1976b) Mechanisms of phase transformations within the miscibility gap of Fe-rich Fe–Al alloys, *Acta Metall.* **24**, 425–437.

Blencoe, J. G. (1977) Computation of thermodynamic mixing parameters for isostructural, binary crystalline solutions using solvus experimental data, *Comput. Geosci.* **3**, 1–18.

Blencoe, J. G. (1979a) The use of thermodynamic excess functions in the Nernst distribution law: discussion. *Amer. Mineral.* **64**, 1122–1128.

Blencoe, J. G. (1979b) An ion-exchange method for evaluating $\Delta\mu_1^\circ$ and $\Delta\mu_2^\circ$ for nonisostructural binary crystalline solutions, *Amer. Geophys. Union Trans.* **60**, 404–405.

Carpenter, M. A. (1980) Mechanisms of exsolution in sodic pyroxenes, *Contrib. Mineral. Petrol.* **71**, 289–300.

Carpenter, M. A. (1981) A "conditional spinodal" within the peristerite miscibility gap of plagioclase feldspars, *Amer. Mineral.* **66**, 553–560.

Donnay, G., and Donnay, J. D. H. (1952) The symmetry change in the high-temperature alkali–feldspar series, *Amer. J. Sci.* **250A**, 115–132.

Ehrenfest, P. (1933) Phasenumwandlunsen im ueblichen und erweiterten sinn, klassifiziert nach den entsprechenden singularitaeten des thermodynamischen potentiales, *Proc. Acad. Sci. Amsterdam* **36**, 153.

Epstein, P. S. (1937) *Textbook of Thermodynamics*. Wiley, New York.

Goldsmith, J. R. (1972) Cadmium dolomite and the system $CdCO_3$–$MgCO_3$, *J. Geology* **80**, 611–626.

Goldsmith, J. R., and Heard, H. C. (1961) Subsolidus relations in the system $CaCO_3$–$MgCO_3$. *J. Geology* **69**, 45–74.

Goldsmith, J. R., and Newton, R. C. (1974) An experimental determination of the alkali feldspar solvus, in *The Feldspars*, edited by W. S. MacKenzie and J. Zussman, Proc. NATO Adv. Study Inst., pp. 337–359. Manchester University Press, Manchester, England.

Gordon, P. (1968) *Principles of Phase Diagrams in Material Systems*. McGraw-Hill, New York.

Hazen, R. M. (1976) Sanidine: predicted and observed monoclinic-to-triclinic reversible transformations at high pressure, *Science* **194**, 105–107.

Hazen, R. M., and Finger, L. W. (1979) Polyhedral tilting: A common type of pure displacive phase transition and its relationship to analcite at high pressure, *Phase Transitions* **1**, 1–22.

Henderson, C. M. B. (1979) An elevated temperature X-ray study of synthetic disordered Na–K alkali feldspars, *Contrib. Mineral. Petrol.* **70**, 71–79.

Henderson, C. M. B., and Roux, J. (1979) Inversions in subpotassic nephelines, *Contrib. Mineral. Petrol.* **61**, 279–298.

Henderson, C. M. B., and Thompson, A. B. (1980) The low-temperature inversion in subpotassic nephelines, *Amer. Mineral.* **65**, 970–980.

Hovis, G. L. (1980) Angular relations of alkali feldspar series and the triclinic-monoclinic displacive transformation, *Amer. Mineral.* **65**, 770–778.

Hovis, G. L., and Waldbaum, D. L. (1977) A solution calorimetric investigation of K–Na mixing in a sanidine–analbite ion-exchange series, *Amer. Mineral.* **62**, 680–686.

Kroll, H. (1971) Feldspäte im System $KAlSi_3O_8$–$NaAlSi_3O_8$–$CaAl_2Si_2O_8$: Al, Si-Verteilung und Gitterparameter, Phasen-Transformationen und Chemismus. Dissertation, Westf Wilhems-Universität, Munster, Germany.

Kroll, H., Bambauer, H. U., and Schirmer, U. (1980) The high albite–monalbite and analbite–monalbite transitions, *Amer. Mineral.* **65**, 1192–1211.

Lagache, M., and Weisbrod, A. (1977) The system: two alkali feldspars–KCl–NaCl–H_2O at moderate to high temperatures and low pressures, *Contrib. Mineral. Petrol.* **62**, 77–101.

Laves, F. (1952) Phase relations of the alkali feldspars. II. The stable and pseudo-stable phase relations in the alkali feldspar system, *J. Geology* **60**, 549–574.

Lindsley, D. H., Davidson, P. M., and Grover, J. E. (1980) Ca–Mg pyroxenes: a solution model that permits coexisting En_{ss} + Pig + Di_{ss}, *Geol. Soc. Amer., Abstr. Progs.* **12**, 472.

Lindsley, D. H., Grover, J. E., and Davidson, P. M. (1981) The thermodynamics of the $Mg_2Si_2O_6$–$CaMgSi_2O_6$ join: a review and a new model, in *Advances in Physical Geochemistry*, edited by R. C. Newton, A. Navrotsky, and B. J. Wood, Vol. 1, Springer-Verlag, New York.

Luth, W. C., and Fenn, P. M. (1973) Calculation of binary solvi with special reference to the sanidine–high albite solvus, *Amer. Mineral.* **58**, 1009–1015.

Luth, W. C., Martin, R. F., and Fenn, P. M. (1974) Peralkaline alkali feldspar solvi, in *The Feldspars*, edited by W. S. MacKenzie and J. Zussman, Proc NATO Adv. Study Inst., pp. 297–312. Manchester University Press, Manchester, England.

Luth, W. C., and Querol-Suñé, F. (1970) An alkali feldspar series, *Contrib. Mineral. Petrol.* **25**, 25–40.

Luth, W. C., and Tuttle, O. F. (1966) The alkali feldspar solvus in the system Na_2O–K_2O–Al_2O_3–SiO_2–H_2O, *Amer. Mineral.* **51**, 1359–1373.

MacKenzie, W. S. (1952) The effect of temperature on the symmetry of high-temperature soda-rich feldspars, *Amer. J. Sci.* **252**A, 319–342.

Merkel, G. A., and Blencoe, J. G. (1978) Calculated activity coefficients and thermodynamic excess properties for high albite–sanidine feldspars at 2 kbar pressure, 600–700°C, *Amer. Geophys. Union Trans.* **59**, 395.

Merkel, G. A., and Blencoe, J. G. (in preparation) Thermodynamic mixing properties of binary analbite-sanidine feldspars.

Müller, G. (1971) Der einfluss der Al, Si-verteilung auf die mischungslücke der alkalifeldspäte, *Contrib. Mineral. Petrol.* **34**, 73–79.

Nukui, A., Nakazawa, H., and Akao, M. (1978) Thermal changes in monoclinic tridymite, *Amer. Mineral.* **63**, 1252–1259.

Okamura, F. P., and Ghose, S. (1975) Analbite → monalbite transition in a heat treated twinned Amelia albite, *Contrib. Mineral. Petrol.* **50**, 211–216.

Orville, P. M. (1963) Alkali ion exchange between vapor and feldspar phases, *Amer. J. Sci.* **261**, 201–237.

Orville, P. M. (1967) Unit-cell parameters of the microcline–low albite and the sanidine–high albite solid solution series, *Amer. Mineral.* **52**, 55–86.

Parsons, I. (1978) Alkali feldspars: which solvus? *Phys. Chem. Minerals* **2**, 199–213.

Perchuk, L. L., and Ryabchikov, I. D. (1968) Mineral equilibria in the system nepheline–alkali feldspar–plagioclase and their petrological significance, *J. Petrology* **9**, 123–167.

Pippard, A. B. (1966) *Elements of Classical Thermodynamics*. Cambridge University Press, Cambridge.

Rao, C. N. R., and Rao, K. J. (1978) *Phase Transitions in Solids*. McGraw-Hill, New York.

Seck, H. A. (1972) The influence of pressure on the alkali feldspar solvus from peraluminous and persilicic materials, *Fortschr. Mineral.* **49**, 31–49.

Smith, P., and Parsons, I. (1974) The alkali–feldspar solvus at 1 kilobar water-vapour pressure, *Mineral. Mag.* **39**, 747–767.

Thompson, A. B., and Perkins, E. H. (1981) Lambda transitions in minerals, in *Advances in Physical Geochemistry*, edited by R. C. Newton, A. Navrotsky, and B. J. Wood, Vol. 1. Springer-Verlag, New York.

Thompson, A. B., and Wennemer, M. (1979) Heat capacities and inversions in tridymite, cristobalite, and tridymite–cristobalite mixed phases, *Amer. Mineral.* **64**, 1018–1026.

Thompson, J. B., Jr., and Hovis, G. L. (1979a) Entropy of mixing in sanidine, *Amer. Mineral.* **64**, 57–65.

Thompson, J. B., Jr., and Hovis, G. L. (1979b) Structural–thermodynamic relations of the alkali feldspars, *Trans. Amer. Crystallogr. Assoc.* **15**, 1–26.

Thompson, J. B., Jr., and Waldbaum, D. R. (1968) Mixing properties of sanidine crystalline solutions: I. Calculations based on ion-exchange data, *Amer. Mineral.* **53**, 1965–1999.

Thompson, J. B., Jr., and Waldbaum, D. R. (1969) Mixing properties of sanidine crystalline solutions: III. Calculations based on two-phase data, *Amer. Mineral.* **54**, 811–838.

Traetteberg, A., and Flood, H. (1972) Alkali ion exchange equilibria between feldspar phases and molten mixtures of potassium and sodium chloride, *Trans. Roy. Inst. Technol. Stockholm* **296**, 608–618.

Willaime, C., Brown, W. L., and Perucaud, M. C. (1974) On the orientation of the thermal and compositional stain ellipsoids in feldspars, *Amer. Mineral.* **59**, 457–464.

Winter, J. K., Okamura, F. P., and Ghose, S. (1979) A high-temperature structural study of high albite, monalbite, and the analbite→monalbite phase transition, *Amer. Mineral.* **64**, 409–423.

Wood, B. J. (1977) Experimental determination of the mixing properties of solid solutions with particular reference to garnet and clinopyroxene solutions, in *Thermodynamics in Geology*, edited by D. G. Fraser, pp. 11–27. Reidel, Dordrecht, The Netherlands.

Wright, T. L., and Stewart, D. B. (1968) X-ray and optical study of alkali feldspar. I. Determination of composition and structural state from refined unit-cell parameters and 2V, *Amer. Mineral.* **53**, 38–87.

Zyrianov, V. N., Perchuk, L. L., and Podlesski, K. K. (1978) Nepheline–alkali feldspar equilibria: I. Experimental data and thermodynamic calculations, *J. Petrology* **19**, 1–44.

Chapter 9

Gibbs Free Energies of Formation for Bayerite, Nordstrandite, $Al(OH)^{2+}$, and $Al(OH)_2^+$, Aluminum Mobility, and the Formation of Bauxites and Laterites

B. S. Hemingway

Introduction

A knowledge of the low-temperature hydrolysis of aqueous aluminum solutions is important because of the role of aluminum in soil formation, because of the use of aluminum in the elimination of colloids and organic matter through flocculation and in other water treatment processes, and to improve the general analytical chemistry of aluminum and the processes for commercial aluminum extraction. Schoen and Roberson (1970) have stated that "our principal gaps in understanding the geochemistry of aluminum arise from the lack of detailed knowledge of the controls on solubility as well as the kinds and amounts of substances in solution. In addition, the aluminous solids that precipitate from supersaturated solutions must be adequately characterized." The purpose of this discussion is the development of a model of the processes that control the low-temperature aqueous aluminum system and, although the problems cited by Schoen and Roberson are not entirely resolved, to review a portion of the vast amount of research gathered during the last two decades that has greatly improved our knowledge of the factors controlling the geochemistry of aluminum.

Extensive research into the factors controlling aluminum solubility and the occurrence of compounds in the system Al–H–O had been reported in the literature (e.g., Kittrick, 1980; May *et al.*, 1980; Violante and Violante, 1980; Kwong and Huang, 1979; Lind and Hem, 1975; Smith and Hem, 1972; Parks, 1972; Ross and Turner, 1971; Schoen and Roberson, 1970; Turner and Ross, 1970; Hem and Roberson, 1967; Barnhisel and Rich, 1965; and Bye and Robinson, 1964). Confusion regarding the interpretation of the experimental data from this system results from the metastability of intermediate phases and the slow rates of reaction (see discussion by Hsu (1967)), from incomplete identification of solid phases (note discussion by Schoen and Roberson (1970)), and from the use of different reference states in calculating the thermodynamic properties of phases or ionic species in this system (note the discussions of Parks (1972) and Hemingway *et al.* (1978)). As recently as 1980,

Violante and Violante (1980) have stated that "it is not clear why nordstrandite and gibbsite, rather than bayerite, are more common in alkaline soils or in
bauxite" and cite the lack of knowledge of the factors controlling the formation of the three aluminum hydroxide polymorphs.

The problems associated with our extrapolation of laboratory experiments
to explain weathering processes arises from our past failure to understand that
the low-temperature chemistry of the aqueous aluminum system (as well as
many other geologic processes) is controlled by a mechanism discussed either
as the law of successive reactions or the Gay-Lussac–Ostwald step rule
(Ostwald, 1897). A model for the low-temperature geochemistry of aqueous
aluminum will be developed based upon the law of successive reactions, and
the kinetics of these low-temperature reactions will be discussed in terms of
the factors that change the normal hydrolysis process and thereby alter the
free-energy differences driving the reactions.

The thermodynamic properties of the aluminum hydroxide phases, bayerite,
and nordstrandite and of the aqueous aluminum species Al^{3+}, $Al(OH)^{2+}$, and
$Al(OH)_2^+$ will be revised based upon the model of the low-temperature
chemistry of aluminum, a reinterpretation of several studies of the solubility of
aluminum hydroxide phases, and upon the recent excellent solubility data
provided by May *et al.* (1979).

The Gay-Lussac–Ostwald Step Rule or Law of Successive Reactions, and the Al–Si–O–H System

In order to establish a framework for the interpretation of low-temperature
experiments in the system Al–Si–O–H, the following observations regarding
the law of successive reactions must be made. The step rule, or law of
successive reactions, holds that in all *processes* in which several relatively
stable states occur it is not the most stable state (the phase having the lowest
free energy) that forms first, but the state nearest to the original state in free
energy. If between the initial and final states there are several intermediate
states of relative stability, one will succeed the next in the order of decreasing
free energy resulting in a stepwise change from the initial to the final state.
Ostwald (1897) formulated the law of successive reactions based upon the
observations that unstable forms frequently arise before the stable form, as
noted by Gay-Lussac.

The law of successive reactions describes the process that operates in the
system Al–Si–O–H at low temperatures (< 373 K). Willstätter and Kraut
(1924) proposed a classification of aluminum hydroxide gels obtained from
dissolved aluminum salts through precipitation and aging. Bye and Robinson
(1964) summarized the results presented in the subsequent 40 years through
the simplified reaction scheme: freshly precipitated amorphous gel → pseudoboehmite (gelatinous) → bayerite (nordstrandite may also be present) →

gibbsite (nordstrandite may also be present). The direction of the arrows indicates the reaction path the system follows spontaneously at temperatures below 343 K in a closed system. Each step is accelerated by raising the pH and may be retarded by impurity anions. Busenberg (1978) has studied the low-temperature alteration of primary aluminosilicates from which Paces (1978) concluded that a similar reaction sequence exists for the aluminosilicates (the direct analogy is made by this author). Paces concluded that a metastable cryptocrystalline aluminosilicate of variable (pH-dependent) composition controlled (reversibly) the concentrations of alumina and silica in natural waters during the irreversible dissolution of primary minerals and the irreversible formation of secondary minerals. Iler (1973) has shown experimentally that alumina and silica combine (reversibly) in water to form a metastable aluminosilicate that maintains a lower solubility of alumina and silica than would exist in the pure end-member systems.

Other examples of the Gay-Lussac–Ostwald step rule may be found in the work of Goldsmith (1949), in which high carnegieite was crystallized below its inversion temperature from a glass of carnegieite composition, and in the results reported by Laves (1952) for the crystallization of sanidine instead of microcline. Also, the work of Hsu and Wang (1980) has shown the same mechanism to be operative in the crystallization of goethite and hematite.

Implicit in the law of successive reactions is the implication that the rate of reaction of each succeeding step is slower than that for the previous step. That is, in the reaction sequence $A \rightarrow B \rightarrow C \rightarrow D$, the reaction of B to form C must be slower than the reaction of A to form B, or B would not be observed. Hückel (1951) noted that factors other than the simple stepwise reduction of free energy may control the formation of individual states. Of importance here are factors that would alter the relative reaction rates (catalyst or inhibitor).

The law of successive reactions has been interpreted (e.g., Hückel, 1951) as meaning that, when a chemical system has several free-energy states, the system must step through all the free-energy states in the order of decreasing free energy. Numerous examples of violations of this interpretation of the "law" have been cited since Ostwald (1897) orginally postulated the step rule. However, this interpretation of the step rule is inconsistent with the original definition of the step rule. The step rule requires a process to have one or more relatively stable states between the initial and final states before the step law is obeyed. The definition of process is the critical element in the application of the step rule.

When we define the process, we place limitations upon the initial and final state of a chemical system and upon subsequent intermediate steps. In order to clarify this point, let us examine one of the commonly cited exceptions to the step rule (e.g., Hückel, 1951, p. 650)—the condensation of phosphorus gas under conditions of rapid cooling. Phosphorus gas heated to near 1273 K and then cooled rapidly forms white phosphorus solid, which is less stable than red phosphorus. When the phosphorus gas is heated above 1273 K (also seen in the literature as strongly heated) and then rapidly cooled, a mixture of red and white solid phosphorus forms. When the same high-temperature gas is held at a temperature below 1273 K and allowed to equilibrate and then rapidly cooled, only white phosphorus forms. The formation of red phosphorus from the gas heated above 1273 K is viewed as a violation of the step law (e.g., by Hückel) as white phosphorus represents a free-energy state intermediate between the phosphorus gas and the solid red phosphorus. The process, however, is the rapid cooling of a gas to form a solid. Above 1273 K, the reaction $2P_2 = P_4$ leads to partial dissociation of the P_4 molecules. Consequently, above 1273 K, two processes occur—P_2 gas going to solid phosphorus and P_4 gas condensing as solid phosphorus. When P_4 gas condenses, white phosphorus represents an intermediate free-energy state between the gas phase and red phosphorus solid, as white phosphorus contains P_4 molecular clusters not found in red phosphorus (Hückel). Between P_2 gas and red phosphorus, there are no intermediate free-energy states in the process of gas condensation. That is, P_4 molecules are not formed and, consequently, white phosphorus is not an intermediate phase. In addition, once red phosphorus has nucleated, some P_4 can be expected to be incorporated into the growing solid phase resulting in red phosphorus forming a larger portion of the solid phase than P_2 represented in the gas phase. Therefore, the formation of red phosphorus from phosphorus gas rapidly cooled from temperatures above 1273 K does not violate the step rule. Proper application of the step rule may be made only if the process is properly understood.

Finally, each step reaction represents a metastable equilibrium, that is, equilibrium can only be reached from one direction, that of lower stability. The state that forms is the stable state under the system in place at that time. In a closed system, the state that forms will be considered a metastable state if a state of lower free energy exists. However, in an open system, that higher free-energy state may be maintained indefinitely in a condition of unstable equilibrium (that is, the system changes so slowly that it appears not to undergo change with time). Any metastable equilibrium in the closed system may be viewed as an unstable equilibrium if the time of observation of the step reaction is shorter than the time interval necessary to complete the reaction.

True equilibrium is attained when the free-energy content of the system is at a minimum for the given values of the variables. A necessary condition of true equilibrium is that the same state can be realized from either direction. No evidence supports the conclusion that any $Al(OH)_3$ phase can be shown to exhibit true equilibrium at low temperatures (less than 345 K). Kittrick (1966) observed that gibbsite initially undersaturated in alkaline solutions quickly became supersaturated and approached equilibrium only from supersaturated conditions. May et al. (1979) attempted to determine the solubility of gibbsite in the acid pH range in two experiments from undersaturation. In the higher pH experiment (about pH 5) the solution supersaturated with respect to gibbsite before equilibrating with gibbsite. In the experiment in which the pH was held near 4, the solution remained undersaturated with respect to gibbsite while the pH drifted, indicating a lack of equilibrium in the experiment. Further comments on this subject will be given in a later section of this paper.

Therefore, we must view all solubility or phase equilibria measurements or observations of $Al(OH)_3$ phases at surface weathering temperatures and above as metastable or unstable equilibria. When we speak of the stable phase in such an experiment, we must understand that the phase is the stable phase only under the conditions of the experiment or the natural weathering environment.

Paces (1978) has formally treated the irreversible processes that transform *primary minerals* into secondary minerals at low temperatures. From this study, Paces had drawn the following conclusions about the low-temperature weathering of aluminosilicate minerals. First, the concentration of a dissolved component in natural waters is not necessarily controlled by chemical equilibrium with a stable phase forming in the solution. Second, the formation and existence of an irreversible metastable or stable phase is not precluded by a reversible reaction (that is, a reaction that controls the concentration of solution components) with a metastable phase. Finally, the formation of a secondary phase can proceed independently of the equilibrium composition of the solution.

The conclusions given by Paces (1978) are simply a restatement of the law of successive reactions. However, Paces' second conclusion is too general. For some solutions, the formation of a secondary phase is excluded by the chemical equilibria controlling the solution composition. Two examples include (1) a solution in which the rate of growth of a metastable phase is significantly greater than the rate of growth of the secondary phase, resulting in an overgrowth of the metastable phase on the secondary phase, thus removing the secondary phase from the system, and (2) a solution that interacts with a component to form a structural unit that is incompatible with the solid structure of the secondary phase, a subject dealt with in some detail in a later section of this paper.

Helgeson (1968 and 1971) considered irreversible reactions in geochemical processes involving aqueous solutions and minerals as a succession of partial equilibrium states, each of which is said to be reversible with respect to the next sequential step and irreversible with respect to the initial state. Paces (1978, p. 1489) criticized Helgeson's (1968) approach of assuming that the evolution of the chemical composition of a natural solution can be calculated from stoichiometric reactions relating primary and secondary phases through their equilibrium constants "unless it is proved that all the secondary solids behave reversibly and the rates of the irreversible dissolution are proportional to the masses of the primary solid." The approach followed by Helgeson differs from the law of successive reactions through the assumption that each step represents a reversible rather than a metastable equilibrium. The consequences of this difference in interpretation can affect the calculated chemical composition of solutions and values of thermodynamic properties for secondary phases calculated from experimental or natural systems, particularly systems at low temperature.

Finally, the number of possible intermediate states that may exist in the aluminum system under study is also controlled by the initial conditions of the system. The importance of the initial conditions should be obvious; however, to complete the discussion, let us note that the initial concentration of aluminum in experimental solutions (systems closed with respect to aluminum), the concentration of aluminum produced by weathering primary and secondary minerals, and the concentration of aluminum produced by a phase or an impurity within a phase studied by solubility techniques limits the number of successive reactions that can take place within this system.

The chemistry of this system changes significantly at moderately higher temperatures as some of the metastable equilibria that complicate the system at lower temperatures are lost. For example, Frink and Peech (1962) obtained a pure gibbsite precipitate at 368 K from a solution in which bayerite and gibbsite coprecipitated at 343 K. Mesmer and Baes (1971) observed direct precipitation of boehmite at 423 K (from solutions that yield $Al(OH)_3$ phases at lower temperatures) under conditions where Hemley, Montoya, Marinenko, and Luce (1980) have shown diaspore to be the stable phase. Russell, Edwards, and Taylor (1955) noted that gibbsite is rapidly transformed to boehmite in basic solutions at temperatures above 373 K. Consequently, we will be unable to extrapolate the results of hydrothermal studies even at moderately low temperatures to explain the processes operative in weathering because the processes studied do not include the important reactions. The work of Hemley *et al.* and Russel *et al.* clearly indicates that diaspore should be the stable aluminum phase under surface weathering conditions.

Comments on Selected Work in the System Al–Si–O–H

In this study, I am interested in the factors that affect the aqueous geochemistry of aluminum during the weathering processes and how these processes are reflected in the natural water and soil systems. Within the context of this broad topic, I must limit this report to an examination of the experimental data collected in the Al–Si–O–H system at low temperatures (< 450 K) and to rather specialized natural water systems, those producing bauxite and laterite soils.

As an example of the problems that arise in this system, let us examine the work of Gayer *et al.* (1958) and the subsequent interpretation of their data by Baes and Mesmer (1976) and Parks (1972). Gayer *et al.* measured the solubility (in the pH range of 4 to 11) of an $Al(OH)_3$ phase prepared by precipitating $Al(OH)_3$ from an aluminum chloride solution by the use of sodium hydroxide. The sample was dried at 323 K for 2 days after several washings to remove sodium. A portion of the sample was ignited to 873 K to determine the H_2O content by weight loss. An X-ray pattern of the sample was taken, but only the three major *d*-spacings were given. Nowhere within the report was the polymorph identified. Baes and Mesmer assumed the phase to be gibbsite and calculated a solubility constant for gibbsite from the high pH data given by Gayer *et al*. Parks observed the X-ray data given by Gayer *et al*. most closely fit the three major *d*-spacings for bayerite. Parks rejected the data in determining the thermodynamic properties of bayerite because there was not sufficient proof that equilibrium was attained in the experiments. However, Parks and Hemingway *et al*. (1978) used the results given by Gayer *et al*. and similar data to calculate the formation constant β_2 (see Sillen and Martell (1964) for a discussion of this nomenclature) and the thermodynamic properties of $Al(OH)_2^+$. The solubility values in the intermediate pH range reported by Gayer *et al*. do not represent the assumed equilibria as we will see in a later section. Therefore, the thermodynamic properties of $Al(OH)_2^+$ calculated by Parks and by Hemingway *et al*. are not consistent with the properties of Al^{3+} and $Al(OH)_4^-$ derived in those studies. Similar questions can be raised regarding the experiments reported by Mesmer and Baes (1971) and Frink and Peech (1962 and 1963) which were used by Baes and Mesmer in their analysis of the monomeric aluminum hydrolysis species. These studies are considered in latter sections.

The purpose of the study performed by May *et al*. (1979) was to define the minimum in gibbsite solubility as a function of the solution pH, a task that had not been accomplished previously, and thereby to derive a consistent set of thermodynamic parameters for the aluminum hydrolysis species Al^{3+}, $Al(OH)^{2+}$, $Al(OH)_2^+$, and $Al(OH)_4^-$. If it can be shown that the data of May *et al*. (1979) do adequately define the equilibrium solubility minimum for a well-defined $Al(OH)_3$ phase as a function of pH, then consistent and more reliable theomodynamic parameters for the monomeric aluminum solution species can be derived.

The solubility results presented by May *et al*. (1979) represent the best values for aluminum solubility in the intermediate pH range. The results describe several metastable equilibria in the system. Hence, if these metastable equilibria can be defined, the thermodynamic properties of the monomeric aluminum species can be calculated. Necessarily, these equilibria must be defined through inference from what is known about the system from the literature because May *et al*. did not unequivocally determine the phase controlling the aluminum solubility in their experiments.

The solubility data obtained by May *et al*. (1979) for a synthetic sample of gibbsite were shown in their Fig. 1 (a schematic plot of their data is given here in Fig. 1) as a plot of the aluminum concentration as a function of pH. The authors noted an inflection in the curve at about pH = 6.7 reflecting a trend toward lower aluminum concentrations than would be predicted from an extrapolation of the trend obtained in the pH range 6 to 6.7. The data obtained for solutions that initially had a pH > 7 show a marked offset from the extrapolated trend. A similar pattern was found from experiments utilizing natural gibbsite samples.

May *et al*. (1979) used solutions buffered by acetic acid covering the pH ranges of 4 to 6.5, *bis-tris*(2, 2-*bis*-(hydroxymethyl)-2, 2′, 2″-nitrilotriethanol) from pH 6 to 7, and *tris*(*tris* (hydroxymethyl) aminomethane) from pH 7 to 9.

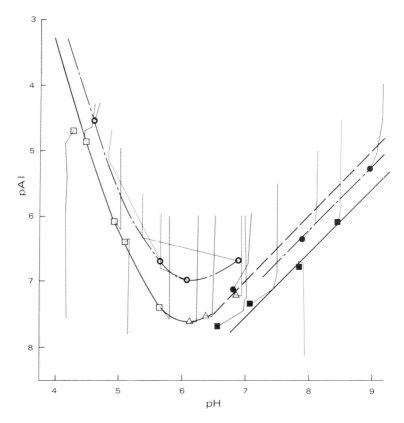

Fig. 1. Aluminum concentrations determined in the pH range of 4 to 9 from the data given by May *et al.* (1979). The solid curves represent the equilibrium aluminum concentrations obtained in the dissolution study of synthetic gibbsite. The dash–dot curves represent the results for equilibration with a natural gibbsite sample. The lines terminating in squares and triangles for the synthetic gibbsite and circles for the natural gibbsite represent the change in aluminum concentration as determined by May *et al.* from analyses at selected time intervals. The filled symbols represent solutions containing a *tris* buffer (see text). The dashed curve represents the estimate of gibbsite solubility in the basic pH region from May *et al.*

The authors, cognizant of the correlation of their change in solubility with their change in buffer chemistry, cited two lines of evidence to suggest that a reaction between the buffers and the aluminum solution species was not responsible for the lower apparent solubility obtained in their work.

Figure 2 has been constructed to show the relationship between several sets of values for the solubility of aluminum phases as a function of the solution pH. The data of May *et al.* (1979) for the natural gibbsite sample clearly blend smoothly with the results of Hem and Roberson (1967) and Raupach (1963) in the region of pH 6 to 6.7. Above a pH of 7, the *synthetic* gibbsite sample describes a separate curve which is consistent with the data of Gayer *et al.* (1958) and Raupach (1963). I propose here that this latter set of data

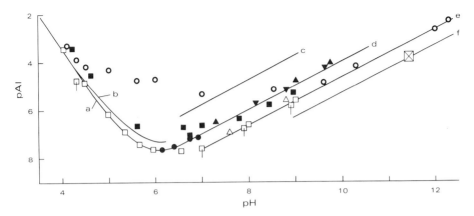

Fig. 2. Aluminum concentrations determined in the pH range of 4 to 12 from data given by Gayer *et al.* (1958), ◯), Raupach (1963, ▲ and △), Kittrick (1966), ⊠), Hem and Roberson (1967, ▼), and May *et al.* (1979). The open squares and filled circles represent the data of May *et al.* for solutions equilibrated with synthetic gibbsite (⊞ denotes equilibration from undersaturated initial conditions). The filled squares are the data of May *et al.* for solutions equilibrated with the natural gibbsite sample. Curves indicating the total of dissolved aluminum in equilibrium with gibbsite from May *et al.* (a and d), Smith and Hem (1972) (b), and Kittrick (f) and with bayerite aged 1 day (c) and 10 days (d) from Hem and Roberson are also shown. The curve representing the equilibrium concentration of $Al(OH)_4^-$ with nordstrandite (e) in the pH range of 7 to 12 is from this study (see text).

represents the equilibrium of $Al(OH)_4^-$ and nordstrandite, a phase less stable than gibbsite.

Nordstrandite was selected as the phase in equilibrium with $Al(OH)_4^-$ in the results presented by May *et al.* (1979), Gayer *et al.* (1958), and Raupach (1963) on the basis of the data reported by Schoen and Roberson (1970) (and supported by the result of Violante and Violante (1980)). Long-term aging of bayerite in basic solutions produced nordstrandite rather than gibbsite. This choice is also consistent with natural occurrences of nordstrandite associated with carbonate rocks as reported by Wall, Wolfenden, Beard, and Dean (1952, Sarawak. Borneo), Goldberg and Loughnan (1977, Sydney Basin, Australia), and Hathaway and Schlanger (1965, Guam) and with calcites in clays (Naray-Szabo and Peter (1967)).

Assuming the foregoing conclusions to be true, one can readily hypothesize that the poorly crystalline material that May *et al.* (1979) observed to be present in the *natural* sample of gibbsite was either bayerite or amorphous $Al(OH)_3$. The buffer present in the basic solutions caused rapid initial precipitation of $Al(OH)_3$, which was, in part, rapidly replaced by the dissolution of the less stable phase in the natural gibbsite sample. According to the results of Schoen and Roberson (1970), the phase precipitated was bayerite. One may further hypothesize that, although rapid precipitation took place in the buffered experiments in the alkaline solutions in which the *synthetic* gibbsite sample was suspended, the synthetic, well-crystallized gibbsite provided aluminum solution species through dissolution at concentrations lower than that required to stabilize bayerite.

The question remains, however: "Why are the results of Kittrick (1966) and May et al. (1979) different for alkaline solutions even though both experiments were conducted on well-crystallized gibbsite?" In particular, how can the solutions of May et al. reach equilibrium from both undersaturation (only apparent undersaturation) and oversaturation with nordstrandite rather than gibbsite if Kittrick's solutions reached equilibrium with gibbsite? May et al. have noted that the data reported by Kittrick do not represent equilibrium experiments. Kittrick plotted his observations as a function of the inverse square root of the elapsed time of the experiment and extrapolated the linear trends to infinite time. Although the procedure followed by Kittrick is not generally accepted today, it remains to be shown that the procedure introduced significant error in Kittrick's results. Kittrick further observed that all his experiments approached equilibrium from oversaturation even though several were initially undersaturated with respect to gibbsite. If Kittrick's results are essentially correct, then two differences in the experimental procedures must be examined. May et al. used Tris as a pH buffer and continuously agitated the sample throughout the experiment. Because the results from the Tris-buffered solutions are consistent with the data of Raupach (1963) and Gayer et al. (1958), one may agree with May et al. that Tris may not adversely affect the results that they have reported. The continuous agitation of the sample in the experiments of May et al. should have led to higher dissolution rates for the gibbsite sample than that found in Kittrick's experiments.

From these observations, one may hypothesize that the aluminum hydroxide polymorph that is stable in the alkaline pH region at any time is dependent upon the concentration of aluminum ions provided to the system as well as upon time-dependent aging and recrystallization. One may further hypothesize that bayerite will slowly be converted to nordstrandite, which, in turn, will be replaced by gibbsite as the concentration of dissolved aluminum species supplied to the system decreases. Because diaspore and water are stable with respect to gibbsite at room temperature, gibbsite should slowly recrystallize to diaspore. In natural systems associated with surface weathering processes, this process may progress too slowly to be of importance. Instead, a lowering of the fugacity of water in the soils during periodic dry spells in the weathering cycle most likely results in the conversion of precipitated $Al(OH)_3$ to diaspore. This topic is discussed in greater detail in a later section of the paper.

Hemingway et al. (1978) reported enthalpy of solution values for some of the gibbsite samples that Kittrick (1966) used to determine the solubility of gibbsite in alkaline solutions. Kittrick's samples had remained intact and stored at room temperature for more than a decade prior to their use by Hemingway et al. Of importance here is the fact that the average of the enthalpies of solution of the gibbsite samples equilibrated in alkaline solutions by Kittrick (-159.66 kJ/mol) was essentially identical to the enthalpy of solution reported by Hemingway and Robie (1977a, -159.58 kJ/mol) for a well-crystallized gibbsite. This result argues against the formation of a phase more stable than gibbsite from basic solutions equilibrated with well-crystallized gibbsite.

The results presented by May *et al.* (1979) for the dissolution of *synthetic* gibbsite in the acidic region are similar to those reported by Smith and Hem (1972). Smith and Hem monitored the changes in the aluminum concentration and pH of solutions, initially supersaturated with respect to crystalline and amorphous oxides and hydroxides of aluminum, which equilibrated with gibbsite that grew in the solution. The solutions studied by Smith and Hem did not contain macrocrystalline gibbsite at the inception of the experiment. With two exceptions, the data of May *et al.* also represent equilibration from supersaturation (refer to Fig. 1). The use of pH buffering (May *et al.*) forced the solutions to equilibrate through the precipitation of $Al(OH)_3$ unlike the unbuffered solutions studied by Smith and Hem which equilibrated through a pH shift at nearly the same concentration of dissolved aluminum.

The major difference between the results of May *et al.* (1979) and Smith and Hem (1972) was the time required to reach equilibrium. The data of May *et al.* show near-equilibrium in the buffered solutions within about 24 hours, whereas the unbuffered solutions of Smith and Hem required nearly half a year to reach equilibrium. When the buffer capacity of the solution was exceeded, as for example in several of the experiments in the intermediate pH range using the natural gibbsite starting material (refer to Fig. 1), equilibration was considerably slower than in the fully buffered experiments. Hence, the shorter apparent equilibration period noted by May *et al.*, as compared with the times reported by Kittrick (1966), Singh (1974), and Smith and Hem, was directly related to the difference in the experimental design (note the discussion of the interaction of fulvic acid and aluminum given in a later section).

The data given by May *et al.* (1979) indicate that pH buffering increases by several orders of magnitude the rate at which the solution aluminum species equilibrate with the aluminum hydroxide phase stable under the experimental conditions through the pH range of 4 to 9. Schoen and Roberson (1970) have shown that both bayerite and microcrystalline gibbsite are initially precipitated in the intermediate pH range, but later the microcrystalline gibbsite is dissolved. Hence, we can reasonably conclude that the *bis-tris*-buffered solubility values obtained for the experiments using the synthetic gibbsite sample in a solution initially oversaturated with respect to gibbsite would equilibrate with bayerite between pH 6 and 7 as is clearly shown in Fig. 2.

Revised Free Energies of Formation for the Aqueous Aluminum Species $Al(OH)^{2+}$ and $Al(OH)_2^+$ and for Bayerite and Nordstrandite

Hsu (1967) proposed the existence of two structural series of hydrolyzed aluminum ions. The first type is based upon the form designated as oxo by

Hsu and representing the linear Al–O–Al linkage. This form is discussed by others (e.g., Moolenaar et al., 1970) as the AlO_2^- ion. The second form is designated as ol linkage by Hsu and represents the polymeric structure for $Al(OH)_4^-$. These ionic structures are related to the structural units in $Al(OH)_3$ phases (ol linkage) and $AlO(OH)$ phases (oxo linkage). Although the conclusion reached by Hsu was derived empirically and was based simply upon an observation of the change in the aluminum phase that precipitated from solutions containing variable amounts of NaCl, the theory is considered here to be valid and shall be explored further.

The observations of Hsu (1967) are discussed in some detail elsewhere in this paper. Briefly, however, Hsu concluded that as the concentration of NaCl was increased in solutions, which became supersaturated (through the addition of OH^-) with respect to $Al(OH)_3$ or $AlO(OH)$ phases, water was lost from the dissolved aluminum species; this water loss resulted in a transformation from ol to oxo linkages in the solution and a change from the precipitation of bayerite to precipitation of pseudoboehmite.

A similar conclusion was reached by Dibrov, Mal'tsev, and Mashovets (1964) in a study of the vapor pressure of aqueous sodium aluminate solutions. These workers concluded that $Al(OH)_4^-$ was the dominant aqueous aluminum species at low aluminum concentrations. As the concentration of NaOH was increased, $Al(OH)_4^-$ was thought to dehydrate to form AlO_2^-. Further studies by Mal'tsev, Malinin, and Mashovets (1965) which correlated a proton chemical shift in the aqueous aluminate solution with the change in vapor pressure observed by Dibrov et al. added support for this conclusion.

Carreira, Maroni, Swaine, and Plumb (1966) examined the infrared and Raman spectra for solutions highly supersaturated in aluminum over the pH range of 8 to 13. They concluded that $Al(OH)_4^-$ was the dominant solution species from pH 8 to 12 and that AlO_2^- was the dominant form at pH 13. Moolenaar et al. (1970) and Lippincott, Psellos, and Tobin (1952) also used Raman data but found evidence of only $Al(OH)_4^-$ in the higher pH solutions that had low alumina concentrations.

No direct evidence shows that the linear oxo linkage is the dominant structure for aluminum ions in basic solutions from which $AlO(OH)$ phases precipitate. The circumstantial evidence is quite strong, however. The presence of the ol linkage structure in the low temperature, low-concentration solutions from which $Al(OH)_3$ phases precipitate appears well established (Moolenaar et al., 1970).

If we infer that solutions from which $AlO(OH)$ phases precipitate directly contain dissolved aluminum species that have a different structure and hence different thermodynamic properties from those species from which $Al(OH)_3$ phases precipitate, then we must conclude that the structure of the dissolved aluminum species must change as the temperature of the system is increased above about 370 K. The temperature at which thermal dehydration would take place cannot be established directly. Indirectly, however, Russell et al. (1955) have shown that gibbsite converts to boehmite at temperatures above 368 K, and Frink and Peech (1962) have shown that gibbsite can be precipitated at the same temperature from NaOH solutions in which aluminum metal was dissolved if a stream of CO_2 is slowly passed through the solution. Glastonbury (1969) has concluded that the structural change is complete at 348 K.

Dibrov et al. (1964) also studied the vapor pressure of aqueous sodium aluminate solutions from 298 to 620 K. They concluded that throughout this temperature range, an equilibrium existed between univalent anions of different degrees of hydration. The reaction given was $Al(OH)_4^- \rightarrow AlO(OH)_2^- + H_2O \rightarrow AlO_2^- + 2H_2O$. Dibrov et al. observed that increasing dilution (of NaOH) favored the tetrahydroxyaluminate ion and that increasing temperature and pH favored the dehydration reactions.

Mal'tsev and Mashovets (1965) measured the specific heat of sodium hydroxide and sodium aluminate solutions for a wide range of concentrations and to 368 K. They concluded that the dependence of the specific heat of the aluminate solution on temperature at any fixed concentration of NaOH and on NaOH concentration at any fixed temperature showed that dehydration and/or a breakdown of the polymeric structure of the aluminate ion took place as either temperature or the concentration of NaOH was increased. Mal'tsev and Mashovets considered this reaction to be largely complete at 348 K.

Calvet, Thibon, Maillard, and Boivinet (1950) measured diffusion coefficients for sodium aluminate solutions from 293 to 323 K. They showed that the average particle size calculated for the aluminate ion decreased by 50% at the higher temperatures of their study. This observation supports the interpretation of Mal'tsev and Mashovets (1965).

Sharma and Kashyap (1972) recorded the Raman spectra of zinc and aluminum ions in solutions of KOH and NaOH at several concentrations and at several temperatures between 293 and 363 K. The spectra for the aluminum ionic species in both KOH and NaOH solutions were similar. Sharma and Kashyap confirmed the presence of the linear AlO_2^- ion in strongly alkaline solutions saturated with aluminum. The change in the spectra from that of the $Al(OH)_4^-$ structure to that defining AlO_2^- was accompanied by a change in the structure of the KOH and NaOH solutions as either the concentration or temperature was increased. This change was interpreted as an increase in the ion-pair association for KOH and NaOH.

Mesmer and Baes (1971) studied the hydrolysis of aluminum in acidic solutions between 335.65 and 422.95 K and in a basic solution at 422.95 K through potentiometric measurements. An extended set of data was collected at 398 K. Measurements at temperatures lower than 373 K were less precise as a result of the slow reaction rates. At temperatures above 423 K, precipitation took place immediately. The phase that precipitated at 423 K was identified as boehmite. Of importance here was the observation by Mesmer and Baes that the maximum value of the average number of hydroxides bound per aluminum which could be studied by the potentiometric method changed abruptly between 335.65 and 373 K. This observation supports that idea that the structure of the aluminum solution species changes in the region of 370 K.

Baes and Mesmer (1976) have reported values of the hydrolysis constants for $Al(OH)^{2+}$, $Al(OH)_2^+$, $Al(OH)_3^0$, and $Al(OH)_4^-$ at 298.15 K based largely on the results of Mesmer and Baes (1971). Because of the limited pH range accessible to Mesmer and Baes, the potential change in the structure of the aluminum solution species at higher temperatures, and the lack of precise data at lower temperatures, the hydrolysis constants reported by Baes and Mesmer must be considered suspect for the ol linkage series, but may apply to the oxo linkage series.

Thus, the best method to determine the low-temperature hydrolysis constants for the aluminum solution species would appear to be the method followed by May et al. (1979). The most internally consistent data set will be derived from a study of the (metastable) equilibrium of a single $Al(OH)_3$ phase throughout the pH region of 4 to 9.

Schoen and Roberson (1970) noted that bayerite formed above pH 5.8 which they interpreted to be the pH of minimum aluminum solubility. This point would not represent the minimum solubility value unless bayerite and gibbsite had equivalent free energies of formation. Because bayerite is less stable than gibbsite, the solubility curve for bayerite should intersect the gibbsite solubility curve above the minimum in gibbsite solubility if bayerite does not precipitate from solutions having a pH < 5.8 and if the mechanism controlling $Al(OH)_3$ precipitation of bayerite in solutions becomes dominant at a pH of about 5.8.

The data provided by May et al. (1979) for the solubility of synthetic gibbsite do not adequately describe the solubility minimum for gibbsite. Even when the data of Kittrick (1966) are combined with the data of May et al. for pH < 6, the two sets do not provide adequate information to allow a calculation of the free energies of formation of the several aluminum solution species.

The data given by May et al. (1979) for the natural gibbsite sample describe a solubility curve subparallel to that obtained from the synthetic gibbsite experiments for pH < 7. Because the initial solutions of the experiments with the natural and synthetic gibbsite samples were each supersaturated with equivalent aluminum concentrations, the subparallel solubility curves indicate that equilibration of the solutions containing natural gibbsite was with a phase less stable than gibbsite. This phase was not identified by May et al., but may be inferred to be nordstrandite from the work of Barnhisel and Rich (1965).

The assumption is further supported by the work of Ross and Turner (1971) in which nordstrandite was identified as the initial phase to precipitate (at 3 weeks) from an aluminum chlorate solution that was rapidly titrated to 80% neutrality with $1N$ NaOH. The solid phase that had precipitated from the solution aged for 8 months yielded an X-ray diffraction pattern similar to that given by Hauschild (1963) for nordstrandite. After 13 months, the X-ray diffraction pattern for the solid phase resembled the pattern for synthetic gibbsite given by Rooksby (1961).

On the assumption that nordstrandite forms in the acid pH region after the dissolution of the *natural* sample and in the alkaline pH region after the dissolution of the *synthetic* gibbsite sample, we may fit the solubility data for these data sets following the procedures of May *et al.* (1979).

Using the value for $*K_{S4}$ obtained from this fit and the free energies of formation of the ancillary phases and $Al(OH)_4^-$ from Hemingway *et al.* (1978), I calculated the free energy of formation of nordstrandite as -1151.5 ± 2 kJ/mol. Because this value is based upon the free energy of formation of $Al(OH)_4^-$, which, in turn, is tied to the free energy of formation of Al^{3+} through the free energy of formation of gibbsite (Hemingway and Robie, 1977b, and Hemingway *et al.*, 1978), we may calculate the free energy of formation of the aluminum species Al^{3+} from the derived value of $*K_{S0}$ for nordstrandite to examine the reasonableness of our assumptions. Using the free energy of formation of nordstrandite calculated above and $*K_{S0} = 5.19 \times 10^8$ derived from the data of May *et al.* (1979), we obtain -489.8 ± 4 kJ/mol for the free energy of formation of Al^{3+}. This value is in good agreement with the value of -489.4 ± 1.4 kJ/mol reported by Hemingway and Robie (1977b), and the arguments are therefore consistent, but not proven.

The free energies of formation of $Al(OH)^{2+}$ and $Al(OH)_2^+$ were also calculated from the fit to the nordstrandite data set and are listed with the free energies of formation of Al^{3+} and $Al(OH)_4^-$ in Table 1. Using the free energy of formation of $Al(OH)_4^-$ from Hemingway *et al.* (1978), we can evaluate the free energies of formation of "bayerite" aged 1 day and aged 10 days from the data of Hem and Roberson (1967) as -1140.4 ± 3 and -1144.5 ± 3 kJ/mol, respectively. A somewhat more negative value may be obtained from the *tris*-buffered natural gibbsite experiments of May *et al.* (1979) as -1149.0 ± 2 kJ/mol. The latter value is assumed to represent the best estimate of the free energy of formation of bayerite.

The values of the solubility constants for gibbsite, nordstrandite, microcrystalline gibbsite, and a precipitate of $Al(OH)_3$ aged 24 hours listed in Table 2 were calculated from the free energies of the phases given above and from the solution species listed in Table 1. The data are also shown graphically in Fig. 3 with some of the solubility results of May *et al.* (1979), Gayer *et al.* (1958), and Smith and Hem (1972). The activities of the solution species and the free

Table 1. Free energies of formation of the solution species Al^{3+}, $Al(OH)^{2+}$, and $Al(OH)_2^+$ (at 298.15 K, 1 bar, and for an ideal $1M$ solution) determined from the solubility data for nordstrandite in the acid and basic region (see text) and from the free energy of $Al(OH)_4^-$ from Hemingway *et al.* (1978).

Solution species	Al^{3+}	$Al(OH)^{2+}$	$Al(OH)_2^+$	$Al(OH)_4^-$
Free Energy of Formation in kJ/mol	-498.8 ± 4.0	-701.4 ± 4.0	-903.4 ± 4.0	-1305.0 ± 1.3

Table 2. Solubility constants and free energies of formation for several Al(OH)$_3$ compounds: bayerite, nordstrandite, gibbsite, microcrystalline gibbsite, and a precipitated Al(OH)$_3$ phase aged 24 hours. Values calculated from constants given in Table 1 and the activity coefficients given by Smith and Hem (1972).

phase	$^*K_{S0}$	$^*K_{S1}$	$^*K_{S2}$	$^*K_{S4}$	$\Delta G^\circ_{f,298}$
24-hour precipitate[a]	3.87×10^{10}	1.52×10^6	1.06	1.95×10^{-13}	-1140.4 ± 3.0
Microcrystalline gibbsite (this study)	8.69×10^9	3.42×10^5	0.238	4.37×10^{-14}	-1144.5 ± 3.0
Microcrystalline gibbsite (Hem and Roberson, 1967)	2.21×10^9	8.67×10^4	6.03×10^{-2}	1.11×10^{-14}	-1147.5 ± 3.0
Bayerite[a]	1.20×10^9	4.73×10^4	3.29×10^{-2}	6.06×10^{-15}	-1149.0 ± 2.0
Nordstrandite[a]	5.19×10^8	1.73×10^4	1.20×10^{-2}	2.21×10^{-15}	-1151.5 ± 2.0
Gibbsite[b]	1.11×10^8	4.38×10^3	3.05×10^{-3}	5.60×10^{-16}	-1154.9 ± 1.2

[a] See discussion in the text.
[b] Free energy from Hemingway *et al.* (1978).

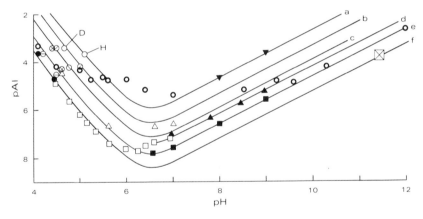

Fig. 3. The total aluminum concentrations in equilibrium with several Al(OH)$_3$ phases (see text discussion) in the pH range of 4 to 12 as derived in this study (see Tables 1 and 2). Curve a, 24-hour precipitate; curve b, microcrystalline gibbsite (this study); curve c, microcrystalline gibbsite (Hem and Roberson, 1967); curve d, bayerite; curve e, nordstrandite; and curve f, gibbsite. The activity coefficients were taken from Smith and Hem (1972). The series of light circles represents the measured aluminum concentration determined for solutions D and H by Smith and Hem. The sample designated by the symbol ○ represents an equilibration time of 1.1 hours, ⊘ = 4 days, ⊖ = 7 days, ⊗ = 26 days, ⊖ = 121 days, and ● = 259 days. The other symbols are as defined in Fig. 2, with the exception that the open and filled triangles represent the data of May *et al.* (1979) for the natural gibbsite experiments. The several data points which lie above curve a in the region of pH 6 most likely represent equilibration with a phase containing mixed anions (e.g., Gastuche and Herbillon, 1962; Fripiat and Pennequin, 1965; and Turner and Ross, 1970, as well as the text).

energies for microcrystalline gibbsite and the 24-hour precipitate are based on the results given by Smith and Hem.

Factors Affecting the Rate of Crystallization of Al(OH)$_3$ and AlO(OH) Phases and of Aluminosilicate Phases

In a previous section we noted that the time required by a solution to reach equilibrium with gibbsite varied from as little as 24 hours for the pH buffered solutions investigated by May *et al.* (1979) to periods greater than a half year as shown by Smith and Hem (1972) and Kittrick (1966). We hypothesized that pH buffering increased the rate of equilibration by several orders of magnitude. In this section, we shall examine some other techniques that appear to result in an increased rate of equilibration in some of the metastable reactions involving Al(OH)$_3$ phases and some factors that decrease the rate of one or more of the metastable reactions, or inhibit the reaction altogether.

Kittrick (1980) has recently applied the immiscible displacement method originated by Mubarak and Olsen (1976) to a solubility study of gibbsite and kaolinite. Kittrick found that using a ratio of 4 ml of solution per gram of mineral produced equilibrium solubility values in several months that required several years for equilibration at a ratio of 20 ml per gram of mineral. Kittrick noted, however, that if the solution that is mixed with the kaolinite sample is far from equilibrium (Kittrick appears to mean that if the pH is not near the equilibrium pH), rapid and correct results cannot be obtained by the immiscible displacement method. The mechanism thus operative in Kittrick's experiments would appear to be similar to that operative in the pH buffered experiments of May *et al.* (1979).

Frink and Peech (1963) studied the hydrolysis of aluminum chloride in solutions having aluminum concentrations ranging from $0.01\,M$ to $0.00001\,M$ that were prepared by diluting an initial stock solution containing $0.1\,M$ aluminum. The conductance and pH of the solutions were monitored for 8 months and little change was observed for most samples (see Fig. 4). Solutions having concentrations greater than $0.005\,M$ were considered by Frink and Peech to have equilibrated immediately after dilution to the final concentration. Solutions more dilute than $0.005\,M$ became more acidic on standing. Frink and Peech indicated that cyclic heating of similar solutions between 298 and 313 K slightly accelerated the decrease in pH for solutions less dilute than $0.0005\,M$, but appeared to have no effect upon the more concentrated solutions. We may conclude from the data given in Fig. 4 that all the initial solutions studied by Frink and Peech were supersaturated with respect to nordstrandite, but that solutions more dilute than $0.0005\,M$ were significantly different in pH from that required for equilibrium with nordstrandite at the given aluminum concentration, and that resolution of the change in pH on

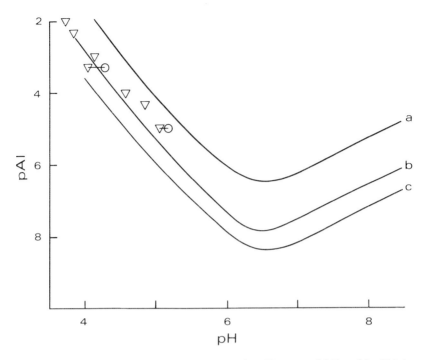

Fig. 4. Aluminum concentrations determined in the pH range of 3.7 to 5 by Frink and Peech (1963). Samples were prepared by diluting $AlCl_3$ solutions to the desired Al concentration. The samples aged 8 months ($\triangledown$) were equivalent to the intial (○, not shown unless significantly different after 8 months) solutions in most cases. The concentration–pH conditions of the solutions equilibrated near a pH of 4 are similar to those found by May *et al.* (1979) for their natural gibbsite sample. The equilibrium concentration curves for microcrystalline gibbsite (a), nordstrandite (b), and gibbsite (c) are shown.

cyclic heating could be observed only in these solutions. An interpretation of the results of Frink and Peech would be analogous to that given for the data of Smith and Hem (1972) in a later section.

Frink and Peech (1962) presented similar results (see Fig. 5) to their later (1963) work (see Fig. 4) described above, but some details of the experimental procedures must be assumed. Probably we can assume that the aluminum chloride solutions used in this study were diluted from an initial stock solution as described by Frink and Peech (1963). Under that assumption, the presence of gibbsite in solutions having a pH of 4 to 5 indicates rapid equilibration (within 3 months) with gibbsite. According to the procedure outlined by Frink and Peech (1962), the gibbsite was added after the solution of aluminum chloride had been diluted to the desired aluminum concentration. In the more concentrated solutions (where pH < 4), the solutions remain close to the solubility expected for nordstrandite. Where additional acid (as HCl) was

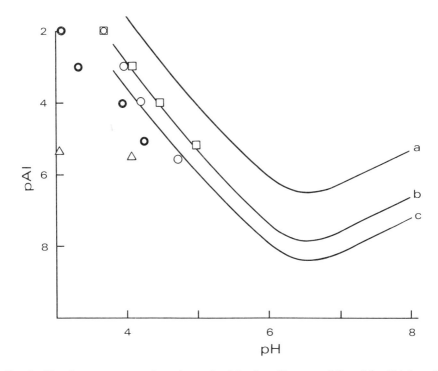

Fig. 5. Aluminum concentrations determined in the pH range of 3 to 5 by Frink and Peech (1962) under varying experimental conditions: $\square$, solution aged 3 months, prepared by dilution of $AlCl_3$ solution like solutions shown in Fig. 4; $\bigcirc$, diluted $AlCl_3$ solution aged 3 months in the presence of gibbsite; $\bigcirc$, diluted $AlCl_3$ solution aged 1 month in the presence of gibbsite with H^+ added as HCl; and $\triangle$, a suspension of gibbsite to which H^+ (as HCl) was added. The addition of HCl delays equilibration of these solutions. The curves are described in Fig. 4.

added to solutions containing gibbsite with or without aluminum chloride, the solutions remained significantly undersaturated with respect to gibbsite, that is, the rate of equilibration was greatly reduced.

In order to properly interpret the results of experiments performed at low temperatures in the system Al–Si–O–H, one needs to know the precise procedures followed in the experiment, as the work of Hsu (1967) clearly demonstrates. Hsu studied the effect of several cations (for which no effect was observed) and anions on the aluminum phase (in this study either bayerite or pseudoboehmite) that precipitated from basic solutions. Hsu found that at a given concentration ratio of NaOH/Al, the higher the content of Cl^-, ClO_4^-, NO_3^-, Br^-, or I^- in the solution, the greater the tendency for pseudoboehmite to precipitate. At a fixed concentration of salt impurity, the higher the NaOH/Al ratio, the greater the tendency for bayerite precipitation. However, once the NaOH and Al solutions were mixed slowly, the addition of other salts had no effect on which phase precipitated (bayerite). When the aluminum chloride solution was rapidly mixed with the sodium hydroxide solution in the absence of other salts, the initial precipitate contained pseudoboehmite which disappeared within 1 week.

The initial appearance of pseudoboehmite as a precipitate from aluminum-bearing solutions to which hydroxyl ions have been added rapidly, followed by the subsequent slow loss of pseudoboehmite as a solid phase coincident with the appearance of an $Al(OH)_3$ phase in the solid phase, may be explained by reference to the work of Carreira et al. (1966). The rapid addition of the hydroxyl ion causes a temporary and local increase in the ratio of OH^- to Al in solution. The temporary high OH^- density causes the rapid formation of the linear AlO_2^- solution species and subsequent precipitation of gelatinous pseudoboehmite. As mixing reduces the local ratio of OH^- to Al, formation of AlO_2^- rapidly ceases and $Al(OH)_4^-$ ions form. The presence of aluminum in the form $Al(OH)_4^-$ inhibits the precipitation of additional pseudoboehmite, allowing instead only $Al(OH)_3$ forms to precipitate. Because pseudoboehmite is poorly crystallized and fine grained, it dissolves reasonably rapidly. Aluminum lost through dissolution is incorporated in additional $Al(OH)_4^-$ ions leading to the loss of pseudoboehmite from the solid precipitating from the solution.

Ross and Turner (1971) made a similar study in the acid pH region. Solutions of $AlCl_3$, $Al(NO_3)_3$, and $Al(ClO_4)_3$ were rapidly titrated to 80% neutrality with vigorous stirring and then maintained at a constant temperature for several years. Nordstrandite rapidly precipitated from the chlorate solution and was replaced after 13 months by gibbsite. An aluminum hydroxynitrate of poor crystallinity originally precipitated from the nitrate solution. After $9\frac{1}{2}$ months, gibbsite crystallized in the solution. No $Al(OH)_3$ phase was observed to precipitate from the chloride solution during the 2-year study (solution was $0.6 M$ Cl^-). Barnhisel and Rich (1965) obtained both gibbsite and nordstrandite from chloride solutions (in the acidic pH region) having a lower concentration of Cl^-. Aluminum oxychloride precipitated from the more concentrated salt solutions studied by Hsu and by Barnhisel and Rich.

Turner and Ross (1970) examined the effect of temperature as well as Cl^- concentration on the formation of precipitates from aluminum chloride solutions that were titrated with NaOH and diluted to obtain the final concentration. The system was examined at 283, 299, and 313 K. At each temperature, the same sequence of events took place; however, the rate of these reactions decreased dramatically at the lower temperatures. Of importance here are the observations that an initial solid phase formed and then decreased to a minimum value as the particle size of the solid decreased correspondingly, and that this process was then followed by a slow increase in the amount of solid formed. The concentration of polynuclear hydroxyaluminum ions followed an inverse relationship to the solid, showing a maximum concentration in the solution where the solid was a minimum.

Gastuche and Herbillon (1962) and Fripiat and Pennequin (1965) found that the first solid phase to form when OH^- ions were added to an aluminum chloride solution contains chlorine. Turner and Ross (1970) found that the first solid to precipitate in their studies was amorphous to X-rays. When the concentration of the polynuclear hydroxyaluminum ions decreased to near zero, gibbsite was detected in the solid (60 days at 313 K and 100 days at 299 K).

Hsu (1967) and Turner and Ross (1970) have concluded that the presence of significant concentrations of impurity anions can retard or prevent the formation of $Al(OH)_3$ phases. Hsu also noted that once the polynuclear hydroxyaluminum ions have formed, the addition of Cl^- or other anions will not retard or prevent the precipitation of $Al(OH)_3$ phases.

Kwong and Huang (1979) have used similar reasoning in describing the effect that low-molecular-weight organic acids (those commonly found in soil and aquatic environments) have on the hydrolysis and precipitation of aluminum. Citric acid, malic acid, tannic acid, aspartic acid, and p-hydroxybenzoic acid in concentrations of $10^{-4}M$ and $10^{-6}M$ have been shown to reduce the amount of precipitate formed when $AlCl_3$ solutions are titrated with NaOH to a pH of 4 to 5. The phase that precipitated was not identified by Kwong and Huang.

Lind and Hem (1975) performed a study similar to that of Kwong and Huang (1979) except that quercetin (a low-molecular-weight organic material having a structure similar to organic coloring material found in natural waters) was the organic phase studied. Lind and Hem observed that in solutions having a pH lower than 6.5 (which precipitated gibbsite without the presence of the organic material), the polymerization of aluminum was strongly inhibited. In solutions near neutrality, boehmite precipitation was confirmed by X-ray diffraction analysis. Bayerite precipitated from the more alkaline solutions. Lind and Hem also observed an increased yield of well-crystallized kaolinite from solutions containing small amounts of quercetin over the yield of similar solutions without added organic material. They concluded that quercetin inhibited aluminum hydroxide precipitation and thus allowed the formation of Al–O–Si bonds at a rate comparable to that needed in the crystallization of the aluminosilicate phase.

We may conclude from the observations given above that the presence of substances which inhibit the rapid polymerization of hydroxyaluminum ions in solution allows the system time to produce a more ordered and, therefore, a more stable or lower free-energy phase as the precipitate. We may further conclude that the $AlO(OH)$ phases should be stable under surface weathering conditions, but that the rapid formation of ol-linked hydroxyaluminum ions in natural waters precludes direct precipitation of diaspore or boehmite under typical weathering conditions. In natural water, the source of aluminum is through the dissolution of primary and secondary minerals. The natural system, thus, has many relatively stable intermediate states through which it may pass in order to reach the lowest energy state.

Some Comments on Bauxite and Laterite Formation, on the Mobility of Aluminum in Natural Waters, and on the Interpretation of Experimental Solubility Measurements

Several models have been proposed to describe the control of the concentration of silica and alumina in natural waters. Helgeson (1968, 1971) has suggested a partial equilibrium between the solution and clay minerals and/or gibbsite; Hem, Roberson, Lind, and Polzer (1973) proposed equilibrium between the solution and halloysite and/or microcrystalline gibbsite; Paces (1978) proposed a reversible equilibrium between the solution and a cryptocrystalline aluminosilicate of varied (pH dependent) composition; and Iler (1973) suggested that the concentration of silica and alumina in solution was controlled by the adsorption of silica and alumina on silicate surfaces. Paces' model was developed to explain the chemistry of 152 natural water samples, whereas the remaining models were developed from laboratory experiments which were designed to simulate portions of the weathering process (e.g., dissolution of feldspars or synthesis of clay minerals). Each model proposes a different free-energy state within the Al–Si–O–H system; the models of Paces and Iler represent the least stable state, and the model of Helgeson proposes the most stable state. Within the context of our earlier discussion of the law of successive reactions, we may conclude that each model should be applicable under a given set of weathering conditions.

I do not intend to examine all the factors which control aluminosilicate deposition. However, I suggest that aluminosilicate deposition proceeds through a series of metastable intermediate states that have controls similar to those already examined for the Al–O–H system. In the absence of rate-modifying factors (such as low pH solutions containing high-molecular-weight organic material), the model proposed by Paces would appear to be the most generally applicable for the weathering of felsic rocks (also see Busenberg (1978)).

In this study, I am primarily interested in the conditions that affect the solubility of aluminum in weathering processes and in natural waters. Laterites are only briefly mentioned here and then only with respect to factors that affect the chemistry of aluminum. As noted earlier, similar processes may be expected in the iron–water system in the low pH region (e.g., Hsu and Wang, 1980).

As solutions bearing dissolved aluminum equilibrate with solids, they vary in aluminum concentration and/or pH. The mechanisms that may be responsible for such changes are recrystallization and cementation without a change in the crystal structure (Ostwald ripening) or dissolution and recrystallization of a more stable phase. Smith and Hem (1972) suggested that the former mechanism was active in their study of the aging of acidified aqueous aluminum solutions. Brosset (1952), Bye and Robinson (1964), and Schoen and Roberson (1970) suggested that the latter mechanism prevailed in neutral or alkaline solutions.

Barnhisel and Rich (1965) presented a simplified "phase diagram" showing the variation of $Al(OH)_3$ polymorphs as a function of pH and of NaCl concentration. For the pure $Al_2O_3-H_2O$ system they found that gibbsite is the only phase to precipitate from solutions having a pH less than about 4.4; gibbsite and nordstrandite precipitate from solutions having a pH in the range of about 4.4 to 5.8; gibbsite, nordstrandite, and bayerite precipitate from solutions having a pH in the range of 5.8 to 7; nordstrandite and bayerite precipitate from solutions having a pH of 7 to 9; and above pH 9, only bayerite precipitates. Nordstrandite precipitated from both lower and higher pH solutions as the concentration of NaCl was increased (see Barnhisel and Rich for a discussion of acid NaCl solutions). Thus, all these phases must be considered capable of controlling the concentration of soluble aluminum under some conditions.

Geiling and Glocker (1943) reported that the freshly precipitated aluminum hydroxide gel (from basic solutions) had an "amorphous liquid structure," which, upon aging, produced diffuse boehmite lines in X-ray photographs. Brosset (1952) noted that the amorphous gel first changed to pure boehmite and then, in succession, to bayerite and to gibbsite in solutions that were equilibrated at 40°C. Bye and Robinson (1964) concluded, as others before them had (see Bye and Robinson for references), that, at room temperature, freshly precipitated aluminum hydroxide crystallized to pseudoboehmite and then recrystallized successively to bayerite and gibbsite (however, recrystallization to gibbsite was not observed). Schoen and Roberson (1970) found that bayerite altered to nordstrandite and not gibbsite in their studies in alkaline solutions and that early formed microcrystalline gibbsite that coprecipitated with bayerite dissolved and recrystallized as bayerite in nearly neutral solutions.

Thus, once we realize that equilibrium is reached in this system through a series of metastable reactions, we can see how various studies have obtained significantly different results for seemingly identical reactions as a result of varying conditions of the experiment (such as the aluminum concentration or the rate of addition of OH^-) which lead to different metastable reactions being observed. In most experiments, for example in the study by Smith and Hem (1972) of the aging of aqueous aluminum solutions, the phase actually controlling the concentration of aluminum species was not properly identified. Smith and Hem identified hexagonal plates shown in their electron micrographs as gibbsite because of their similarity to electron micrographs of gibbsite published by Schoen and Roberson (1970). However, numerous rectangular plates, which resemble nordstrandite, also appear on Smith and Hem's electron micrographs. Consequently, even in these fairly acidic solutions, aging may proceed through the slow dissolution of less stable phases and precipitation of the next most stable phase.

Smith and Hem (1972) suggested that the mechanism of Ostwald ripening was responsible for changes seen in the solid phase grown in their experiments. In fact, the data of Smith and Hem can more easily be interpreted as representing a solution that is undergoing equilibration with an initial poorly ordered solid followed by equilibration with nordstrandite and finally with gibbsite. These successive equilibrations may be inferred from the data presented in Fig. 3 for the total aluminum concentration in solutions D and H as reported by Smith and Hem. The data of Hem and Roberson (1967) may also represent successive equilibrations as described above.

The data used by Hem and Roberson (1967) to obtain the solubility product constant for microcrystalline gibbsite fall in a pH concentration range between the presumed limiting curve for nordstrandite and an inflection in the data for solution H (at the point given for the solution aged 4 days). The data of Gayer et al. (1958) also show an inflection at this point suggesting that microcrystalline gibbsite (or some other as-yet-unidentified aluminum hydroxide phase) may be defined by the data set near this inflection rather than that given by Hem and Roberson. The data used by Hem and Roberson to determine the equilibrium constant for microcrystalline gibbsite scatter excessively, suggesting that rather than representing an equilibrium assemblage, the data represent a slow time-dependent reaction involving the dissolution of a less stable initially precipitated phase (here assumed to be microcrystalline gibbsite) and the precipitation of the next metastable phase, which in this experiment appears to be nordstrandite. On the basis of these assumptions and the data given earlier, the Gibbs free energy of formation of microcrystalline gibbsite was calculated as -1144.5 ± 4.0 kJ/mol. However, the initially precipitated phase may be the phase identified as pseudoboehmite in the literature (e.g., Bye and Robinson, 1964).

The data given by Smith and Hem (1972) for several aged solutions show a break in the slope of the curves representing the change in total aluminum (referenced Al a + b) in the solution as a function of the solution pH between the samples taken at about 26 days and at 121 days (see Fig. 3). The total aluminum concentration remains fairly constant for about the first month as the pH of the solution decreases. From the sample taken at 46 days (not shown) to the sample taken at 259 days, both the pH and aluminum concentration decrease. The inflection in the curve corresponds with the stability limit derived for nordstrandite from the data of May et al. (1979) and with a change in the equilibration rate of the solution.

The solutions studied by Smith and Hem (1972) remained significantly supersaturated with respect to gibbsite for 3 months, and the supersaturation during the first 7 to 26 days remained an order of magnitude greater than the equilibrium value for gibbsite. The solutions studied by May et al. (1979) equilibrated with gibbsite or nordstrandite within several hours at the same conditions of pH and initial supersaturation. This result has important implications regarding aluminum mobility and the formation of bauxite and laterites.

Wollast (1967) studied the kinetics of feldspar dissolution in solutions buffered to pH values between 4 and 10. Wollast used NaOH and potassium biphthalate from pH 4 to 6 and NaOH and H_3BO_3 from pH 8 to 10 as the pH buffers. These buffers had no effect upon the rate at which the dissolved alumina equilibrated with Al(OH)$_3$ phases. Similar results were obtained in the study of Busenberg and Clemency (1976) where the solutions in which the feldspars were suspended were acidified but not buffered. Thus, buffering with OH$^-$ will not affect the rate of equilibration of dissolved aluminum with Al(OH)$_3$ phases in the pH range of 4 to 9.

Two hypotheses regarding bauxite formation have been proposed. Microcrystalline gibbsite found as a replacement of feldspar in which the crystal outline and cleavage was preserved led Young and Stephen (1965) and Sherman, Cady, Ikawa, and Blumsberg (1967) to conclude that primary aluminosilicate weathered directly to gibbsite. Other workers (e.g., Bates, 1962; Allen, 1952) concluded that gibbsite formed from an intermediate clay mineral, and some workers have reported substantial aluminum mobility.

Hsu (1977) examined the literature on this topic and noted that authors proposing the direct weathering of primary aluminosilicates to gibbsite worked in areas characterized by high rainfall, good drainage, tropical to subtropical temperatures, and basic to intermediate rocks, whereas those proposing an intermediate clay phase worked in areas of poor drainage, but otherwise similar conditions. Wolfenden (1961), for example, found a thick layer of kaolinite between basic rock and bauxite where a high ground-water table prevented drainage, and Gordon, Tracey, and Ellis (1958) reported kaolinite formation in the ground-water table below Arkansas bauxites.

Patterson and Roberson (1961) found the pH of ground water taken from wells that drained the bauxite soils of Kauai, Hawaii, to range from 4.0 to 5.9, whereas wells in the underlying fresh basalt produced waters having a pH range from 7.6 to 7.8. Valeton (1972) noted that decomposition of staghorn fern plants supplied some Hawaiian bauxite soils with organic acids. Schnitzer and Desjardins (1969) collected natural rainwater leachate from between the Ae and Bhf horizon of a humic podzol and measured a pH of 3.9. They found that 87% of the organic leachate was fulvic acid.

Schnitzer and Kodama (1977) have reviewed the literature concerning the interaction of soil humic substances with soil minerals and solution species. Of the several humic fractions, only fulvic acid is soluble in the low pH range commonly associated with bauxite soils or lateritic soils. Fulvic acid forms stable complexes with aluminum in aqueous solutions (Schnitzer and Hansen (1970) provided stability constants for several metal–fulvic acid complexes).

The interaction of fulvic acid with soluble aluminum as well as with aluminum hydroxide (reaction with boehmite has not been shown) phases is pH dependent, and the greatest interaction takes place in the region of pH 2 to 5 (Evans and Russell, 1959; Schnitzer and Skinner, 1965). Above pH 5, fulvic acid interaction with Fe becomes dominant with respect to the interaction with Al.

Linares and Huertas (1971) precipitated gibbsite from solutions of pH 4 to 6, which contained aluminum–fulvic acid complexes, in time periods of less than 1 month. They hypothesized that the fulvic acid complex was important in the formation of the sixfold coordination of aluminum necessary for the formation of gibbsite and kaolinite (De Kimpe, Gastuche, and Brindley, 1961).

Schnitzer and Hansen (1970) determined the stability constant for aluminum–fulvic acid complexes at a pH of 2.25. They found that the stability constant decreased at a fixed pH when the ionic strength of the solution was increased. They also found that in solutions of low ionic strength, an increase in pH resulted in a lowering of the ratio of fulvic acid/aluminum in solution and subsequently led to precipitation of aluminum hydroxide.

Wada (1977), in a review of the weathering of volcanic ash in humid, temperate climatic zones, suggested that the activity of aluminum released by the weathering processes is suppressed by aluminum–humus complexes that fix aluminum in the soils and lead to the formation of opaline silica rather than aluminosilicate phases. Wada further suggested that the balance between the supply of organic matter and the rate of aluminum production through weathering would control the mineral formation in soils.

La Iglesia Fernandez and Martin Vivaldi (1973) confirmed the catalytic effect of fulvic acid in solutions having a pH near 4 in the formation of kaolinite. Aluminosilicates were precipitated from room-temperature solutions in which the pH was fixed at several values between 3 and 7. After 65 days, the yield of well-crystallized kaolinite (described also as a T polytype close to dickite) was highest from the solution held at pH = 4. The yield of well-crystallized material decreased greatly at higher pH values although appreciable amorphous aluminosilicate did precipitate.

Finally, Evans and Russell (1959) found no tendency for fulvic acid or humic acid to be adsorbed on boehmite in solutions having a pH of 3.2. Interaction with gibbsite was observed in solutions having pH values ranging from 3.2 to 7. Linares and Huertas (1971) observed only gibbsite and bayerite precipitates from solutions of pH < 6 that contained fulvic acid and low silica concentrations. These data suggest that the presence of fulvic acid in soils promotes $Al(OH)_3$ precipitation at the expense of $AlO(OH)$ phases.

Thus, both laboratory (May et al., 1979; and Linares and Huertas, 1971) and field (Wada, 1977) observations support the important role that some organic compounds play in suppressing aluminum mobility in the pH range of 3 to 5 and in the formation of bauxites and lateric soils. Furthermore, the local presence of organic compounds and solutions having a pH range of 3 to 5 in developing soils could lead to the formation of small amounts of gibbsite in most soils at any time during their development.

The role of climate, drainage, and rainfall may be viewed in terms of the effect of each upon the quantity of organic material contributed to the soil and upon the pH and ionic strength (as well as composition) of the ground water interacting with the developing soil. High rainfall and temperate to tropical climates lead to high accumulations of organic debris. A well-drained soil coupled with high rainfall leads to solutions having low ionic strength and low pH. As the residence time of ground water in a soil increases, the ionic strength and pH increase, yielding conditions more favorable to the growth of aluminosilicate phases such as allophane, imogolite, halloysite, or kaolinite.

Variations in the frequency of rainfall and degree of drainage will affect the activity of water in the soil and consequently the distribution of gibbsite, boehmite, and diaspore. Local variations in dissolved aluminum concentrations arising from factors such as the mixing of ground water from different sources (e.g., drainage regimes) or inhomogeneous distribution of organic material may lead to the formation of $Al(OH)_3$ phases less stable than gibbsite.

Conclusions

The thermodynamic properties of the phases bayerite, nordstrandite, and microcrystalline gibbsite and of the solution species $Al(OH)^{2+}$ and $Al(OH)_2^+$ presented in this report are based upon assumptions regarding equilibria inferred from the work presented in several studies. Although the inferences appear reasonable, the lack of adequate phase characterization in studies of this system requires further substantiation of these findings.

I have assumed that the results presented by May et al. (1979) may be interpreted as representing several metastable equilibria, one of which spans the pH range of 4 to 9 and, thus, properly describes the minimum solubility of a single $Al(OH)_3$ phase for the first time. These data, together with the results presented by Hemingway et al. (1978), allow the calculation of an internally consistent set of thermodynamic properties for the monomeric aluminum solution species Al^{3+}, $Al(OH)^{2+}$, $Al(OH)_2^+$, and $Al(OH)_4^-$.

Verification of the $Al(OH)_3$ phase as nordstrandite should be possible through identification of the phase that precipitates from solutions prepared in the manner followed by Frink and Peech (1962, 1963) or through a modification of the procedure followed by May et al. (1979), that is, an organically buffered solution (without suspended solid $Al(OH)_3$) to which aluminum and OH^- are added at a rate sufficient to keep the solution *slightly* supersaturated with respect to the equilibrium solubility assumed to represent nordstrandite.

Some high-molecular-weight organic materials have a catalytic role in the formation of gibbsite and kaolinite at low pH, whereas low-molecular-weight organic solutes may have an inhibiting effect on $Al(OH)_3$ precipitation.

The weathering of primary aluminosilicate minerals proceeds through a series of metastable (irreversible) reactions that progress spontaneously at low temperatures toward the lowest free-energy state. Models that fail to consider factors affecting the relative rates of these metastable reactions or the role of organic compounds (e.g., Norton, 1973), or that assume that the metastable reactions are reversible (e.g., Helgeson, 1968, 1971) must be viewed as incomplete.

Finally, in order to properly interpret the role of aluminum in metasomatic alteration, the structure and thermodynamic properties of the aluminum solution species must be defined at temperatures above 373 K.

Acknowledgments

I wish to thank my U.S. Geological Survey colleagues, D. K. Nordstrom and J. D. Rimstidt, for several valuable discussions. I also wish to thank J. J. Hemley, J. D. Rimstidt, and E. Busenberg for their critical review of this manuscript.

References

Allen, V. T. (1952) Petrographic relations in some typical bauxite and diaspore deposits, *Geol. Soc. Amer. Bull.* **63**, 649–688.

Baes, C. F., Jr., and Mesmer, R. E. (1976) *The Hydrolysis of Cations*. Interscience, New York.

Barnhisel, R. I., and Rich, C. I. (1965) Gibbsite, bayerite, and norstrandite formation as affected by anions, pH and mineral surfaces, *Soil Sci. Soc. Amer. Proc.* **29**, 531–534.

Bates, T. F. (1962) Halloysite and gibbsite formation in Hawaii, *Clays Clay Mineral.* **9**, 315–328.

Brosset, C. (1952) On the reaction of the aluminum ion with water, *Acta Chem. Scand.* **6**, 910–940.

Busenberg, E. (1978) The products of the interaction of feldspars with aqueous solutions at 25°C, *Geochim. Cosmochim. Acta* **42**, 1679–1686.

Busenberg, E., and Clemency, C. V. (1976) The dissolution kinetics of feldspars at 25°C and 1 atm CO_2 partial pressure, *Geochim. Cosmochim. Acta* **40**, 41–49.

Bye, G. C., and Robinson, J. G. (1964) Crystallization processes in aluminum hydroxide gels, *Kolloid-Z.Z. Polym.* **198**, 53–60.

Calvet, E., Thibon, H., Maillard, A., and Boivinet, P. (1950) Sodium aluminate solutions and the decomposition of these solutions, *Soc. Chim. France Bull.*, 1308–1312.

Carreira, L. A., Maroni, V. A., Swaine, J. W., Jr., and Plumb, R. C. (1966) Raman and infrared spectra and structures of the aluminate ions, *J. Chem. Phys.* **45**, 2216–2220.

De Kimpe, C., Gastuche, M. C., and Brindley, G. W. (1961) Ionic coordination in alumino-silicic gels in relation to clay mineral formation, *Amer. Mineral.* **46**, 1370–1381.

Dibrov, I. A., Mal'tsev, G. Z., and Mashovets, V. P. (1964) Vapor pressure of sodium hydroxide and sodium aluminate solutions of a wide concentration range at 25–350°, *Zh. Prikl. Khim.* **37**, 1920–1929.

Evans, L. T., and Russell, E. W. (1959) The adsorption of humic and fulvic acids by clays, *J. Soil Sci.* **10**, 119–132.

Frink, C. R., and Peech, M. (1962) The solubility of gibbsite in aqueous solutions and soil extracts, *Soil Sci. Soc. Amer. Proc.* **26**, 346–347.

Frink, C. R., and Peech, M. (1963) Hydrolysis of the aluminum ion in dilute aqueous solutions, *Inorg. Chem.* **2**, 473–478.

Fripiat, J. J., and Pennequin, M. (1965) Modification of the composition and molecular weight of dialysis-purified iron and aluminum hydroxides, *Soc. Chim. France Bull.*, 1655–1660.

Gastuche, M. C., and Herbillon, A. (1962) Etude des gels d'alumine: cristallisation en milieu desionise, *Soc. Chim. France Bull.*, 1404–1412.

Gayer, K. H., Thompson, L. C., and Zajicek, O. T. (1958) The solubility of aluminum hydroxide in acidic and basic media at 25°C, *Can. J. Chem.* **36**, 1268–1271.

Geiling, S., and Glocker, R. (1943) Atomic arrangement in Al(OH)₃ gel, *Z. Elektrochem.* **49**, 269–273.

Glastonbury, J. R. (1969) Nature of sodium aluminate solutions, *Chem. Ind. (London)* **5**, 121–125.

Goldberg, R., and Loughnan, F. C. (1977) Dawsonite, alumohydrocalcite, nordstrandite and gorceixite in Permian marine strata of the Sydney Basin, Australia, *Sedimentology* **24**, 565–579.

Goldsmith, J. R. (1949) Some aspects of the system NaAlSiO₄–CaO–Al₂O₃, *J. Geology* **59**, 19–31.

Gordon, M., Jr., Tracey, J. I., Jr., and Ellis, M. W. (1958) Geology of the Arkansas bauxite region, *U.S. Geol. Survey Prof. Paper No. 299*.

Hathaway, J. C., and Schlanger, S. O. (1965) Nordstrandite (Al₂O₃ · 3H₂O) from Guam, *Amer. Mineral.* **50**, 1029–1037.

Hauschild, U. (1963) Uber nordstrandite, γAl(OH)₃, *Z. Anorg. Allg. Chem.* **324**, 15–30.

Helgeson, H. C. (1968) Evaluation of irreversible reactions in geochemical processes involving minerals and aqueous solutions—I. Thermodynamic relations, *Geochim. Cosmochim. Acta* **32**, 853–877.

Helgeson, H. C. (1971) Kinetics of mass transfer among silicates and aqueous solutions, *Geochim. Cosmochim. Acta* **35**, 421–469.

Hem, J. D., and Roberson, C. E. (1967) Form and stability of aluminum hydroxide complexes in dilute solution, *U.S. Geol. Surv. Water Supply Paper No. 1827-A*.

Hem, J. D., Roberson, C. E., Laird, C. J., and Polzer, W. L. (1973) Chemical interactions of aluminum with aqueous silica at 25°C, *U.S. Geol. Survey Water-Supply Paper No. 1827-E*.

Hemingway, B. S., and Robie, R. A. (1977a) Enthalpies of formation of low albite (NaAlSi₃O₈), gibbsite (Al(OH)₃), and NaAlO₂; revised values for $\Delta H^{\circ}_{f,298}$ and $\Delta G^{\circ}_{f,298}$ of some aluminosilicate minerals, *U.S. Geol. Surv. J. Res.* **5**, 413–429.

Hemingway, B. S., and Robie, R. A. (1977b) The entropy and Gibbs free energy of formation of the aluminum ion, *Geochim. Cosmochim. Acta* **41**, 1402–1404.

Hemingway, B. S., Robie, R. A., and Kittrick, J. A. (1978) Revised values for the Gibbs free energy of formation of [Al(OH)₄⁻aq], diaspore, boehmite and bayerite at 298.15 K and 1 bar, the thermodynamic properties of kaolinite to 800 K and 1 bar, and the heats of solution of several gibbsite samples, *Geochim. Cosmochim. Acta* **42**, 1533–1543.

Hemley, J. J., Montoya, J. W., Marinenko, J. W., and Luce, R. W. (1980) Equilibria in the system Al₂O₃–SiO₂–H₂O and some general implications for alteration/ mineralization processes, *Econ. Geol.* **75**, 210–228.

Hsu, P. H. (1967) Effects of salts on the formation of bayerite versus pseudoboehmite, *Soil Sci.* **103**, 101–110.

Hsu, P. H. (1977) Aluminum hydroxides and oxyhydroxides, in *Minerals in Soil Environments*, edited by J. B. Dixon and S. B. Weed, Chap. 4, pp. 99–143. Soil Science Society of America, Madison, Wisc.

Hsu, P. H., and Wang, M. K. (1980) Crystallization of goethite and hematite at 70°C, *Soil Sci. Soc. Amer. J.* **44**, 143–149.

Hückel, W. (1951) *Structural Chemistry of Inorganic Compounds*. Elsevier, Amsterdam.

La Iglesia Fernandez, A., and Martin Vivaldi, J. L. (1973) A contribution to the synthesis of kaolinite. Internat. Clay Conf., 1972, Division de Ciencias, C.S.I.C. Madrid, Spain, Proc., pp. 173–185.

Iler, R. K. (1973) Effect of adsorbed alumina on the solubility of amorphous silica in water, *J. Colloid Interface Sci.* **43**, 399–408.

Kittrick, J. A. (1966) The free energy of formation of gibbsite and $Al(OH)_4^-$ from solubility measurements, *Soil Sci. Soc. Amer. Proc.* **30**, 595–598.

Kittrick, J. A. (1980) Gibbsite and kaolinite solubilities by immiscible displacement of equilibrium solutions, *Soil Sci. Soc. Amer. J.* **44**, 139–142.

Kwong, K. F. Ng Kee, and Huang, T. M. (1979) The relative influence of low-molecular-weight, complexing organic acids on the hydrolysis and precipitation of aluminum, *Soil Sci.* **128**, 337–342.

Laves, F. (1952) Phase relations of the alkali feldspars. I. Introductory remarks, *J. Geology* **60**, 436–450.

Linares, J., and Huertas, F. (1971) Kaolinite: synthesis at room temperature, *Science* **171**, 896–897.

Lind, C. J., and Hem, J. D. (1975) Effects of organic solutes on chemical reactions of aluminum, *U.S. Geol. Surv. Water-Supply Paper No.* 1827-G.

Lippincott, E. R., Psellos, J. A., and Tobin, M. C. (1952) Raman spectra and structures of aluminate and zincate ions, *J. Chem. Phys.* **20**, 536.

Mal'tsev, G. Z., and Mashovets, V. P. (1965) Heat capacity of sodium aluminate solutions at 25–90°, *Zh. Prikl. Khim.* **38**, 92–99.

Mal'tsev, G. Z., Malinin, G. V., and Mashovets, V. P. (1965) Structure of aluminate solutions, *Zh. Strukt. Khim.* **6**, 378–383.

May, H. M., Helmke, P. A., and Jackson, M. L. (1979) Gibbsite solubility and thermodynamic properties of hydroxy-aluminum ions in aqueous solutions at 25°C, *Geochim. Cosmochim. Acta* **43**, 861–868.

Mesmer, R. E., and Baes, C. F., Jr. (1971) Acidity measurements at elevated temperatures. V. Aluminum ion hydrolysis, *Inorg. Chem.* **10**, 2290–2296.

Moolenaar, R. J., Evans, J. C., and McKeener, L. D. (1970) The structure of the aluminate ion in solutions at high pH, *J. Phys. Chem.* **74**, 3629–3636.

Mubarak, A., and Olsen, R. A. (1976) An improved technique for measuring soil pH. *Soil Sci. Soc. Amer. J.* **40**, 880–882.

Naray-Szabo, I., and Peter, E. (1967) Nachweis von Nordstrandit und Bayerit in ungarischen Zeigeltonen, *Acta Geol. Hung.* **11**, 375–377.

Norton, S. A. (1973) Laterite and bauxite formation, *Econ. Geol.* **68**, 353–361.

Ostwald, W. (1897) Studien über die bildung und umwandlung fester köper, *Z. Physik. Chem.* **22**, 289–330.

Paces, T. (1978) Reversible control of aqueous aluminum and silica during the irreversible evolution of natural waters, *Geochim. Cosmochim. Acta* **42**, 1487–1493.

Parks, G. A. (1972) Free energies of formation and aqueous solubilities of aluminum hydroxides and oxide hydroxides at 25°C, *Amer. Mineral.* **57**, 1163–1189.

Patterson, S. H., and Roberson, C. E. (1961) Weathered basalt in the eastern part of Kauai, Hawaii, *U.S. Geol. Surv. Prof. Paper No.* 424-C, C195–C198.

Raupach, M. (1963) Solubility of simple aluminum compounds expected in soils—I. Hydroxides and oxyhydroxides, *Austr. J. Soil Res.* **1**, 28–35.

Rooksby, H. P. (1961) Oxides and hydroxides of aluminum and iron, in *The X-Ray Identification and Crystal Structures of Clay Minerals*, edited by G. Brown, pp. 354–392. Mineral Society, London.

Ross, G. J., and Turner, R. C. (1971) Effect of different anions on the crystallization of aluminum hydroxide in partially neutralized aqueous aluminum salt systems, *Soil Sci. Soc. Amer. Proc.* **35**, 389–392.

Russell, A. S., Edwards, J. D., and Taylor, C. S. (1955) Solubility of hydrated aluminas in NaOH solutions, *Am. Inst. Mining Metall. Eng. Trans., J. Metals.* **203**, 1123–1128.

Schnitzer, M., and Desjardins, J. G. (1969) Chemical characteristics of a natural soil leachate from a humic podzol, *Can. J. Soil Sci.* **49**, 151–158.

Schnitzer, M., and Hansen, E. H. (1970) Organo-metallic interactions in soils: 8. An evaluation of methods for the determination of stability constants for metal–fulvic acid complexes, *Soil Sci.* **109**, 333–340.

Schnitzer, M., and Kodama, H. (1977) Reactions of minerals with soil humic substances, in *Minerals in Soil Environments*, edited by J. B. Dixon and S. B. Weed, Chap. 21, pp. 741–770. Soil Science Society of America, Madison, Wisc.

Schnitzer, M., and Skinner, S. I. (1965) Organo-metallic interactions in soils: 4. Carboxyl and hydroxyl groups in organic matter and metal retention, *Soil Sci.* **99**, 278–284.

Schoen, R., and Roberson, C. E. (1970) Structures of aluminum hydroxide and geochemical implications, *Amer. Mineral.* **55**, 43–77.

Sharma, S. K., and Kashyap, S. C. (1972) Ionic interactions in alkali metal hydroxide solutions–A Raman spectral investigation, *Inorg. Nucl. Chem.* **34**, 3623–3630.

Sherman, G. D., Cady, J. G., Ikawa, H., and Blumsberg, N. E. (1967) Genesis of the bauxitic Hailu soils, *Hawaii Agric. Exp. Stn. Tech. Bull. No.* 56.

Sillén, L. G., and Martell, A. E. (1964) Stability constants of metal ion complexes, *Chem. Soc. (London) Spec. Pub.* 17.

Singh, S. S. (1974) The solubility product of gibbsite at 15, 25, and 35°C, *Soil Sci. Soc. Amer. Proc.* **38**, 415–417.

Smith, R. W., and Hem, J. D. (1972) Effect of aging on aluminum hydroxide complexes in dilute aqueous solutions, *U.S. Geol. Surv. Water-Supply Paper No.* 1827-D.

Turner, R. C., and Ross, G. J. (1970) Conditions in solution during the formation of gibbsite in dilute Al salt solutions. 4. Effect of Cl concentration and temperature and a proposed mechanism for gibbsite formation, *Can. J. Chem.* **48**, 723–729.

Valeton, I. (1972) *Bauxites*. Elsevier, Amsterdam.

Violante, A., and Violante, P. (1980) Influence of pH, concentration, and chelating power of organic anions on the synthesis of aluminum hydroxides and oxyhydroxides, *Clays Clay Minerals* **28**, 425–435.

Wada, K. (1977) Allophane and imogolite, in *Minerals in Soil Environments*, edited by J. B. Dixon and S. B. Weed, Chap. 16, pp. 603–638. Soil Science Society of America, Madison, Wisc.

Wall, J. R. D., Wolfenden, E. B., Beard, E. H., and Dean, T. (1952) Norstrandite in soil from West Sarawak, Borneo, *Nature* **196**, 261–265.

Willstätter, R., and Kraut, H. (1924) Hydrates and hydrogels. V. The hydroxides and their hydrates in different alumina gels, *Ber.* **57B**, 1082–1901.

Wolfenden, E. B. (1961) Bauxite in Sarawak, *Econ. Geol.* **56**, 972–981.

Wollast, R. (1967) Kinetics of the alteration of K–feldspar in buffered solutions at low temperature, *Geochim. Cosmochim. Acta* **31**, 635–648.

Young, A., and Stephen, I. (1965) Rock weathering and soil formation on high-altitude plateaus of Malawi, *J. Soil Sci.* **16**, 322–333.

Chapter 10
Hydrostatic Compression of Perovskite-Type MgSiO₃

T. Yagi, H. K. Mao, and P. M. Bell

Introduction

Recent laboratory data on high-pressure and high-temperature experiments of various silicates strongly suggest that silicate minerals having the perovskite-type structure are dominant phases in the earth's lower mantle (e.g., Liu, 1976; Mao *et al.*, 1977). Discovery of the existence of silicate perovskites has naturally preceded knowledge of their physical properties, and thus no experimental data have been reported so far on their elastic properties.

In the present study $MgSiO_3$ perovskite was synthesized in the MBC (megabar pressure cell; Mao and Bell, 1978) using laser heating. The sample was then subjected to hydrostatic pressures of 40–75 kbars in two sets of experiments. One set of experiments was designed as a study of the axial ratio $a:b:c$, as a function of pressure. The equivalent cubic perovskite has $a = b = c = 1 = 1.4142$. The enstatite–perovskite (EP) has $a = b = c = 1 = 1.0320 : 1.4439$. Silicate perovskite has lower symmetry (orthorhombic) than many compounds with the perovskite structure (cubic) that are stable at 1 bar, although it has been suggested that silicate perovskites may become cubic at high pressure. Data on the behavior of the axial ratio of the enstatite–perovskite may be important in interpreting the response of this structure type to pressure.

Assuming that this perovskite structure is stable in the mantle, the bulk modulus could be an important factor to be used in assessing the general elastic properties of the deep earth. The second group of experiments included determining the volume compression curve in the same pressure range (40–75 kbars), by utilizing the axial ratio data and diffraction data obtained in a separate set of experiments that contained a mixture of $MgSiO_3$–perovskite with either MgO or Pd metal as internal standards.

The ultimate value of the present data rests on their accuracy. The sources of experimental error and the resulting uncertainties are evaluated.

Experimental

Sample Preparation

The $MgSiO_3$-perovskite phase was synthesized in the MBC from clinoenstatite ($MgSiO_3$) and from forsterite (Mg_2SiO_4) starting materials by the laser heating technique. The details of these synthesis experiments are reported elsewhere (Yagi et al., 1978a). The synthesis conditions were approximately 400 kbars and 1000°C. Quenched samples are almost pure $MgSiO_3$ perovskite (from clinoenstatite) or the mixture of $MgSiO_3$ perovskite and cubic MgO (from forsterite) with a minor amount of unreacted starting material and a trace of finely divided platinum powder that was added as an absorber of the laser radiation. The MgO forms an intimate mixture with the EP phase, so it could be used as an X-ray diffraction internal standard (Mao and Bell, 1979).

Unit cell dimensions of the $MgSiO_3$ perovskite and MgO determined at 1 bar were compared with previous data (Yagi et al., 1978a) and found to be identical.

X-Ray Diffraction Techniques

The disk-shaped quenched samples (approximately 100 μm in diameter $\times$ 10 μm thick) from two synthesis runs were placed in the sample chamber of the MBC in a fluid mixture of methanol and ethanol that served as a hydrostatic pressure transmitting fluid. Small crystals of ruby were also included in the sample chamber for pressure determination.

Monochromatized Mo-$K\alpha_1$ X-radiation from a microfocus X-ray source (focal spot on the target was 1.0×1.0 mm) was collimated by a special precision slit system. The resulting X-ray beam focused incident on the sample had a rectangular cross section of approximately 100 μm $\times$ 50 μm; the diffraction linewidth recorded on film was less than 100 μm for a sample-to-film distance of approximately 50 mm. At X-ray generator power settings of 50 kV and 4 mA, the exposure time was 200 to 400 hours.

The sample pressure was measured before and after X-ray exposure by the ruby fluorescence technique. In this pressure region the pressure calibration scale of Mao et al. (1978) is indistinguishable from that of Piermarini et al. (1975).

The objectives of the experiments were to determine the axial ratios of enstatite–perovskite as a function of pressure and to determine the volume compressibility in the same range. Thus, the two separate sets of experiments were conducted in order to optimize each type of measurement. The axial ratio study was made with the EP phase synthesized from enstatite from which an average of 25 diffraction lines could be utilized. Only lines that could be indexed without ambiguity were utilized. No internal standard was required in these experiments.

In the second set of experiments either the EP plus MgO mixture synthesized from forsterite or a mixture of EP plus Pd metal was used, with the MgO or Pd as internal standards. Diffraction data on the internal standards were used in conjunction with pressure values measured by the ruby fluorescence scale and volume equations of state for each material (MgO, Anderson and Andreatch, 1966; Pd, Carter *et al.*, 1971) to deduce the absolute film-to-sample distance to ±20 μm and film shrinkage. The diffraction lines (200) and (220) for MgO, or (111) and (200) for Pd, were measured with three to nine lines of the EP phase.

Once $a:b:c$ is determined as a function of pressure, the specific volume $\overline{V}$ becomes a single parameter function of each d-value, and thus $\overline{V}$ was calculated from each d-value. In overlapping lines such as (002) + (110) and (004) + (220), the volume was calculated so that the weighted average of two d-values became equal to the observed d-value. The calculated intensity ratio at one bar [8:29 for (002):(110) and 32:56 for (004):(220); Yagi *et al.*, 1978a] was the weighting factor. The linear compression of the structure turned out to be less than 1% in the present pressure region and the compression was essentially isotropic, indicating that this assumption was reasonable.

The position of the diffraction lines on the X-ray films was measured visually using a micrometer. Each line was read three times or more and only the lines which could be read within an accuracy of ±10 μm were used for the analysis. Experiments were made between 40 and 90 kbars, but the pressure became nonhydrostatic between 75 and 90 kbars. Each line became broad and some lines overlapped with each other, making accurate measurements impossible.

Results

The variation of the b/a and c/a ratio determined in the present experiments is summarized in Table 1. An example of the observed and calculated d-values in this study is listed in Table 2. In order to determine the orthorhombic unit

Table 1. Unit cell parameters of MgSiO$_3$–EP (enstatite–perovskite) at high pressure, 25°C.

P (kbars)	0.001	74.5	84.5
a (A)	4.780(1)	4.729(3)	4.722(3)
b (A)	4.933(1)	4.892(3)	4.891(3)
c (A)	6.902(1)	6.853(6)	6.849(8)
V (A^3)	162.75(3)	158.5(1)	158.2(2)
b/a	1.0320(4)	1.035(1)	1.036(1)
c/a	1.4439(5)	1.449(2)	1.450(3)

Table 2. Observed and calculated *d*-values of $MgSiO_3$–EP (enstatite–perovskite) at $P = 74.5$ kbars and $T = 23°C$.

hkl	d (cal)	d (obs)	I[a]	hkl	d (cal)	d (obs)	I
002	3.427	3.404	s	122	1.835	1.834	mw
110	3.400			004	1.713	1.700	s
111	3.046	3.042	w	220	1.700		
020	2.446	2.446	s	221	1.650	1.648	mw
112	2.413	2.412	vs	123	1.574	1.578	w
200	2.364	2.365	m	130	1.542	1.542	mw
120	2.173	2.176	w	114	1.530	1.521	m
210	2.129	2.129	mw	222	1.523		
121	2.071	2.073	m	131	1.504	1.503	m
103	2.057	2.059	w	132	1.406	1.405	w
211	2.033	2.034	m	024	1.403		
022	1.991	1.990	w	204	1.387	1.387	w
202	1.946	1.944	w	312	1.374	1.376	w
113	1.896	1.892	w				

[a] *I*, intensity; s, strong; vs, very strong; m, medium; w, weak; mw, medium weak.

cell, 19 lines, free from overlap, were used for least-squares refinement. All the observed and calculated *d*-values are in agreement within an error of $\pm 0.15\%$.

From Table 1 it can be seen that the variation of both b/a and c/a is small, increasing slightly with pressure. Based on this observation, the following equations were proposed:

$$b/a = 1.032 + 0.04P \quad \text{(Mbar)},$$
$$c/a = 1.444 + 0.07P \quad \text{(Mbar)}.$$

This relation indicates that the *a*-axis is more compressible than either the *b*- or *c*-axis. If orthorhombic perovskite were to shift to an ideal cubic perovskite, *a*, *b*, and *c* must satisfy the relation $a = b = c/\sqrt{2}$. With the increase of pressure, c/a approaches this value, but b/a and c/a depart from the cubic value. Therefore, it is not conclusive whether or not degree of distortion of orthorhombic perovskite decreases under pressure.

In Table 3 all the experimental data used for the determination of the isothermal compression curve of $MgSiO_3$ perovskite are summarized and the result of the volume compression is shown in Fig. 1. A Birch–Murnaghan equation of state was fitted to these data assuming various values for dK_0/dP. If $dK_0/dP = 4$, the best fit obtained is $K_0 = 2.58$ Mbars and K_0 varies from 2.62 to 2.55 Mbars, assuming dK_0/dP between 3 and 5. The bulk sound velocity calculated from these values is 8.0 km/second.

Error Analysis

The three most probable sources of error in the present experimental procedure are considered as follows.

Table 3. Experimental data (plotted in Fig. 1).

Run No.	P (kbars)	b/a	c/a	hkl	d (obs)	V (calc)	V/V$_0$
8C-24	71(1)	1.035	1.449	002 + 110	3.410	159.1	
				112	2.415	158.9	
				004 + 220	1.702	157.8	
						158.6(5)	0.9744(39)
8C-25	73(1)	1.035	1.449	020	2.451	159.3	
				112	2.415	158.9	
				004 + 220	1.700	157.2	
						158.5(8)	0.9737(49)
8C-28	72(2)	1.035	1.449	002 + 110	3.404	158.3	
				020	2.449	159.0	0.9737(4)
				112	2.412	158.3	
				004 + 220	1.705	158.6	
				131	1.504	158.4	
						158.5(1)	0.9739(9)
8C-29	44(1)	1.034	1.447	002 + 110	3.418	160.2	
				020	2.455	160.3	
				112	2.424	160.7	
				004 + 220	1.713	160.9	
						160.5(2)	0.9862(11)
8C-34	60(1)	1.034	1.448	002 + 110	3.409	159.0	
				020	2.452	160.0	
				112	2.414	158.7	
						159.1(9)	0.9776(22)
8C-36	44(1)	1.034	1.447	112	2.420	159.9	0.9825(31)
8C-37	68(2)	1.035	1.449	002 + 110	3.409	159.0	
				020	2.451	159.4	
				112	2.411	158.1	
				200	2.368	159.2	
				122	1.834	158.3	
				004 + 220	1.704	158.3	
				221	1.651	158.8	
				131	1.504	158.4	
				312	1.377	159.5	
						158.8(2)	0.9755(12)
8C-38	41(1)	1.034	1.447	002 + 110	3.419	160.3	
				020	2.460	161.3	
				112	2.422	160.3	
				004 + 220	1.709	159.7	
				312	1.382	161.0	
						160.5(3)	0.9863(19)
8C-39	51(2)	1.034	1.448	002 + 110	3.414	159.7	
				112	2.414	158.7	
				004 + 220	1.706	158.9	
				312	1.380	160.3	
						159(4)	0.9794(27)

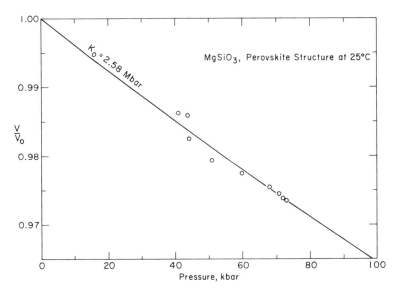

Fig. 1. Volume compression of MgSiO₃–EP (enstatite–perovskite) at high pressure and 25°C.

General

In the present experiments, all the pressures are measured by the ruby fluorescence technique. The uncertainty of this scale reflects the uncertainty of Decker's equation of state of NaCl (Decker, 1971) in which the reported error is 1.7% at pressures below 100 kbars. Therefore, a systematic error of 1.7% in the pressure calibration would cause an error of ±0.05 Mbar in calculation of bulk modulus.

In high pressure *in situ* X-ray diffraction experiments, nonhydrostatic pressure components could cause systematic errors in the compression curves (e.g., Sato *et al.*, 1975; Wilburn *et al.*, 1978). The type of systematic error depends on not only the type of apparatus but also the experimental technique, and thus it could be difficult to obtain meaningful corrections. To avoid this problem, samples were placed in the sample chamber with ethanol–methanol.

The positions of the diffraction lines on the X-ray film were measured visually by micrometer. The reproducibility of the measurements of line was ±10 μm, which corresponds to ±0.013° for a sample-to-film distance of 50 mm. For example, an uncertainty of ±10 μm in the internal standard line (200) corresponds to an uncertainty of ±20 μm in sample-to-film distance and ±0.13 Mbar in bulk modulus at 70 kbars. If, however, this uncertainty causes a 0.1% uncertainty in *d*-value, and if the bulk modulus is calculated from a single line, there is an uncertainty of 0.30 Mbar. These errors are random and the uncertainty was reduced by taking the average of many lines, hence there is a significant scatter. In a preliminary report of this study (Yagi *et al.*, 1978b), a slightly higher value was reported for the bulk modulus of MgSiO₃, based on fewer observations.

Axial Ratio Determination

Sample-to-film distance was calculated from the diffraction pattern and the equation of state of MgO or Pd mixed with specimen. Values of the bulk modulus and pressure derivative of MgO were selected from ultrasonic data (Anderson and Andreatch, 1966; Chang and Barsch, 1969). Different equations of state, such as the Murnaghan or second-order Birch–Murnaghan equations, gave identical values within an uncertainty of $\pm 0.2\%$. The uncertainty of a parameter and the form of the equation of state combined produce an effect of only ± 0.02 Mbar in the final result.

The elastic properties of Pd are not as well known, but in this material the correction to the isotherm is exceedingly small, so the isothermal compression could be reliably calculated from shock compression data (Carter et al., 1971). Results obtained from the two internal standards (Pd and MgO) show no systematic discrepancy.

In the present study, variation of b/a and c/a with pressure was determined first, and then the volume was calculated from each observed d-value assuming these ratios, so the calculated volume is sensitive to the choice of ratio. At 1 bar, 31 reflections were used for least-squares refinement, with the result that b/a and c/a are 1.0320 ± 0.0004 and 1.4439 ± 0.0005, respectively. At 74.5 kbars, 19 nonoverlapping lines were used and the values 1.0345 ± 0.0013 and 1.4491 ± 0.0022 were obtained, and at 84.5 kbars the result from 13 nonoverlapping reflections gave the values 1.0358 ± 0.0013 and 1.4504 ± 0.0028 for b/a and c/a, respectively. In these runs the sample-to-film distance and film shrinkage are uncertain, but this error is systematic and tends to cancel in calculation of the ratio $a:b:c$. If the uncertainty of the camera constant were assumed to be $\pm 0.3\%$, larger than ever observed, the resultant uncertainty in $a:b:c$ would be only $\pm 0.005\%$, which is negligible. Therefore, an uncertainty of $\pm 0.1\%$ was assumed for these ratios, which could cause an error of approximately ± 0.08 Mbar in the bulk modulus.

In experimental run 8C-37, nine diffraction lines were observed other than those of the internal standard. Direct least-squares calculation yielded a unit cell volume of 158.7(1) Å³. The volume calculated from each diffraction line using the given axial ratio yielded an average value of 158.8(2) Å³.

Volume Determination

The specific volumes calculated from the (020) line give somewhat larger values than from the (112) line. This difference could be caused by the systematic error in c/a and/or b/a ratio, but, more likely, this error is caused by the systematic error in reading the line position because the two lines are only 0.2 mm apart and the (112) line is relatively intense. This difference tends to cancel when the values of two lines are average.

In the present analysis, overlapping lines, (002) + (110) and (004) + (220), were also used for the calculation assuming that the intensity ratios are constant with pressure. A slight change of the intensity ratio in these lines would not affect the results.

The total possible systematic error results in a 0.15-Mbar error in bulk modulus. In addition, there is a maximum uncertainty of $0.3 \pm 0.13 = 0.43$ Mbar due to the measurement error, but because 36 data points are used, this uncertainty is reduced to 0.07 Mbar. All errors combined result in an uncertainty of ± 0.22 Mbar in the present bulk modulus.

Systematics of the Elastic Properties of Perovskite-Type Compounds

Lieberman et al. (1977) studied the elastic properties of aluminate, titanate, stannate, and germanate compounds with the perovskite structure by the ultrasonic method. Systematics, such as the known relationship that the modulus K_s is inversely proportional to the volume or $v_\phi M = $ const. (v_ϕ is bulk sound velocity and $\overline{M}$ is the mean atomic weight), in isostructural compounds hold reasonably well in various perovskite-type compounds. Based on systematics, the values $K_s = 2.5 \pm 0.3$ Mbars and $v_\phi = 7.9 \pm 0.4$ km/second were predicted for $MgSiO_3$ perovskite. The present experimental results yielded the value $K_T = 2.6 \pm 0.2$ Mbars and $v_\phi = 8.0$ km/second for enstatite perovskite in reasonable agreement with these predictions.

The fact that the elasticity systematics appear to hold for a variety of perovskite studies suggests that the bulk elastic properties are insensitive to distortion of the cation site. Evidently, the relatively short O–O distances in silicate perovskite (2.48 Å, minimum) do not appear to affect the bulk elastic properties either.

Summary

The bulk modulus of orthorhombic perovskite-type $MgSiO_3$, measured by *in situ* high-pressure X-ray diffraction study in the pressure range up to 80 kbars, is $K_0 = 2.6 \pm 0.2$ Mbars, assuming dK_0/dP between 3 and 5. This value can be used as a boundary condition in models of the earth's lower mantle.

Comparison of the elastic properties of various compounds with perovskite structure indicate that the proposed systematics between bulk modulus and molar volume and between bulk sound velocity and mean atomic weight are valid for this silicate perovskite.

References

Anderson, O. L., and Andreatch, P., Jr. (1966) Pressure derivatives of elastic constants of single-crystal MgO at 23 and $-195.8°C$, *J. Amer. Ceram. Soc.*, **49**, 404–409.

Carter, W. J., Marsh, S. P., Fritz, J. N., and McQueen, R. G. (1971) The equation of state of selected materials for high-pressure reference, in *Accurate Characterization of the High Pressure Environment*, edited by E. C. Lloyd, NBS Special Publication 326, pp. 147–158. U.S. GPO, Washington, D.C.

Chang, Z. P., and Barsch, G. R. (1969) Pressure dependence of the elastic constants of single-crystalline magnesium oxide, *J. Geophys. Res.* **74**, 3291–3294.

Decker, D. L. (1971) High-pressure equation of state for NaCl, KCl, and CsCl, *J. Appl. Phys.* **42**, 3239–3244.

Liebermann, R. C., Jones, L. E. A., and Ringwood, A. E. (1977) Elasticity of aluminate, titanate, stannate, and germanate compounds with the perovskite structure, *Phys. Earth Planet. Inter.* **14**, 165–178.

Liu, L. G. (1976) Orthorhombic silicate phase observed in olivine, pyroxene, and garnet at high pressures and temperatures, *Phys. Earth Planet. Inter.* **11**, 289–298.

Mao, H.-K., Yagi, T., and Bell, P. M. (1977) Mineralogy of the earth's deep mantle: quenching experiments at high pressure and temperature, *Carnegie Inst. Washington Yearbook* **76**, 502–504.

Mao, H.-K., and Bell, P. M. (1978) Design and varieties of the megabar cell. *Carnegie Inst. Washington Yearbook* **77**, 904–908.

Mao, H.-K., and Bell, P. M. (1979) Equations of state of MgO and ε-Fe under static pressure conditions, *J. Geophy. Res.* **84**, 4533–4536.

Mao, H.-K., Bell, P. M., Shaner, J. W., and Steinberg, D. J. (1978) Specific volume measurements of Cu, Mo, Pd, and Ag, and calibration of the ruby R_1 fluorescence pressure gauge from 0.06 to 1 Mbar, *J. Appl. Phys.* **49**, 3276–3283.

Piermarini, G. J., Block, S., Barnett, J. D., and Forman, R. A. (1975) Calibration of the pressure dependence of the R_1 ruby fluorescence line to 195 kbar, *J. Appl. Phys.* **46**, 2774–2780.

Sato, Y., Yagi, T., Ida, Y., and Akimoto, S. (1975) Hysteresis in the pressure-volume relations and stress inhomogeneity in composite materials, *High Temp.-High Pressure* **32**, 315–323.

Wilburn, D. R., Bassett, W. A., Sato, Y., and Akimoto, S. (1978) X-ray diffraction, compression studies of hematite under hydrostatic, isothermal conditions, *J. Geophys. Res.* **83**, 3509–3512.

Yagi, T., Mao, H.-K., and Bell, P. M. (1978a) Structure and crystal chemistry of perovskite-type MgSiO₃, *Phys. Chem. Minerals* **3**, 97–110.

Yagi, T., Mao, H.-K., and Bell, P. M. (1978b) Isothermal compression of perovskite-type MgSiO₃, *Carnegie Inst. Washington Yearbook* **77**, 835–837.

Chapter 11

Quantitative Spectra and Optics of Some Meteoritic and Terrestrial Titanian Clinopyroxenes

R. G. J. Strens,* H. K. Mao, and P. M. Bell

Introduction

Of all mineral spectra, those of titanian clinopyroxenes are among the most difficult to measure, interpret, and assign. Many overlapping (and often unresolvable) single-ion d–d (Fe^{2+}, Fe^{3+}, and Ti^{3+}) and intermetallic charge-transfer ($Fe^{2+} \rightarrow Fe^{3+}$, $Fe^{2+} \rightarrow Ti^{4+}$, $Ti^{3+} \rightarrow Ti^{4+}$) bands contribute to a broad absorption envelope stretching from 420 to beyond 800 nm, which is superimposed on the tail of a strong ligand-metal charge-transfer band centered in the near ultraviolet (Manning and Nickel, 1969; Burns et al., 1976). The measurement and interpretation of these spectra is further complicated by large differences between the orientations of the principal indices and principal absorbances in the (010) plane (Dowty, 1978) and by the occurrence of both spin-allowed (Smith, 1978a) and spin-forbidden (Ferguson et al., 1966) d–d transitions of exchange-coupled ion-pairs. These pair transitions have the energy, width, width/energy ratio, and pressure dependence of single-ion d–d bands, but the polarization-, composition-, and temperature-dependence of charge-transfer absorptions. The result is that such bands have often been assigned by different authors to both charge-transfer and d–d origins, depending on the type of measurement made. Examples are the 725- and 1100-nm bands in tourmaline (Smith, 1978b) and the 620-nm band in the Allende fassaite, assigned to $Ti^{3+} \rightarrow Ti^{4+}$ charge transfer by Dowty and Clark (1973) and as a d–d transition of Ti^{3+} by Burns and Huggins (1973) and Mao and Bell (1974).

Two promising approaches to these problems of interpretation and assignment are the use of perturbation methods such as the measurement of spectra as a function of temperature or pressure and the application of quantitative methods, including the analysis of polarization dependence of intensity and comparison of width/energy ratios and oscillator strengths of absorption bands. We use both approaches to assign the spectra of two meteoritic titanian fassaites, Allende and Angra dos Reis, each of which contains only two major transition metal ions. Both have been previously studied in this laboratory

*Deceased.

(Mao and Bell, 1974; Mao *et al.*, 1977), and we now report new experimental data for the Angra dos Reis material and a reinterpretation of the spectra of the Allende mineral. The Allende fassaite contains major Ti^{3+} and Ti^{4+}, the iron content being negligible (Dowty and Clark, 1973; Burns and Huggins, 1973; Mao and Bell, 1974). The Angra dos Reis fassaite contains major Fe^{2+} and Ti^{4+} with little Fe^{3+} or Ti^{3+} (Hazen and Finger, 1977; Mao *et al.*, 1977). The resulting spectra are easily resolved and assigned, providing reliable quantitative data for d–d bands of Fe^{2+} and Ti^{3+} and for $Ti^{3+} \rightarrow Ti^{4+}$ and $Fe^{2+} \rightarrow Ti^{4+}$ charge-transfer absorption.

The optical properties of titanian clinopyroxenes also present many interesting problems, and the information on band energies, widths, and absorbances, taken with the equations derived by Strens and Freer (1978) which relate optical properties to the spectra, enables these problems to be treated quantitatively.

Spectral Measurements on Monoclinic Crystals

If we consider only electric dipole transitions, then the absorption properties of crystals of orthorhombic or higher symmetry are completely specified by spectra measured with the electric vector (E) of the incident light parallel to each of the crystal axes in turn, i.e., by measurement of the ϵ and ω or α, β, and γ spectra. The crystal axes and the principle axes of the absorption figure and the indicatrix coincide in these crystals.

In monoclinic crystals such as the titanian clinopyroxenes, one principal absorbance (a_3) and one principal index (n_3) must coincide in direction with the unique (b) axis. The other two principal indices (n_1, n_2) lie in the (010) plane at angles and $(90 + \phi)°$ to the c-axis, where ϕ is the extinction angle measured in the obtuse angle β. The other two principal absorbances (a_1, a_2) lie at angles θ and $(90 + \theta)$ to c, and thus make angles $(\phi - \theta)$ with the principal index directions. Provided ϕ and θ or $(\phi - \theta)$ are known, it does not matter whether the principal absorbances are measured directly (as in this paper) or derived from the spectra measured with the electric vector parallel to each of the two principal indices in the (010) plane (as in most previous work). The choice is a matter of convenience, the values being interconvertible using

$$E \parallel n_1: \qquad a' = a_1 \cos^2(\phi - \theta) + a_2 \sin^2(\phi - \theta),$$

$$E \parallel n_2: \qquad a'' = a_1 \sin^2(\phi - \theta) + a_2 \cos^2(\phi - \theta). \qquad (1)$$

Most titanian clinopyroxenes crystallize in space group $C2/c$, with the transition metal ions responsible for absorption in and near the visible region concentrated in the octahedral M(1) and eight-coordinated M(2) positions (Hazen and Finger, 1977; Dowty and Clark, 1973). Since M(1) and M(2) positions have identical x-coordinates, and both ion-pair and charge-transfer transitions are polarized along the metal–metal vectors (Robbins and Strens,

1968; Smith and Strens, 1976; Smith, 1978a, b), the intermetallic charge-transfer and ion-pair bands have principal absorbance directions a^*, b, c, with zero absorbance for $E \| a^*$, and $\theta = 0$. In the P2/n pyroxenes (omphacites) there are two types of M(1) position, one enriched in (Mg, Fe^{2+}) and the other in (Al, Fe^{3+}). The M(2) positions are occupied by calcium and alkalis (Clark et al., 1968). Charge-transfer and ion-pair transitions are again polarized in (100).

In C2/c pyroxenes, the M(1) and M(2) polyhedra have point group symmetry 2, so that the z electronic axes of transition metal ions in these sites lie along $b = \beta$ (the diad). The x and y electronic axes lie near γ and α, respectively (Demspey, 1976), with $(\phi - \theta) \sim 7°$ so that about 98% of x (y) polarized absorption appears in the γ (α) spectrum, with 2% in α (γ).

The most convenient orientations for measuring $d-d$ spectra are therefore $E \| \alpha, \beta, \gamma$, using (010) and (h01) sections, while the best orientations for measuring ion-pair or charge-transfer spectra are $E \| a^*, b, c$, using (100) and a^*b sections.

Experimental

Preliminary microscopic examination of fragments of Angra dos Reis fassaite mounted on a spindle stage, and particularly of those with the unique axis $b = \beta$ as the rotation axis, showed the orientation of the remaining two principal absorption axes to be indistinguishable within experimental error from a^* and c, with absorption for white light being $c > b \gg a^*$. Two sections bc and a^*b were prepared by orienting fragments on the spindle stage, mounting in epoxy resin, grinding and polishing one side, remounting in Lakeside cement, and grinding and polishing the reverse side.

Using the apparatus described by Bell and Mao (1972), spectra were obtained with $E \| b$, $E \| c$ and $E : b = 45°$ using the (100) section, and with $E \| b$, $E \| a^*$ and at four intermediate orientations using the a^*b section. The areas sampled ranged up to 50 μm square, and the divergence of the sample beam was approximately 5°, corrections for divergence being negligible (Goldman and Rossman, 1979).

Graphs of the logarithm of the optical density against photon energy for the three orientations showed the background absorption (absorption edge) to obey Urbach's rule $d \ln D / dv = $ constant (Davydov, 1968), and subtraction of this baseline from the spectra enabled true wavelengths, widths, and absorbances to be measured. Oscillator strengths (f) were calculated by the methods of Robbins and Strens (1972) and Smith and Strens (1976).

Hazen et al. (1977) reported high-pressure spectra of a (110) cleavage flake of Angra dos Reis fassaite, but as this orientation yields impure (γ') spectra, the measurements were repeated for $E \| b$ on a third section at pressures of 1 atm and 48 kbars.

Description and Orientation Dependence
of the Spectra

The measured spectra of the Angra dos Reis fassaite normalized to the 139-μm thickness of the $a*b$ section, are shown in Fig. 1, in which the observed optical density $D = \log_{10}(I_0/I_{tr})$ is plotted against wavenumber ($\bar{\nu}$ in μm^{-1}, 1 μm^{-1} = 10,000 cm^{-1}) and wavelength (λ in nm). The $E \parallel b$ spectra measured at 0 and 48 kbars are shown in Fig. 2.

All three spectra (Fig. 1) consist of discrete absorption bands centered near 0.5, 1, and 2 μm superimposed on an absorption edge rising towards the ultraviolet. The $E \parallel b$ and $E \parallel c$ spectra are dominated by an intense broad absorption at 0.5 μm, with a width $w \simeq 200$ nm, which is assigned to $Fe^{2+} \rightarrow Ti^{4+}$ charge transfer. Absorption by Fe^{2+} occurs near 1 μm, with a broad weak band for $E \parallel a*$, a weak doublet for $E \parallel c$, and a sharp and moderately strong band at 1026 nm in the $E \parallel b$ spectrum. A pair of broad weak bands in the $E \parallel a*$ spectrum may represent absorption by small

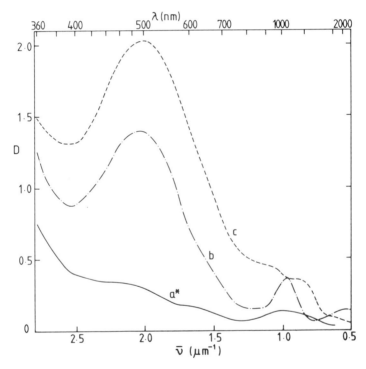

Fig. 1. Polarized spectra of the Angra dos Reis fassaite measured at 1 atm and room temperature with $E \parallel a*$, b, c. The strong 0.5-μm (500-nm) $Fe^{2+} \rightarrow Ti^{4+}$ charge-transfer band is polarized in (100), but d–d transitions of Fe^{2+} in M(1) occur near 1 μm (1000 nm) in all polarizations. The spectra have been normalized to the 139-μm thickness of the (100) section.

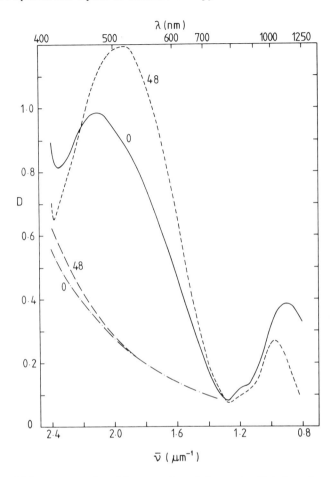

Fig. 2. The $E \parallel b$ spectrum of an 86-μm section of Angra dos Reis fassaite measured at room temperature and pressures of 1 atm and 48 kbars. Low- and high-pressure baselines are shown. Both d–d and charge-transfer bands sharpen as atomic vibration amplitudes are reduced by increasing pressure. The absorption edge and the 0.5-μm $Fe^{2+} \rightarrow Ti^{4+}$ band both move to lower energy, and the d–d band at 1026 nm moves to higher energy, with increasing pressure. Reduced distortion of the M(1) octahedron causes a reduction in area of the 1-μm band at high pressure.

amounts of Ti^{3+} (as in the Allende fassaite); these bands may also occur in the $E \parallel b$ and $E \parallel c$ spectra, where they would be obscured by the strong 0.5-μm band. The band at 2 μm in the $E \parallel b$ spectrum may represent absorption by Fe^{2+} in the distorted M(2) position. The arguments for these assignments are presented later.

The 1-atm and 48-kbar $E \parallel b$ spectra (Fig. 2) show a small shift of the absorption edge to lower energies, as expected for a ligand-metal charge-transfer absorption. The 0.5-μm band shifts to lower energies and narrows, improving its resolution and showing a slight increase in intensity. The

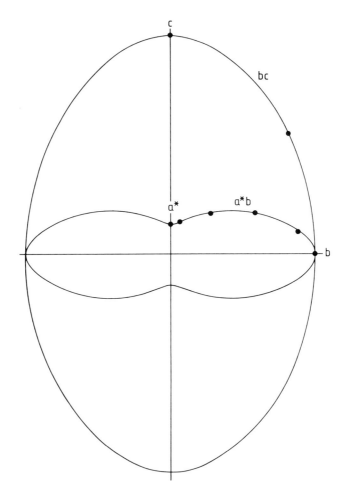

Fig. 3. Observed (●) and calculated (—) orientation dependence of optical density of $a*b$ and bc sections of the Angra dos Reis fassaite at 500 nm, using Eq. (2b) with values of $D(a*) = 0.29$, $D(b) = 1.39$, $D(c) = 2.03$ taken from Fig. 1.

1026-nm band shifts to higher energies, narrows, and (unusually for a d–d band) loses intensity with increasing pressure.

The observed angular variation of absorption in the (100) and $a*b$ sections is compared in Fig. 3 with theoretical behavior calculated using

$$a(\phi') = \ln\left(\exp(-a'\cos^2\phi') + \exp(-a''\sin^2\phi')\right)^{-1}, \qquad (2a)$$

where a' and a'' are the linear absorption coefficients $a = (\ln 10)D/t$ for the two allowed vibration directions in the plane considered, and ϕ' is the angle between the E-vector and one of those directions. The calculated variation of absorption at 0.5 μm in the optical axial plane (010) is shown in Fig. 4. Absorbances for $E \parallel \alpha$ and γ were first calculated using Eq. (1), and the

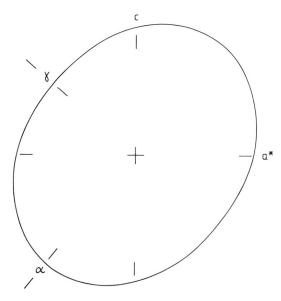

Fig. 4. Calculated variation of optical density in the (010) plane of the Angra dos Reis fassaite at 500 nm, using Eq. (1) with $D' = 1.31$, $D'' = 1.01$, $\phi = 50°$, $\theta = 0°$. Despite the extreme anisotropy of the principal absorbances in this plane $(D(c)/D(a^*) = 7)$, isotropic behavior is approached as $(\phi - \theta) \rightarrow 45°$.

orientation dependence in (010) was then found using Eq. (2a). The results confirm the visual observation that absorption of white light varies little in the (010) plane and agree with the results of instrumental measurements by Mao *et al.* (1977). In terms of optical densities D, D', and D'', Eq. (2a) becomes

$$D(\phi') = \log_{10}(10^{-D'} \cos^2\phi' + 10^{-D''} \sin^2\phi')^{-1}. \tag{2b}$$

Used in this form, G. R. Rossman (personal communication, 1977) found excellent agreement between observed and calculated absorption at 1049 nm in the (010) plane of olivine at ten orientations between $\gamma = a$ and $\beta = c$.

Assignment of Spectra

The strength of an absorption band is proportional to the product of chromophore concentration $[c]$ and the oscillator strength (f), the latter being a measure of the probability $(0 < f < 1)$ that a transition will be caused by an incident photon. A gross approximation is that spin- and symmetry-allowed single-ion d–d bands have oscillator stengths of $f \sim 10^{-4}$, and intermetallic charge-transfer bands and d–d transitions of exchange-coupled ion-pairs have

Table 1. Estimated single-ion concentrations (site populations $[C_1]$, $[C_2]$), donor × acceptor products ($[C_{11}] \times 10^4$) and ion-pair concentrations ($[C_{11}] \times 10^4$) in selected titanian clinopyroxenes.

Chromophore		A. dos Reis	Allende	Schivas
$[C_1]$	Fe^{2+}	0.205	0	0.229
	Fe^{3+}	< 0.005	0	0.128
	Ti^{3+}	< 0.012	0.34	0
	Ti^{4+}	< 0.059	0.14	0.166
	Cr^{3+}	0.005	0.001	0.005
$[C_2]$	Fe^{2+}	0.018	0	0
$[C_{11}] \times 10^4$	$Fe^{2+} \cdot Fe^{3+}$	< 10	0	293
	$Fe^{2+} \cdot Ti^{4+}$	121	0	380
	$Ti^{3+} \cdot Ti^{4+}$	6	476	0
	$Fe^{3+} \cdot Fe^{3+}$	< 0.25	0	164
	$Ti^{3+} \cdot Ti^{3+}$	< 1.4	1156	0

$f \sim 10^{-3}$, resulting in linear absorption coefficients $a \sim 1$ mm^{-1} for $[c] \sim 0.01$ and 0.001, respectively.

The integrated absorbance (area under absorption band) depends on the square of the transition moment $I^{1/2} \alpha \langle \psi_g \mu \psi_e \rangle$, where ψ_g, ψ_e are the wave functions of the ground and excited states, and μ is the dipole moment operator. For a charge-transfer transition, $\mu \simeq \Delta z e r_{da}$, where Δz is the difference in formal charge of the donor and acceptor ions, r_{da} is their separation, and e is the unit of electronic charge. Thus, $Fe^{2+} \rightarrow Ti^{4+}$ bands ($\Delta z = 2$) should be about four times as strong as $Fe^{2+} \rightarrow Fe^{3+}$ or $Ti^{3+} \rightarrow Ti^{4+}$ bands ($\Delta z = 1$), other things being equal.

Single-ion concentrations $[c_1] > 0.001$ and ion-pair concentrations $[c_{ij}] > 0.0001$ of selected meteoritic and terrestrial titanian clinopyroxenes are given in Table 1. Consideration of these alone suggests that the spectrum of the Angra dos Reis fassaite will be dominated by single-ion d–d transitions of Fe^{2+} in the M(1) and M(2) positions and by $Fe^{2+} \rightarrow Ti^{4+}$ charge-transfer absorption. In Allende, d–d transitions of Ti^{3+} may be intensified by formation of exchange-coupled Ti_2^{3+} pairs, and $Ti^{3+} \rightarrow Ti^{4+}$ charge transfer should be important. Terrestrial titanian clinopyroxenes (with abundant Fe^{3+} but little Ti^{3+}) could show marked intensification of both Fe^{2+} and Fe^{3+} d–d bands due to the formation of exchange-coupled Fe^{2+}–Fe^{3+} and Fe_2^{3+} pairs, with $Fe^{2+} \rightarrow Fe^{3+}$ charge-transfer absorption, in addition to the $Fe^{2+} \rightarrow Ti^{4+}$ charge-transfer band seen in the Angra dos Reis fassaite.

The discrimination of single-ion d–d, ion-pair and multiple-ion d–d and charge-transfer bands has been considered by Smith and Strens (1976) and Smith (1978a). Single- and multiple-ion d–d bands are similar in energy, w/λ ratio (typically 9–18%) and pressure-dependence, but differ in composition-, temperature-, and polarization-dependence. Multiple-ion d–d and charge-transfer bands resemble each other in temperature-, polarization-, and composition-dependence, but differ in w/λ ratio (typically 20–40% for charge-transfer bands) and pressure-dependence. Quite detailed work is therefore required to distinguish single-ion from multiple-ion d–d transitions, and multiple-ion d–d from charge-transfer transitions.

Angra dos Reis

The 0.5-μm Band

The chromophore concentrations (Table 1), taken with the great width (w/λ = 36–39% in Table 2), the pressure dependence ($d\lambda/dP$ positive), and the lack of absorption for $E \parallel a^*$ mark this as an $Fe^{2+} \to Ti^{4+}$ charge-transfer

Table 2. Major absorption bands in the Angra dos Reis fassaite at 1 atm and 298 K.

Band	Property	$E \parallel a^*$	$E \parallel b$	$E \parallel c$
absorption	a_{500} (mm^{-1})	3.7	5.2	9.1
edge	slope	2.2[a]	2.2	2.2
0.5 μm	λ (nm)	—	504	513
	w (nm)	—	185	209
	a (mm^{-1})	—	17.7	24.8
	$\bar{\nu}$ (μm^{-1})	—	1.984	1.950
	w/λ (%)	—	36.7	39
1 μm	λ (nm)	—	—	952
	w (nm)	—	—	210
	a (mm^{-1})	—	—	2.6
	$\bar{\nu}$ (μm^{-1})	—	—	1.05
	w/λ (%)	—	—	22
1 μm	λ (nm)	1042	1022	1176
	w (nm)	454	160	224
	a (mm^{-1})	1.6	4.9	3.0
	$\bar{\nu}$ (μm^{-1})	0.960	0.978	0.850
	w/λ (%)	43[b]	15.6	19
2 μm	λ (nm)	—	1818	—
	a (mm^{-1})	—	1.7	—
	$\bar{\nu}$ (μm^{-1})	—	0.550	—

[a] Change in photon energy (eV) associated with a tenfold change in absorbance.
[b] Unresolved multiple absorption.

transition. The d–d transitions of Fe^{2+} occur at longer wavelengths (~ 1 and 2 μm), and no other chromophore is present in sufficient concentrations to account for the strong absorption. The transition is within the range (440–600 nm) found by Smith and Strens (1976) for charge transfer between Fe^{2+} and Ti^{4+} in edge-sharing octahedra.

One anomalous feature of the 0.5-μm absorption that must be explained before the assignment is accepted as conclusive is the ratio of absorbances $a(b)/a(c)$ of 0.712, compared with the theoretical value of 0.389 for charge transfer between M(1) positions calculated from the structure data of Hazen and Finger (1977). Assuming the transition to be polarized along the donor–acceptor vector, and allowing for M(1)–M(2) as well as M(1)–M(1) interactions, we write for the integrated absorbance I

$$I \propto \sum_{ij} (f_{ij}c_{ij})(g_{ij}l_{ij}^2) \tag{3}$$

where f_{ij} is the oscillator strength of a transition involving a donor (d_i) in site i and an acceptor (a_j) in site j; g_{ij} is the number of ij contacts per unit cell, l_{ij}^2 is the square of the direction cosine of the donor–acceptor vector along the direction considered, and the concentration product $[c_{ij}] = [d_i][a_j]$. Using data from Table 3 the intensity ratio becomes

$$0.712 = \frac{I(b)}{I(c)} = \frac{4.496(f_{11}c_{11}) + 6.66(f_{21,21'}c_{21,21'})}{11.514(f_{11}c_{11}) + 5.34(f_{21,21'}c_{21,21'})}. \tag{4}$$

Solving Eq. (4) we find $(f_{21,21'}c_{21,21'})/(f_{11}c_{11}) = 1.305$. If $f_{11} = f_{21} = f_{21'}$ and $c_{11} = 121 \times 10^{-3}$, then $c_{21} = c_{21'} = 158 \times 10^{-4}$, i.e., 88% of the Fe^{2+} ions in M(2) are associated with Ti^{4+} in M(1), compared with 6% for a random distribution of Fe^{2+} and Ti^{4+}. Formation of Fe^{2+}–Ti^{4+} pairs has been noted in other minerals by Smith and Strens (1976), Smith (1977), and Ferguson and Fielding (1971, 1972).

Alternative explanations are that Fe^{2+} and Ti^{4+} in M(1) avoid each other, reducing c_{11} below the expected value, that Ti^{4+} is randomly distributed over

Table 3. Donor–acceptor vectors in the Angra dos Reis fassaite (structure data and site populations from Hazen and Finger (1977), see also Table 1).

$i \to j$	r_{ij} (Å)	g_{ij}	a^*	b	c	$10^4 \times [C_{ij}]$	$E \parallel b$	$E \parallel c$
				l_{ij}^2			$g_{ij}l_{ij}^2$	
$1 \leftrightarrow 1$	3.114	16	0	0.280	0.720	121	4.49	11.51
$1 \to 2$	3.233	8	0	0.333	0.667	11	2.66	5.34
$1 \to 2'$	3.513	4	0	1	0	11	4	0
$2 \to 1'$	3.513	4	0	1	0	0	4	0
$2 \to 1$	3.233	8	0	0.333	0.667	0	2.66	5.34
$2 \leftrightarrow 2$	4.832	16	0	0.637	0.363	0	10.19	2.90

M(1) and M(2), which seems unlikely on grounds of size and charge, or that $f_{21} \simeq 15 f_{11}$. Robbins and Strens (1972) noted marked differences ($f \simeq 30 \times 10^{-4}$ and 80×10^{-4}) between oscillator strengths for different types of donor–acceptor contact in biotite, but a $15:1$ ratio is unexpectedly large. We conclude that $Fe^{2+} - Ti^{4+}$ pairs are probably formed in more than random concentration in the Angra dos Reis fassaite. Whatever the cause, $f_{11} = 40 \times 10^{-4}$, compared with $\geqslant 57 \times 10^{-4}$, $\geqslant 81 \times 10^{-4}$, $\geqslant 45 \times 10^{-4}$, and $\geqslant 20 \times 10^{-4}$ found by Smith and Strens (1976) for $Fe^{2+} \rightarrow Ti^{4+}$ charge transfer in corundum (ω), corundum (ϵ), kyanite, and andalusite, respectively.

1- and 2-μm Bands

Using a modification of the methods of Wood and Strens (1972), Dempsey (1976) has calculated the energies, widths, polarizations, and approximate relative intensities of the $d–d$ bands of Fe^{2+} in the M(1) and M(2) positions of diopside.

For M(1) the interpretation is complicated by the near-degeneracy of both the (xy, xz) and $(x^2 - y^2, z^2)$ orbitals and by relaxation of the structure around the Fe^{2+}. A single band or Jahn–Teller split doublet centered at 1020 nm will occur in β (xy ground state) or α, γ spectra (xz ground state), or in both if the energy separation is comparable with kT as suggested by the calculations, the w/λ ratio being 15%. For M(2), transitions are predicted at about 1180 and 1850 nm, either β or $\alpha + \gamma$ polarized according to the choice of axes adopted. Assuming that these calculations apply with only slight modifications to the Angra dos Reis mineral, we assign the bands near 1 μm in both β and α, γ spectra to $d–d$ transitions of Fe^{2+} in M(1), with some contribution from M(2) possible near 1200 nm, and the band at 1818 nm to a $d–d$ transition of Fe^{2+} in the distorted M(2) position.

The high-pressure spectrum (Fig. 2) shows the 1026-nm band moving to shorter wavelengths ($d\lambda/dP = -0.2$ nm/kbar) and narrowing, this behavior being characteristic of $d–d$ transitions as the crystal field and the vibration amplitude decrease with pressure. The intensity decrease is unusual, but understandable in terms of Hazen and Finger's (1977) structure determination, which shows M(1) and M(2) octahedra becoming less distorted with increasing pressure.

The Allende Fassaite

Exploratory spectra of the Allende fassaite were recorded by Dowty and Clark (1973) and by Burns and Huggins (1973), but this account is based on the spectra published by Mao and Bell (1974).

The green color and strong pleochroism of the Allende fassaite are attributable to a well-resolved band centered at 620 nm and polarized in (100) with $a(c) > a(b)$. Synthetic $NaTiSi_2O_6$, which presumably contains little Ti^{4+}, is similarly colored, suggesting assignment to a $d–d$ transition of Ti^{3+} rather than to $Ti^{3+} \rightarrow Ti^{4+}$ charge transfer. This assignment is supported by the low w/λ ratio (13%), and the pressure dependence $(d\lambda/dP$ negative), but the polarization dependence, as emphasized by Dowty and Clark, is that of a charge-transfer band. This is exactly the behavior expected of a d–d transition of an exchange-coupled Ti_2^{3+} ion-pair, and this is the assignment we adopt, finding $f_{11} = 8 \times 10^{-4}$. Recalculating the observed absorbances to the a^*, b, c orientation, we find polarization ratio $a(b)/a(c)$ of 0.32, compared with a theoretical value of 0.28.

What appears to be a d–d band of Ti^{3+} at 485 nm does not have the same polarization dependence, and it may be a single-ion band. The broad absorption near 700 nm would then represent the $Ti^{3+} \rightarrow Ti^{4+}$ charge-transfer band, for which we find $f_{11} = 14 \times 10^{-3}$. Recalculating the observed absorbances to the a^*, b, c orientation, we find a polarization ratio $a(b)/a(c)$ of 0.28, which is the theoretical value for M(1)–M(1) charge-transfer.

Lunar and Terrestrial $C2/c$ Pyroxenes

The spectra of the great majority of lunar and terrestrial titanian clinopyroxenes consist of a broad absorption envelope stretching from about 420 to 800 nm and is composed of several ill-resolved overlapping bands, which is superimposed on an absorption edge rising into the ultraviolet (Burns et al., 1976). This complexity results from the presence of three major transition metal ions (Fe^{2+}, Fe^{3+}, Ti^{4+} in terrestrial minerals, Fe^{2+}, Ti^{3+}, Ti^{4+} in lunar), rather than the two present in the Allende and Angra dos Reis fassaites, and renders quantitative studies extremely difficult, as the overlap is so great that curve resolution becomes unreliable.

Despite these difficulties, a recent review by Burns et al. (1976) raises some interesting points. It appears from the spectra of four titanaugites differing greatly in Fe^{2+}/Fe^{3+} ratio (Burns et al., 1976, Fig. 4) that the intensity of the $^6A_{1g}$–$^4A_{1g}$ transitions of both octahedral and tetrahedral Fe^{3+} near 445 nm increases more than linearly with Fe^{3+} content, suggesting intensification of the absorption by formation of Fe_2^{3+} pairs. The same spectra suggest that the oscillator strength of the $Fe^{2+} \rightarrow Fe^{3+}$ transition near 550 nm is well below that of the $Fe^{2+} \rightarrow Ti^{4+}$ transition in the same minerals.

The spectra of the iron-rich but titanium-poor clinopyroxenes from Kolbeinsey (Iceland) recorded by Bell and Mao (1972a) tend to confirm these deductions, although the interpretation is complicated by the presence of substantial amounts of tetrahedral ferric iron. The 450-nm band of Fe^{3+} is extremely strong, possibly due to intensification by pair formation, although an alternative explanation is increased p–d mixing in the tetrahedral site. There is little sign of $Fe^{2+} \rightarrow Fe^{3+}$ charge transfer, despite a concentration product exceeding 10^{-2}, which is comparable with the $Fe^{2+} \cdot Ti^{4+}$ product responsible for the strong absorption in the Angra dos Reis mineral.

Molecular oribtal calculations by Burns et al. (1976) also confirm the dependence of the energies of Fe^{2+}–Ti^{4+} charge-transfer bands on donor–

acceptor distance deduced from limited experimental data by Smith and Strens (1976). The increase in energy with distance was attributed to the effects of increasing oxygen electron density between the cations. The calculated change as r_{ij} increases from $r_{11} = 3.14$ to $r_{21'} = 3.51$ Å is 0.37 μm^{-1}, which may contribute to the great width of the $Fe^{2+} \to Ti^{4+}$ band ($w/\lambda = 36-39\%$) in Angra dos Reis and other minerals.

Blue Omphacites

Abu-Eid (1976) measured the spectrum of one of a suite of blue omphacites from central Labrador described by Curtis and Gittins (1979). The spectrum (thickness and orientation not specified) shows that the color is attributable to absorption of red and yellow light by a broad band centered at 665 nm (1 atm) and moving to 695 nm at 40 kbars. The optical properties (Curtis and Gittins, 1979) are α dark blue, β azure, γ colorless, $\gamma : c = \phi = 74°$ in obtuse β, OAP (010). As $\cos^2 \phi = 0.08$, only 8% of the intensity of a band polarized in (100) would appear in the γ spectrum (colorless), with 92% in α (dark blue). Comparison of α (dark blue) with β (azure) suggests $a(c) > a(b)$.

The polarization in or near (100), the width ($w/\lambda = 24\%$), and the positive $d\lambda/dP$ are all consistent with a charge-transfer origin, and Curtis and Gittins note that the color is related to Ti content, and that there is evidence for replacement of $Fe^{3+} + Al^{3+}$ by $Fe^{2+} + Ti^{4+}$ rather than the $Fe^{2+} \to Fe^{3+}$ charge transfer suggested by Abu-Eid (1976). Comparison of chromophore concentrations shows abundant $Fe^{2+} - Fe^{3+}$ and $Fe^{2+} - Ti^{4+}$ centers in Abu-Eid's sample, but dark blue and colorless to pale blue omphacites described by Curtis and Gittins (Table 4) differ little in the $Fe^{2+} \times Fe^{3+}$ product, making assignment to $Fe^{2+} \to Fe^{3+}$ charge transfer improbable. By contrast, the $Fe^{2+} \times Fe^{3+}$ product in the dark blue material is 6.4 times that in the colorless to pale blue sample, providing good evidence for $Fe^{2+} \to Ti^{4+}$ charge transfer.

In the absence of data on thickness and orientation, it is not possible to

Table 4. Composition dependence of color in blue omphacites from central Labrador (Abu-Eid, 1976; Curtis and Gittins, 1979).

Concentration	Chromophore	Abu-Eid "Blue"	CG (a) Colorless	CG (b) Dark Blue
$[C_1]$	Fe^{2+}	0.330	0.288	0.345
	Fe^{3+}	0.140	0.023	0.033
	Ti^{4+}	0.099	0.023	0.123
$10^4[C_{11}]$	$Fe^{2+} \cdot Ti^{4+}$	327	66	424
	$Fe^{2+} \cdot Fe^{3+}$	462	66	114
	$Fe^{3+} \cdot Fe^{3+}$	196	5	11

calculate the oscillator strength or principal absorbances of the 665-nm band in omphacite, or to relate spectra to structure. However, the charge transfer is probably between Fe^{2+} and Ti^{4+} in the Mg- and Al-rich octahedra which share edges to form chains lying in the (100) plane. The Mg–Al vectors make angles of about $\pm 32°$ with c, so that the calculated $b:c$ intensity ratio is $0.28:0.72$. The donor–acceptor distance averages 3.11 Å, which is shorter than all but r_{11} in fassaite (Table 2). This may account for the difference in wavelength (665 nm versus 504–513 nm).

Optics of Titanian Clinopyroxenes

Dowty and Clark (1973) sought to relate the pleochroism and dispersion of titanian clinopyroxenes to the presence of strong charge-transfer bands in or near the visible region and polarized in the (100) plane. The derivation by Strens and Freer (1978) of equations relating refractive index and extinction angle to the wavelength, width, and absorbance of bands in the spectrum of a material enables Dowty and Clark's hypothesis to be tested quantitatively, and thus helps to elucidate the optics of titanian clinopyroxenes.

Dispersion of Indices, Extinction Angle, and $2V$

The dispersion of refractive index in a material that absorbs strongly in or near the visible, but otherwise behaves as a normal dielectric (transparent insulator) is described by Eq. (9) of Strens and Freer (1978):

$$n^2 = 1 + S + H + k^2, \tag{5}$$

where S and H are Sellmeier and Helmholtz dispersion terms, respectively. For the materials considered here, $n^2 \sim 3$, $S \sim 2$, $H \sim 10^{-4}$, and the square of the dimensionless absorption coefficient $k^2 \sim 10^{-8}$ and may be neglected. The Sellmeier dispersion term is given by

$$S = AL = IL/\pi^2, \tag{6}$$

where A is Sellmeier's constant, $L = \lambda^2/(\lambda^2 - \lambda_0^2)$, I and λ_0 are the integrated absorbance and characteristic wavelength of the absorption band system causing the dispersion, and λ is the wavelength of observation. The dispersion of indices in the Meiches and Stoffel titanaugites (Holzner, 1934) yields $A = 1.86$–1.93, $\lambda_0 = 134$–144 nm. The Helmholtz dispersion term gives the change in n^2 caused by the perturbing band(s) in or near the visible:

$$H_i = \frac{a_i}{2\pi} \frac{w_i \lambda^2}{\left(\lambda^2 - \lambda_i^2\right) + w_i^2 \lambda^2 / \left(\lambda^2 - \lambda_i^2\right)}, \tag{7}$$

where a_i is the linear absorption coefficient to base e, and w_i and λ_i are the full width at half-height and the wavelength at band maximum. The error in experimentally determined values of n^2 is typically 10^{-3}, whereas $H \sim 10^{-4}$, so that in practice we use Eq. (5) by inserting estimates of $(1 + S)$, e.g., the squares of the observed indices, and finding the changes in these values caused by the absorption band. The positions and magnitudes of old and new maximum and minimum values of n^2 in the (010) plane give the changes in extinction angle and principal indices and the change in optic axial angle is derived by substituting perturbed and unperturbed values of n^2 into

$$\cos^2 V_\alpha = \frac{\gamma^2(\beta^2 - \alpha^2)}{\beta^2(\gamma^2 - \alpha^2)} . \tag{8}$$

An approximation to Eq. (8) is

$$\cos^2 V_\alpha \simeq \frac{\beta - \alpha}{\gamma - \alpha} \quad \text{or} \quad \cos^2 V_\gamma \simeq \frac{\gamma - \beta}{\gamma - \alpha} \tag{9}$$

from which it will be seen that if one partial birefringence is small compared with $(\gamma - \alpha)$, $2V$ will be significantly changed by a very small change in any of the indices. In many titanaugites $\beta - \alpha \simeq 0.003$, $\gamma - \alpha \simeq 0.03$, and values of H of 10^{-4} can cause marked dispersion of optical axial angle. Similarly, changes in extinction angle are most marked when the difference between the two indices in the (010) section is small, and the perturbing band is polarized at a large angle to the extinction direction.

Allende

For sodium D light (589 nm), Dowty and Clark reported $\alpha = 1.747$, $\beta = 1.750$, $\gamma = 1.762$ (all ± 0.005), $2V_\gamma = 64°$, $\gamma : c = 58°$, OAP (010). One optic axis (A) is normal to (100), the other (B) makes an angle of $26°$ with the c-axis in the obtuse angle β. To be consistent with the observed values of α, γ, and $2V$, β should be near 1.751, and we adopt this value in calculations.

   In white light, Dowty and Clark noted three unusual features. First, the purest red color (absence of green) is observed for $E \parallel a^*$, rather than for any of the principal vibration directions. This simply reflects the absence of the 620-nm absorption in this polarization and is analogous to the behavior of the Angra dos Reis fassaite, in which the purest yellow color is seen for $E \parallel a^*$. Second, sections not containing the unique axis fail to extinguish and show anomalous interference colors, in which respect they resemble the Schivas titanaugite (Dixon and Kennedy, 1934). This behavior is caused by dispersion of the extinction angle. Third, one optic axis (B) is strongly dispersed, particularly for red light.

   The perturbations of the principal indices, extinction angle, and optic axial angle caused by the 620-nm band were calculated using Eqs. (5), (7), and (8), with the optical constants quoted above and band parameters reported in Table 5. The resulting changes in H ($\times 10^6$), n ($\times 10^6$), and $\gamma : c$, $2V_\gamma$ and $B : c$ are given in Table 6 for two visible wavelengths (540 and 700 nm) at which the

Table 5. Parameters of 495-, 620-, and 705-μm bands in Allende (Mao and Bell, 1974) and Angra dos Reis fassaites (this work).

Band	Property	Allende			Angra dos Reis $a*$
495 nm	λ (nm)	—	—	495	476
	w (nm)	—	—	70	100
	a (mm^{-1})	13	13	32	1.5
	$\bar{\nu}$ (μm^{-1})	—	—	2.02	2.10
	w/λ (%)	—	—	14.0	21
620 nm	λ (nm)	620	620	—	629
	w (nm)	78	80	—	115
	a (mm^{-1})	46	38	34	0.6
	$\bar{\nu}$ (μm^{-1})	1.613	1.613	—	1.59
	w/λ (%)	12.6	12.9	—	18
705 nm	λ (nm)	705	705	—	—
	w (nm)	127	> 100	—	—
	a (mm^{-1})	34	19	15	—
	$\bar{\nu}$ (μm^{-1})	1.42	1.42	—	—
	w/λ (%)	18	> 14	—	—

Note: Not all band parameters are resolvable.

mineral is least absorbing. Changes in index are in the fourth decimal place (up to 0.0008), with dispersion of 1° to 3° in $\gamma:c$, $2V$ and $B:c$. These are probably sufficient to cause the observed effects although other sources of dispersion may well be important, particularly the 485- and 705-nm bands and the ultraviolet absorption.

Table 6. Perturbations of optical properties of the Allende and Schivas fassaites by 620-nm and 504–513-nm bands (Tables 3 and 4).

		Allende		Schrivas	
		540(g)	700(r)	400(v)	700(r)
10^6 H	$a*$	0	0	0	0
	b	− 739	+ 1727	− 1717	+ 2704
	c	− 1229	+ 3636	− 2392	+ 3767
10^6 Δn	α	− 255	+ 733	− 197	+ 293
	β	− 211	+ 493	− 493	+ 776
	γ	− 95	+ 305	− 494	+ 779
Angles	$\gamma : c$	+ 0.6°	− 1.8°	+ 0.8°	− 1.3°
	$2V_\gamma$	0	− 1.2°	− 7°	− 9°
	$B : c$	+ 0.6°	− 2.4°	—	—

Schivas

The optical peculiarities of this unusual uniaxial titanaugite described from Schivas, Aberdeenshire by Dixon and Kennedy (1934) are of two types: those resulting from the uniaxial nature and those arising from the presence of strong polarized absorption bands in and near the visible.

The uniaxial nature ($\alpha = \beta = \omega = 1.741$, $\gamma = \epsilon = 1.762$, $\gamma : c = 32°$) is probably caused by the reduction in ($\beta - \alpha$) that occurs in clinopyroxenes as (Al, Ti, Fe) contents increase, Schivas being unusually rich in these elements (Al_2O_3 14.3, TiO_2 5.7, Fe_2O_3 4.4, FeO 7.1%). The Berlin blue interference colors seen in sections normal to optic axis resemble those of other minerals in which the dispersion curves for two principal indices approach closely or cross (compare melilite, apophyllite, some iron-poor clinozoisites, and zoisites).

The failure of sections not containing the unique axis to extinguish in white light, the abrupt change in interference colors on passing through extinction, and the dispersion of the optic axis are reminiscent of the behavior of the Allende fassaite, while the intense pleochroism $\alpha = \beta = \omega =$ plum, $\gamma = \epsilon$ = light yellow resembles that of the Angra dos Reis fassaite.

The perturbation of the optical properties of the Schivas titanaugite by the 0.5-μm band was calculated, assuming the absorption coefficient to be 380/121 times that in Angra dos Reis (Table 1). The results are reported in Table 6 for two wavelengths at which the mineral should be least absorbing. Comparatively small changes in extinction angle are predicted, although changes in $2V$ are large because of the pseudouniaxial character. It seems that other bands, probably the ligand → metal charge-transfer absorptions of Fe^{3+} and Ti^{4+} in the near-ultraviolet, are causing the dispersion of extinction angle.

Discussion and Conclusions

In contrast with the materials studied by solid-state physicists and chemists, minerals commonly possess low point group symmetry, and extremely complex chemistry and structure. These features result in complicated spectra that are often difficult to measure and interpret. Two promising approaches to these problems are the use of perturbation methods such as measurement of spectra as a function of temperature or pressure (Smith and Strens, 1976; Mao and Bell, 1974) and the application of quantitative methods such as direct calculation of band energies and widths from structure data (Wood and Strens, 1972; Demspey, 1976), the quantitative study of observational data including polarization and composition dependence, derivation of width/energy or width/wavelength ratios, and the calculation of oscillator strengths. By a careful choice of measurements, it is usually possible to distinguish single-ion from multiple-ion d–d transitions and multiple-ion d–d from charge-transfer transitions and to assign most resolvable bands even in complex spectra to specified transitions of particular chromophores.

Application of these methods to the spectra of the Angra dos Reis fassaite confirms the assignment of the broad band at 0.5 μm to charge transfer between Fe^{2+} ions in $M(1)$ and $M(2)$ and Ti^{4+} in $M(1)$ and suggests formation of $Fe^{2+}-Ti^{4+}$ pairs in more than random concentration. The oscillator strength $f_{11} = 0.0040$. Bands near 1 μm are assigned to single-ion $d-d$ transitions of Fe^{2+} in $M(1)$, and that near 2 μm to a similar transition of Fe^{2+} in the distorted $M(2)$ position. The high-pressure spectrum shows both the absorption edge and the Fe–Ti charge-transfer band shifting to lower energies, while the 1-μm band moves to higher energies as the crystal field strengthens due to the reduction in atomic vibration amplitude and weakens due to the reduced distortion of the $M(1)$ position.

Reinterpretation of the spectra of the Allende titanian fassaite suggests that the 458-nm band is a single-ion $d-d$ transition of Ti^{3+}, the 620-nm band is a $d-d$ transition of exchange-coupled Ti_2^{3+} pairs, and the 705-nm band arises from $Ti^{3+} \rightarrow Ti^{4+}$ charge transfer, for which $f_{11} = 0.0014$. Published spectra of terrestrial ferrian clinopyroxenes suggest that the oscillator strength of the $Fe^{2+} \rightarrow Fe^{3+}$ charge transfer band is similar to that of the $Ti^{3+} \rightarrow Ti^{4+}$ band in Allende, both being about one-third as strong as the $Fe^{2+} \rightarrow Ti^{4+}$ band in Angra dos Reis ($f_{11} = 0.0040$). This difference probably reflects the larger charge difference ($2e$ rather than $1e$), which should double the transition moment and quadruple the oscillator strength of the $Fe^{2+} \rightarrow Ti^{4+}$ band.

The color of blue titanian omphacites is attributed to a $Fe^{2+} \rightarrow Ti^{4+}$ charge-transfer band centered at 665 nm, rather than to $Fe^{2+} \rightarrow Fe^{3+}$ charge transfer as previously suggested by Abu-Eid (1976).

Consideration of the polarization of single-ion $d-d$, multiple-ion $d-d$, and charge-transfer bands suggests that in clinopyroxenes the single-ion bands have principal absorbance directions near the indicatrix axes and are conveniently measured using α, β, and γ spectra, while both types of pair bond are polarized in (100) and are best studied using a^*, b, and c spectra. However, spectra measured on either set of axes may be transformed to the other using Eq. (1), and the variation of absorption with orientation in a principal section is given by Eqs. (2).

Although mineral spectra and optics have usually been regarded as separate subjects, the optical properties of a mineral are completely determined if the absorption spectrum is known for all wavelengths. The quantitative link is provided by dispersion equations which relate the orientation and magnitude of the indicatrix axes to the wavelengths, absorption coefficients, and widths of bands in the spectrum (Strens and Freer, 1978). Dowty and Clark's hypothesis that the optics of titanian clinopyroxenes are significantly modified by the presence in and near the visible region of strong charge-transfer bands polarized in (100) is broadly confirmed by substituting the observed band parameters into the dispersion equations. However, the ultraviolet absorption also plays a major part in determining the properties.

Acknowledgments

One of us (R.G.J.S.) thanks the Carnegie Institution of Washington for financial assistance while working in Washington, and Dr. H. S. Yoder Jr. for facilitating the use of apparatus at the Geophysical Laboratory. We are also grateful to G. Amthauer, R. G. Burns, and G. Smith for critically reviewing this manuscript.

References

Abu-Eid, R. M. (1976) Absorption spectra of transitional metal bearing minerals at high pressures, in *The Physics and Chemistry of Minerals and Rocks*, edited by R. G. J. Strens, pp. 641–675, Wiley, London.

Bell, P. M., and Mao, H. K. (1972a) Crystal field determinations of Fe^{3+}, *Carnegie Inst. Washington Yearbook* **71**, 531–534.

Bell, P. M., and Mao, H. K. (1972b) Apparatus for measurement of crystal-field spectra of single crystals, *Carnegie Inst. Washington Yearbook* **71**, 608–611.

Burns, R. G., and Huggins, F. E. (1973) Visible-region absorption spectra of a Ti^{3+} fassaite from the Allende meteorite: a discussion, *Amer. Mineral.* **58**, 955–961.

Burns, R. G., Parkin, K. M., Loeffler, B. M., Leung, I. S., and Abu-Eid, R. M. (1976) Further characterization of spectral features attributable to titanium on the moon, *Proc. 7th Lunar Sci. Conf.* 2561–2578.

Clark, J. R., and Papike, J. J. (1968) Crystal-chemical characterization of omphacites, *Amer. Mineral.* **53**, 840–868.

Curtis, L. W., and Gittins, J. (1979) Aluminous and titaniferous clinopyroxenes from regionally metamorphosed agpaitic rocks from central Labrador, *J. Petrology* **20**, 165–186.

Davydov, A. S. (1968) Theory of Urbach's rule, *Phys. Status Solidi* **27**, 51–56.

Dempsey, M. J. (1976) Spectra, structure and ordering behaviour of some mixed crystals, Ph.D. Thesis, University of Newcastle upon Tyne.

Dixon, B. E., and Kennedy, W. Q. R. (1934) Optically uniaxial titanaugite from Aberdeenshire, *Z. Kristallogr.* **86**, 112–120.

Dowty, E., and Clark, J. R. (1973) Crystal structure refinement and optical properties of a Ti^{3+} fassaite from the Allende meteorite, *Amer. Mineral.* **58**, 230–242.

Dowty, E., and Clark, J. R. (1978) Absorption optics of low-symmetry crystals, *Phys. Chem. Minerals* **3**, 173–181.

Ferguson, J., and Fielding, P. E. (1972) The origins of the colours of yellow, green and blue sapphires, *Austr. J. Chem.* **25**, 1371–1385.

Ferguson, J., Guggenheim, H. J., and Tanabe, Y. (1966) Absorption of light by pairs of exchange-coupled manganese and nickel ions in cubic perovskite fluorides, *J. Chem. Phys.*, **45**, 1134–1141.

Goldman, D. S., and Rossman, G. R. (1979) Determination of quantitative cation distribution in orthopyroxenes from electronic absorption spectra, *Phys. Chem. Minerals* **4**, 43–53.

Hazen, R. M., Bell, P. M., and Mao, H. K. (1977) Polarized absorption spectra of the Angra dos Reis fassaite to 52 kbar, *Carnegie Inst. Washington Yearbook* **76**, 515–516.

Hazen, R. M., and Finger, L. W. (1977) Crystal structure and compositional variation of Angra dos Reis fassaite, *Earth Planet Sci. Lett.* **35**, 357–362.

Holzner, J. (1934) Beiträge zur Chemie und Optik sanduhrförmiger Titanaugite, *Z. Kristallogr.* **87**, 1–45.

Manning, P. G., and Nickel, E. H. (1979) A spectral study of the origin of colour and pleochroism of a titanaugite from Kaiserstuhl and of a riebeckite from St. Peter's Dome, Colorado, *Can. Mineral.* **10**, 71–83.

Mao, H. K., Bell, P. M., and Virgo, D. (1977) Crystal field spectra of fassaite from the Angra dos Reis meteorite, *Earth Planet. Sci. Lett.* **35**, 352–356.

Mao, H. K., and Bell, P. M. (1974) Crystal-field effects of trivalent titanium in fassaite from the Puerto de Allende meteorite, *Carnegie Inst. Washington Yearbook* **73**, 488–492.

Prewitt, C. T., Shannon, R. D., and White, W. B. (1972) Synthesis of a pyroxene containing trivalent titanium, *Contrib. Mineral. Petrol.* **35**, 77–79.

Robbins, D. W., and Strens, R. G. J. (1968) Polarisation dependence and oscillator strengths of metal-metal charge-transfer bands in iron (II, III) silicate minerals, *Chem. Commun.* 508–509.

Robbins, D. W., and Strens, R. G. J. (1972) Charge transfer in ferromagnesian silicates: the polarised electronic spectra of trioctahedral micas, *Mineral. Mag.* **38**, 551–563.

Smith, G. (1977) Low temperature optical studies of metal-metal charge-transfer transitions in various minerals, *Can. Mineral.* **15**, 500–507.

Smith, G. (1978a) Evidence for absorption by exchange-coupled Fe^{2+}–Fe^{3+} pairs in the near infrared spectra of minerals, *Phys. Chem. Minerals* **3**, 375–383.

Smith, G. (1978b) A reassessment of the role of iron in the 5000–30000 cm^{-1} range of the electronic absorption spectra of tourmaline, *Phys. Chem. Minerals* **3**, 343–373.

Strens, R. G. J., and Freer, R. (1978) The physical basis of mineral optics: I. Classical theory, *Mineral. Mag.* **42**, 19–30.

Wood, B. J., and Strens, R. G. J. (1972) Calculation of crystal field splittings in distorted coordination polyhedra, *Mineral. Mag.* **38**, 909–917.

Index